Functional Food and Superfood: Health and Nutrition

Functional Food and Superfood: Health and Nutrition

Editor: Dorothy Green

STATES
ACADEMIC PRESS
www.statesacademicpress.com

States Academic Press,
109 South 5th Street,
Brooklyn, NY 11249, USA

Visit us on the World Wide Web at:
www.statesacademicpress.com

ISBN: 978-1-63989-220-4 (Hardback)

Cataloging-in-Publication Data

Functional food and superfood : health and nutrition / edited by Dorothy Green.
p. cm.
Includes bibliographical references and index.
ISBN 978-1-63989-220-4
1. Functional foods. 2. Food--Nutritional aspects. 3. Nutrition. 4. Diet.
5. Health. I. Green, Dorothy.

QP144.F85 F86 2022
613.2--dc23

Table of Contents

Preface

Any food which is described to have some additional positive effect on health beyond basic nutrition is referred to as a functional food. It also comprises foods which may have been bred purposely in a particular food variety such as purple or gold potatoes. Many functional foods claim to have physiological benefits. Berries, spinach, almonds, barley, sardines, etc., are some of the common examples of functional foods. Most plant based foods and also some fish and dairy foods which are nutritionally dense and are good for a person's health are termed as superfoods. Superfoods also have various health benefits. They are a rich source of antioxidants, healthy fats and large quantities of fiber. They are believed to be helpful in preventing diabetes and digestive problems. Everyday superfoods include yogurt, olive oil, legumes and tomatoes. This book explores all the important aspects of superfoods in the present day scenario. It presents researches and studies performed by experts across the globe. As this field is emerging at a rapid pace, the contents of this book will help the readers understand the health and nutritional aspects of functional foods and superfoods.

This book unites the global concepts and researches in an organized manner for a comprehensive understanding of the subject. It is a ripe text for all researchers, students, scientists or anyone else who is interested in acquiring a better knowledge of this dynamic field.

I extend my sincere thanks to the contributors for such eloquent research chapters. Finally, I thank my family for being a source of support and help.

Editor

Preface

1

A Systematic Review of the Antioxidant Activity of Apiculture Products in Brazil

Karuane Saturnino da Silva Araújo,
Dark Luzia dos Santos Neto and
Sandra Maria Botelho Mariano

Abstract

The use of substances with antioxidant ability can be very important in the therapeutic prevention of diseases related to increased oxidative stress, such as cancer, heart disease and aging. Brazil has a great diversity of vegetation from which bees can collect resins. This study is a systematic bibliographical review on different electronic scientific databases through descriptors of antioxidant activity of bee products from Brazil. The identification of the articles and their inclusion occurred between the months January to MayW 2016. The bibliographic research was conducted in the following electronic databases: (1) Scientific electronic library online—SciELO; (2) Public library of science—PLOS Medicine; and (3) ScienceDirect. The articles selected were the ones that contained antioxidant activity evaluation of bee products from Brazil during 2011 and 2016. It was possible to observe a large number of articles published in this topic, but a compilation of data from all of these studies was necessary. Given there is a great diversity of vegetation in Brazil, a standardization process of the bee products in Brazil was conducted and by means of this process it was possible to draw a profile of the main antioxidants found in apiculture products in Brazil.

Keywords: phenolic compounds, DPPH, HPLC

1. Introduction

The use of substances with antioxidant ability can be very important in the therapeutic prevention of diseases related to increased oxidative stress, such as cancer, heart disease and aging. Brazil has a great diversity of vegetation from which bees can collect resins. This gives place to large chemical diversity among apiculture products collected in different regions and seasons.

This chemical diversity becomes important due to the derived biological properties. Given that apiculture products have great chemical diversity they must be chemically standardized before use to ensure quality, efficacy and safety and thus it is possible to correlate the type of product with its therapeutic application.

The word propolis is derived from the Greek word pro that means in defense and polis that means city or community; thus propolis means in community defense. It is a material in resin form with complex and varied chemical composition, which is gathered from various species and parts of plants (sprouts and exudates) and results in a substance of different colors and consistencies [1–3]. Propolis production in Brazil is estimated at around 100 tons per year. A large part of this percentage is for export, either in a raw form as in manufactured goods, reaching high prices in foreign trade and representing an important source of income [4, 5]. Among the apiculture products, propolis has been subject of pharmacological studies due to its antimicrobial [6], antiinflammatory [7] and antioxidant properties [8], among others. This biological potential is due to a synergism that occurs amongst its various constituents. Propolis is an important therapeutic alternative from an economic point of view because of its pharmacological efficacy since it is easy to obtain and presents pharmaceutical properties. The use of substances with antioxidant ability can be very important in the prevention and treatment of diseases related to increased oxidative stress, such as cancer, heart disease and aging [9].

Brazil has a great diversity of vegetation from which bees can collect resins. In addition, diversity increases when products are collected in different regions and seasons. Due to this chemical diversity, there has been intense research over the past decades to classify the different types of propolis and bee products found in Brazil [1, 10, 11]. Propolis needs to be chemically standardized before use to ensure quality, efficacy and safety. This way, it is possible to correlate the type of propolis and its therapeutic application, an essential task for a growing market and more demanding throughout the v one of the world's biggest suppliers of bee products [3, 11–13].

In this context, it is important to evaluate the published scientific studies about the antioxidant activity of bee products from Brazil.

2. Research methods

2.1. Study design

Study of systematic literature reviews on different scientific electronic databases through descriptors related to the antioxidant activity of bee products from Brazil. The identification of articles and their inclusion occurred between the months January to May 2016.

2.2. Eletronic databases

The bibliographic research was conducted in the following electronic databases:

1. Scientific Electronic Library Online—SciELO;

2. Public Library Of Science—PLOS Medicine;

3. ScienceDirect.

Additional information was obtained from manual search based on the references listed in the articles included in the review.

2.3. Search strategy

The searches were conducted through cataloged descriptors in Descriptor Health Sciences—DHS and Medical Subject Headings—MeSH, in Portuguese and English, contained in the title or summary of the studies. The combination of terms used together or separately, in the respective databases (SciELO, PLOS Medicine, ScienceDirect) was as follows:

- antioxidants;
- antioxidant response elements;
- phenolic compounds.

2.4. Selection and analysis of publications

For the selection of articles, a personal study was created with the following information: author, year, title, development period of the study, Province, city and research area, study design, descriptor used to locate the publication, objective and main results. An inclusion criterion was used in which the selected articles had to be original, published in international and national journals in English or Portuguese, published between 2011 and 2016 and indexed on one of the databases previously cited. Articles that contained evaluation data of the antioxidant activity of apiculture products from Brazil were selected for review.

2.5. Sampling

About 445.265 scientific articles about antioxidant activity in Brazil and in the world searched in the databases with the descriptors previously mentioned were identified. Of the total number of published articles between the years 2011 and 2016, 77 studies were related to the theme of antioxidant activity of bee products. About 60 articles were excluded for being repeated, in other languages or for not containing evaluation of products from Brazil. For this literature review, 17 studies were selected.

3. Results

For review, only articles containing analyses of evaluation of the antioxidant activity of bee products from Brazil, between 2011 and 2016, were selected. As shown in **Figure 1**, the number of published works on bee products is relevant. However, it is clear that the rate of work on this issue in Brazil is low.

The results obtained with the application of the described search strategy are presented in the logical framework of the study (**Figure 1**).

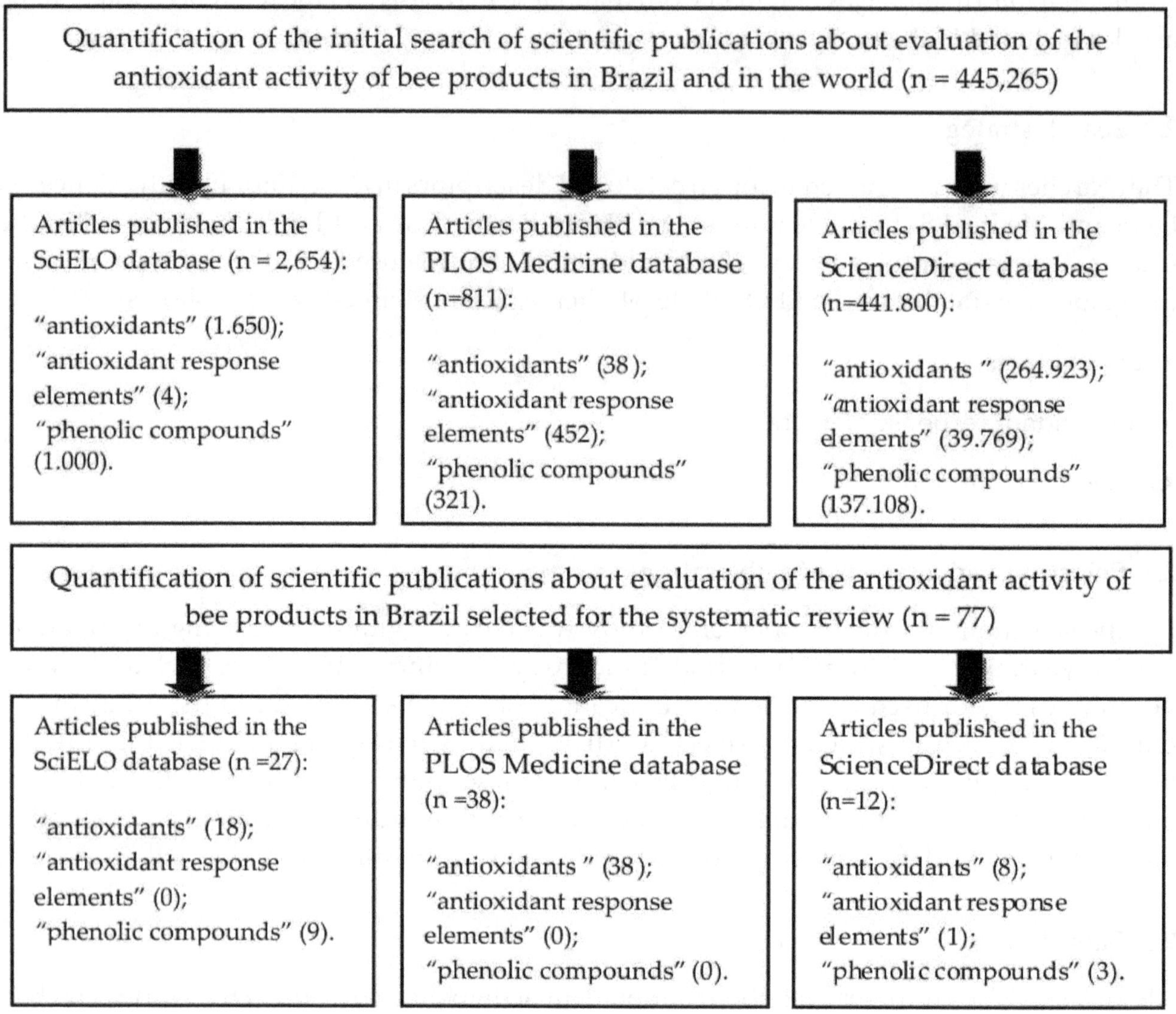

Figure 1. Logical framework of the systematic review, studies about evaluation of the antioxidant activity of bee products in Brazil between the years 2011 and 2016.

Table 1 presents a summary of the studies assessed in this study.

Given the large number of articles published in this topic was observed, a compilation of data from all of these studies was necessary. We can highlight the following identified substances from the main antioxidant substances of apiculture products from Brazil: triterpene compounds of the cycloartane, ursane and oleanane types as the main compounds, in addition to phenolic acids, protocatechuic and gallic acids. It is also important to highlight the high antioxidant potential of geopropolis from Brazil. The works presented here identified the following substances with antioxidant capacity: gallic acid, ellagic acid, catechin, gallocatechin, hesperidin, kaempferol, luteolin, morin, naringin, naringenin and rutin. Analysis of honey also shows the presence of antioxidant substances, especially gallic acid.

No.	Reference	Objectives	Main results/conclusions
01	[14]	Investigate the chemical composition of geopropolis produced by *M. fasciculata* collected from beehives of two phytogeographical regions of Maranhão and evaluate its antioxidant activity.	Geopropolis collected in Palmeirândia contained triterpene compounds of the cycloartane, ursane and oleanane type as the main compounds, in addition to phenolic acids, protocatechuic and gallic acid. In contrast, geopropolis collected in Fernando Falcão contained high concentrations of phenolic acids (gallic acid and ellagic acid) and exhibited high antioxidant activity, suggesting that the high levels of phenolic acids are responsible for the antioxidant property of this geopropolis. The chemical composition and antioxidant activity contribute to the identity and quality of the types of geopropolis produced by *M. fasciculata* collected in two phytogeographical regions of the Maranhão State, northeastern Brazil.
02	[15]	The objective of this study was to determine, among other analyses, the antioxidant capacity of propolis of stingless native bees (Meliponinae) of Tocantins, North of Brazil (State of Tocantins).	Propolis samples collected in the two regions of Tocantins presented physical and chemical characteristics that fit within Brazilian legislation for propolis quality. The flavonoid luteolin was restricted to the samples collected in Santa Maria of Tocantins, while, naringerin and rutin were restricted to the samples collected in Novo Acordo, demonstrating substantial differences between the pasture of the studied regions. And also evidence a high concentration of phenol compounds and good antioxidant capacity.
03	[16]	The objective of this study was to evaluate the effect of cassava starch coatings incorporated with propolis on the content of phytochemicals of nutritional interest and on the antioxidant activity of strawberries stored under refrigeration for 16 days.	The coating with 66% of propolis promoted higher Vitamin C content than fruits submitted to the other treatments at 8 and 12 days of storage.
04	[17]	Identify the antioxidant capacity of propolis samples collected by Apis mellifera L. bees, in four Brazilian regions and compare them using the coupled oxidation method of b-carotene system/linoleic acid.	From the propolis samples collected and analyzed by the coupled oxidation of b-carotene system/ linoleic acid, only two showed antioxidant capacity below 60%, both in the northeast. Samples of the southeast had the highest antioxidant capacity, followed by from the midwest, south and northeast. Pollen analysis revealed the predominant presence of Eucalyptus pollen, mainly from the southeast and south and a greater variety of pollen types in northeast samples.
05	[18]	Isolate phenolic substances of the extract in methanol of geopropolis, from species of *M. interrupta*, Mi5, grown in the National Institute of Amazonian Research—INPA by the Bees Research Group—GPA and MI6, grown in Ramal do Brasileirinho community in Manaus.	Analyses of geopropolis extracts of these species showed promising results, presenting antioxidant activities that are important for body maintenance and disease prevention, as well as being used as food (nutraceutical) by local people. From fractionation of the methanol extracts of geopropolis from *M. interrupta* with higher antioxidant activity was possible to isolate four flavonoid.

No.	Reference	Objectives	Main results/conclusions
06	[1]	This study aimed at verifying whether there is positive correlation between the identity and quality criteria for propolis and the biological activities exhibited by the extracts of propolis.	It was established that G12 propolis has a high content of total flavonoid and phenolic compounds, which gives it an excellent quality according to the Brazilian legislation.
07	[19]	Identify the chromatographic profile of phenolic acids and flavonoids using solid phase extraction (SPE) and liquid chromatography reversed-phase (RP-HPLC) and conduct study of antioxidant activity in honeys from three different species of bees *A. mellifera* (Africanized) exotic species, *Melipona flavolineata* (yellow uruçu) and *M. fasciculata* (gray uruçu) native species. The honey samples were obtained from four Pará State municipalities.	From 36 analyzed samples of honey, the presence of 13 phenolic compounds and three unidentified compounds (Ph1, Ph2 and Ph3) was observed. For most of the honeys, major compounds identified as gallic acid and quercetin were analyzed. The AC and ACP methods were able to distinguish the phenolic composition of the analyzed honeys for the species studied. This factor that indicates possible selectivity of the species in relation to the botanical origin of honeys. In general, analyzed honeys showed significant antioxidant activity, especially the darker honeys which also had higher levels of polyphenol.
08	[20]	The objectives of the present study were to produce and evaluate a propolis extract for use in skin care products.	Reasonable stability was noted and the preferred formula was the one that used combination of propolis extract with tocopheryl acetate.
09	[21]	Comparison between the antioxidant activities of ethanolic extracts of propolis prepared using hydrous ethanol with different ethanol/water contents and different propolis concentrations.	It was noted that the DPPH• and FRAP values are dependent on the propolis concentration and the water/ethanol proportion used in the extraction. This correlation was also observed in phenolic contents and flavonoid contents.
10	[22]	The present work aimed at performing a bioassay-guided fractionation of red propolis samples from Igarassu (Pernambuco, Brazil) in order to determine the main constituents associated with its antimicrobial activity, especially against *Candida sp.*	The botanical origin of propolis samples is difficult to ascertain on the basis of only one palynological analysis and a more definite confirmation depends of analysis comparing the chemical profile of the samples with the chemical profile of resins and extracts from plants found near the hives. It should be stressed that red propolis has been suggested as being the only propolis type derived from a plant from the leguminosae family (*D. ecastaphyllum*), rich in inisoflavones such as genistein and formononetin. Although flavonoids exhibit pleiotropic activity affecting several different targets and the synergistic effects cannot be discarded, our results suggest that the isoflavone formononetin is responsible at least partially for the antimicrobial activity of red propolis.
11	[23]	This work aims at optimization the extraction process of bioactive compounds, evaluate antioxidant activity and also conduct the chemical characterization of propolis, using the high performance liquid chromatography technique.	It was concluded that the analyzed propolis has a promising phenolic content and antioxidant activity. Three phenolic acids derived from hydroxycinnamic acid were identified, common in Brazilian propolis.

No.	Reference	Objectives	Main results/conclusions
12	[24]	This paper aims to identify the pollen types and quantify the total phenolic compounds in propolis samples produced in the arid territory of Bahia—Brazil	The levels of the total phenolic compounds found in the propolis samples from the territory of the arid region of Alagoinhas fulfill the standards of the Brazilian legislation. We recommend that more analyses should be conducted in order to obtain more data that corroborate the information contained here.
13	[25]	We evaluated the effects of phenolic compounds from three propolis-based products with different concentrations of propolis and levels of alcohol on feed intake, digestibility (ruminal and intestinal) and blood parameters in lactating dairy cows.	The propolis-based products have positive effects on protein metabolism in the rumen, without interfering with any other parameter evaluated. The propolis concentration and alcoholic level used in this study influences the amounts of flavonoids and phenolic acids in the propolis-based products, which may interfere with the observed effects on ruminal metabolism and digestive parameters.
14	[26]	This study assessed the polyphenolic profile and the antioxidant and antibacterial activities of monofloral honeys produced by Meliponini in the Brazilian semiarid region.	Honeys from Ziziphus joazeiro Mart. (juazeiro) and Croton heliotropiifolius Kunth (white velame) showed the highest total phenolic contents (TPCs) and the greatest antioxidant activity in assays with DPPH and ABTS + radicals. Malícia's honeys showed the greatest quantities of myricetin, quercetin and kaempferol among the studied honeys.
15	[27]	This study aimed to assess the fatty acid composition of milk, the antioxidant quality of milk and blood lipoperoxidation of dairy cows whose diet was supplemented with flaxseed oil containing a propolis-based product (PBP) with or without vitamin E.	Under the studied conditions, the improvements of milk fat quality, of the oxidative properties of milk and the blood's resistance to oxidation, were reached with PBP supplementation and E vitamin.
16	[28]	The aim of this study was to evaluate antioxidant properties of lyophilized bee pollen extract (LBP), to determine the phenolic profile by liquid chromatography and to evaluate the effect of LBP on the oxidative stability of pork meat sausage.	The LBP (lyophilized bee pollen extract) extract exhibited strong antioxidative effects in pork sausage, probably due to high antioxidant activity and the presence of the phenolic compounds in bee pollen; which has potential to be used in pork sausage.
17	[29]	Evaluation of antioxidant activity as well as the determination of phenolic compounds and antimicrobial activity, among other analyses of honey samples from south Brazil.	In the analyzed samples, the bioactive compounds found in a larger amount were the phenolic compounds. With respect to antimicrobial activity, we can highlight the relevance against Gram-positive microorganisms.

Table 1. Reference, objective, main results and conclusions, studies about evaluation of the antioxidant activity of bee products in Brazil between the years 2011 and 2016.

In addition, with respect to methods of extraction and preparation of propolis extracts, the composition of propolis extracts varies with the concentration of propolis, especially with the water/ethanol content of hydrous ethanol used in the extraction. Oldoni et al. observed the influence of standardization of extraction. The results of their study confirmed that the optimization of the extraction conditions is important to obtain extracts that are rich in phenolic compounds and antioxidant activity of propolis [23].

Since propolis has great chemical diversity, it needs to be chemically standardized before use to ensure quality, efficacy and safety and this way, it is possible to correlate the type of propolis and its therapeutic application, an essential task for a growing market and more demanding throughout the world. Brazil is a pioneer in these practices, since it is considered one of the world's biggest suppliers of bee products [3, 11–13].

Brazilian propolis were classified in the most prevalent types, resulting in 12 groups or types based on their geographical origin, chemical composition and plant origin: five in the south, one in the southeast and six in the northeast. A new type of propolis from a mangrove region from the State of Alagoas had its botanical origin identified as Dalbergiaecastophyllum, a species of legume and was ranked as the 13th type of northeast Brazilian propolis. These findings confirm the great Brazilian biodiversity that has become the subject of several scientific research studies throughout the world [3, 12, 13, 30, 31].

In this context, it is important to note that the evaluation of propolis samples from different geographical and climatic regions means that there are variations in the chemical composition and, therefore, biological activities may differ. Thus, when performing pharmacological studies, we should not indiscriminately compare propolis samples from different regions, neither assign a proven activity in a given sample to other samples from different regions. For this reason, publications on evidence of biological activities should include the physical and chemical characterization of the used propolis [32].

Moreover, there is smaller variation in the chemical composition of propolis found in temperate regions of the planet, where its main bioactive compounds are flavonoids: apigenin, quercetin, hesperetin, rutin, luteolin, genistein, daidzein, anthocyanidin, kaempferol, among others. Although flavonoids are the most extensively studied components of propolis, they are not the only ones responsible for their pharmacological properties. Several other compounds have been linked to its medicinal properties [12, 13, 24].

4. Conclusion: key results

In order to assist in the standardization process of apiculture products which are greatly diverse due to the variety of vegetation that exists in Brazil, it was possible to draw a profile of the main antioxidant substances of apiculture products from Brazil. It was possible to identify several substances of antioxidant capacity in bee products from Brazil, mainly in propolis and geopropolis, propolis extracts and honey.

Standardization has not yet been observed in scientific research and legislation in Brazil. Therefore, the types and quantification of substances with antioxidant capacity become too variable, thus, interfering in a more general analysis of the country's propolis. In this review, we observed that the chemical substances in propolis are determined by the flora of the region.

Additionally, since Brazil has great flora diversity, it is complex to follow a standardization process for the substances found in propolis.

Author details

Karuane Saturnino da Silva Araújo[1]*, Dark Luzia dos Santos Neto[2] and Sandra Maria Botelho Mariano[2]

*Address all correspondence to: karuane@gmail.com

1 Federal University of Maranhão (UFMA), Imperatriz, Brazil

2 Federal University of Tocantins (UFT), Palmas, Brazil

References

[1] Cabral, I. S. R.; Oldoni, T. L. C.; Alencar, S. M.; Rosalen, P. L.; Ikegaki, M. The correlation between the phenolic composition and biological activities of two varieties of Brazilian propolis (G6 and G12). Braz J Pharm Sci. 2012;**48**(3):557–564.

[2] Lima, B.; Tapia, A.; Luna, L.; Fabani, M. P.; Schmeda-Hirschmann, G.; Podio, N. S.; Wunderlin, D. A.; Feresin, G. E. MainFlavonoids, DPPH Activity and Metal 2009. Content allow determination of the geographical origin of propolis from the province of San Juan (Argentina). J Agric Food Chem. 2009;**57**(7):2691–2698.

[3] Lustosa, S. R.; Galindo, A. B.; Nunes, L. C. C.; Randau, K. P.; Rolim-Neto, P. J. Própolis: atualizações sobre a química e a farmacologia. Braz J Pharmacog. 2008;**18**(4):447–454.

[4] Tiemi Inoue, H.; De Sousa, E. A.; Oliveira Orsi, R.; Cunha Funari, S. R.; Carelli Barreto, L. M. R.; Da Silva Dib, A. P. Propolis production by differents methods. Arch. Latinoam. Prod. Anim. 2007;**15**(2):65–69.

[5] Toledo, V. A. A. Comparative study of biological and wax production parameters and royal jelly in Apismellifera bee colonies, Africanized, carnica, Italians and hybrids. College of Agrarian and Veterinarian Sciences - Universidade Estadual Paulista. 1997.

[6] Simões, C. C.; Araújo, D. B.; Araújo, R. P. C. Study, in vitro and ex vivo, of the action of different concentrations of propolis extracts against microorganisms present in human saliva. Braz J Pharmacog. 2008;**18**(4):549–556.

[7] Franchin, M.; Cunha, M. G.; Denny, C.; Napimoga, M. H.; Cunha, T. M.; Bueno-Silva, B.; et al. Bioactive fraction of geopropolis from meliponascutellaris decreases neutrophils migration in the inflammatory process: involvement of nitric oxide pathway. EvidBasedComplementAltern Med. 2013; **2013**:1-10.

[8] Perchyonok, V. T.; Zhang, S.; Grobler, S. R.; Oberholzer, T. G. Insights into and relative effect of chitosan H, chitosan H propolis, chitosan H propolisnystatin and chitosan H nystatin on dentine bond strength. Eur J Dent. 2013;7(4):412-418.

[9] Alencar, S. M.; Aguiar, C. L.; Guzmán, J. P.; Park, Y. K. Composição química de Baccharis dracunculifolia. Ciênc Rural. 2005;**35**(4):909–915.

[10] Cabral, I. S. R.; Oldoni, T. L. C.; Prado, A.; Bezerra, R. M. N.; Alencar, S. M.; Ikegaki, M.; Rosalen, P. L. Phenolic composition, antibacterial and antioxidant activities of brazilian red própolis.Quím Nova. 2009;**32**(6):1523 -1527.

[11] Cunha, L. C.; Alves, L. D. S.; Santana, L. C. L. R.; Nunes, G. B. L.; Neto, P. J. R. Propolis against trypanosomatids of medical importance: a therapeutical perspective for Chagas disease and Leishmaniose. Rev Patol Trop. 2011;**40**(2):105 -124.

[12] Cunha, I. B. S.; Salomão, K.; Shimizu, M.; Bankova, V. S.; Custódio, A. R.; Castro, S. L.; Marcucci, M. C. Antitrypanosomal activity of Brazilian propolis from Apismellifera. ChemPharm Bull (Tokyo). 2004;**52**(5):602-604.

[13] Pereira, A. S.; Seixas, F. R. M. S.; Aquino-Neto, F. R. Propolis: 100 yearsofresearchand future perspectives.Quím Nova. 2002;**25**:321 -326.

[14] Batista, M. C. A.; Abreu, B. V. B.; Dutra, R. P.; Cunha, M. S.; Amaral, F. M. M.; Torres, L. M. B.; Ribeiro, M. N. S. Chemical composition and antioxidant activity of geopropolis produced by Melipona fasciculata (Meliponinae) in flooded fields and cerrado areas of *Maranhão State*, northeastern Brazil. Acta Amaz. 2016;**46**(3):315–322.

[15] Araújo, K. S. S.; Júnior, J. F. S.; Sato, M. O.; Finco, F. D. B. A.; Soares, I. M.; Barbosa, R. S.; Alvim, T. C.; Ascêncio, S. D.; Mariano; S. M. B. Physicochemical properties and antioxidant capacity of propolis of stingless bees (Meliponinae) and Apis from two regions of Tocantins, Brazil. Acta Amaz. 2016;**46**(1):61–68.

[16] Thomas, A. B.; Nassur, R. C. M. R.; Boas, A. C. V.; Lima, L. C. O. Edible Cassava starch coating incorporated with propolis on bioactive compounds in strawberries. Sci Agrotechnol. 2016;**40**(1):87–96.

[17] De-Melo, A. A. M.; Matsuda, A. H.; Freitas, A. S.; Barth, O. M.; Almeida-Muradian, L. B. Antioxidant capacity of propolis. Trop Livest Res. 2014;**44**(3):341–348.

[18] Silva, E. C. C.; Muniz, M. P.; Nunomura, R. C. S.; Nunomura, S. M.; Zilse, G. A. C. Phenolic constituents and antioxidant activity of geopropolis of two species of Amazonic stingless bees. New Chem. 2013;**36**(5):628–633.

[19] Oliveira, P. S.; Müller, R. C. S.; Dantas, K. G. F.; Alves, C. N.; Vasconcelos, M. A. M.; Venturieri, G. C. Phenolic acids, flavonoids and antioxidant activity of honeys of *Melipona fasciculata, M. flavolineata* (Apidae, Meliponini) E Apis mellifera (Apidae, Apini) Amazon. New Chem. 2012;**35**(9):1728–1732.

[20] Gonçalves, G. M. S.; Srebernich, S. M.; Souza, J. A. M. Stability and sensory assessment of emulsions containing propolis extract and/or tocopheryl acetate. Braz J Pharm Sci. 2011;**47**(3):585–592.

[21] Cottica, S. M.; Sawaya, A. C. H. F.; Eberlin, M. N.; Franco, S. L.; Zeoulae, L. M.; Visentainera, J. V. Antioxidant activity and composition of propolis obtained by different methods of extraction. J Braz Chem. 2011;**22**(5):922–935.

[22] Neves, M. V. M.; Silva, T. M. S.; Lima, E. O.; Cunha, E. V. L.; Oliveira, E. J. Isoflavone formononetin from red propolis acts as a fungicide against Candida sp. Braz J Microbiol. 2016;**47**(2016):159–166.

[23] Oldoni, T. L. C.; Oliveira, S. C. ; Andolfatto, S.; Karling, M.; Calegari, M. A.; Sado, R. Y.; Maia, F. M. C.; Alencar, S. M.; Lima, V. A. Chemical characterization and optimization of the extraction process of bioactive compounds from propolis produced by selected bees Apis melífera. J Braz Chem Soc. 2015;**26**(10):2054–2062.

[24] Matos, V. R.; Alencar, S. M.; Santos, F. A. R. Pollen types and levels of total phenolic compounds in propolis produced by Apis mellifera L. (Apidae) in an area of the Semiarid Region of Bahia, Brazil. An Braz Acad Sci. 2014;**86**(1):407–418.

[25] Aguiar, S. C.; Paula, E. M.; Yoshimura, E. H.; Santos, W. B. R.; Machado, E.; Valero, M. V.; Santos, G. T.; Zeoula, L. M. Effects of phenolic compounds in propolis on digestive and ruminal parameters in dairy cows. R Bras Zootec. 2014;**43**(4):197–206.

[26] Sousa, J. M.; Souza, E. L.; Marques, G.; Meireles, B.; Cordeiro, Â. T. M.; Gullón, B.; Pintado, M. M.; Magnani, M. Polyphenolic profile and antioxidant and antibacterial activities of monofloral honeys produced by Meliponini in the Brazilian semiarid region. Food Res Int. 2016;**84**:61–68.

[27] Santos, N. W.; Yoshimura, E. H.; Machado, E.; Matumoto-Pintro, P. T.; Montanher, P. F.; Visentainer, J. V.; Santos, G. T.; Zeoula, L. M. Antioxidant effects of a propolis extract and vitamin E, in blood and milk of dairy cows fed with diet containing flaxseed oil. Livest Sci. 2016;**191**:132–138.

[28] Almeida, J. F.; Reis, A. S.; Heldt, L. F. S.; Pereira, D.; Bianchin, M.; Moura, C.; Plata-Oviedo, M. V.; Haminiuk, C. W. I.; Ribeiro, I. S.; Luz, C. F. P.; Carpes, S. T. Lyophilized pollen extract of bees: a natural antioxidant source to prevent lipid oxidation in refrigerated sausages. LWT – Food Sci Technol. 2016; **76**(B):299–305.

[29] Bueno-Costa, F. M.; Zambiazi, R. C.; Bohmer, B. W.; Chaves, F. C.; Silva, W. P.; Zanusso, J. T.; Dutra, I. Antibacterial and antioxidant activity of honeys from the state of Rio Grande do Sul, Brazil. LWT – Food Sci Technol. 2016;**65**:333–340.

[30] Daugsch, A.; Moraes, C. S.; Fort, P.; Pacheco, E.; Lima, I. B.; Abreu, J. Á.; Park, Y. K. Própolis Vermelha e sua origem botânica. APACAME Mensagem Doce. 2006;**89**:2–15.

[31] Park, Y. K.; Ikegaki, M.; Alencar, S. M.; Moura, F. F. Evaluation of Brazilian propolis by both physicochemical methods and biological activity. Honey Bee Sci. 2000;**21**(2):85–90.

[32] Bankova, V. Chemical diversity of propolis and the problem of standardization. J Ethnopharmacol. 2005;**100**(1–2):114–117.

2

Bee Products as Functional Food

Banu Yucel, Erkan Topal and Mustafa Kosoglu

Abstract

The studies that reveal the impact of the bee products on overall health are accompanied by new researches every year, and the importance of these researches are gradually on the rise. Bee products that are used as food and food supplements and drug concentrations in the historic process are drawing the attention with their marvellous characteristic features. The search for nourishment of the body on behalf of healthy living is currently being searched by many people. Therefore, the consumption of products that protect the health appears as the primary preference of people. In the light of this recent tendency, food sector is now offering well-supported products that are suitable for this preference. At this point, bee products such as honey, pollen, bee bread, royal jelly and propolis gain importance as functional food with their nutritious features that help in protecting the health. In this article, within the consideration of the researches that evaluate bee products as functional food, we aim to introduce the prominence of bee products in our nourishment and overall health.

Keywords: Bee products, functional food, health, prebiotic, probiotic

1. Introduction

At the present time, alterations in food production and consumption occur constantly. Standardized food production gradually gives its place to mass tailor-made production. Throughout this process, consumer needs also change and progress. In every aspect of life, the search for comfort increases its importance. Functional foods come to the light as a result of the search for comfort to fulfil certain needs. Functional food products become prominent within the several product groups particularly due to their direct effect on human health [1].

Functional foods are a highly important sector of the food market that grow rapidly. Within this concept, the interest towards the pro- and prebiotic foods increases by degrees. Foods that beyond their basic nutritional features make positive contribution to our health are named as 'functional foods'. The term functional food underlines the positive relation of nutrition with health [2–5]. The International Food Information Council (IFIC) defines functional foods as 'foods and food components that may provide benefits beyond basic nutrition'. According to the International Life Science Institute (ILSI), the concept of functional foods refers to the 'biologically active components in foods that have the potential to optimize physical and mental well being' [6–13].

The most important target groups of the functional foods are women and elders. The new conception assimilates Hippocrates, the founder of the medicine, nearly 2500 years old philosophy 'Let food be thy medicine and medicine be thy food', and therefore pays more attention to the healthy nutrition. Nowadays, due to reasons such as the increase in the treatment expenses, labour loss, length of life and the portion of the old people inside the population and people's desire of living a quality life, people start to expect more from the food items [14].

According to the data of the research that is made to reveal the consumers' awareness, acceptance and approach towards the functional foods, mineral water, whole-grain diet biscuits and whole-grain breakfast cereals are the most frequently consumed functional foods in the order. The functional food increases the level of the beneficial bacteria in the gut, helps lose weight and supports child development. These three features of the functional food are emphasized by customers as 'the top 3 factors that convince them to consume functional foods' in the research [15].

Education level has advanced hand in hand with the economic development; therefore, people's demand to healthy foods has increased alongside with these progressions. The concept of functional food occurred for the first time in Japan in the 1980s. Functional foods are put on display for customers as an active food trend at the beginning of the 1990s in America and in the middle of 1990s in the countries of Europe.

According to the Foods for Specified Health (FOSHU) affirmed in Japan in 1991, it is stipulated that in order for a product to get functional food licence it is required to meet with certain criteria. Some of these criteria are stated in relation to bee products. Functional foods should help the improvement of the quality of nourishment, and protection and preservation of health, the compound of the product should not differ from the similar type of foods' nutritional substance components within normal conditions, the product should be a well-used food in daily diets, as all bee products have [16–19].

2. Bee products as functional food

Bee products are accepted as 'functional food' by adding them in other food products to increase their nutritional value or used alone with its natural and rich nutritional content and high bioactive components. Honey, pollen, bee bread, royal jelly and propolis have high nutritional value and beneficial effects that affect the human health in a positive manner. Bee products are rich in proteins, simple sugars, essential amino acids and monounsaturated fatty acids. These features strengthen the immunity, help the body fight actively with bacteria and stimulate the quality tissue regeneration, and consequently protect overall body health and treat it [20] (see **Figure 1**).

Figure 1. Honey bee on combs and bee products on sale *(photographed by M. Kosoglu and E. Topal).*

Phenolic compounds	Honey (μg/100 g)	Pollen (μg/100 g)	Bee bread (μg/100 g)
Gallic acid	3.05	3.68	6.17
Protocatechuic acid	7.08	N.D.	91.25
Chlorogenic acid	N.D.	N.D.	N.D.
p-OH benzoic acid	8.98	N.D.	N.D.
Caffeic acid	79.90	N.D.	N.D.
Syringic acid	38.17	4.80	N.D.
Catechin	N.D.	N.D.	N.D.
Vanillic acid	41.40	5.30	N.D.
Epicatechin	N.D.	N.D.	N.D.
Ferulic acid	N.D.	6.62	28.38
Ellagic acid	N.D.	N.D.	N.D.
Rutin	56.66	50.80	217.20
P-coumaric acid	N.D.	N.D.	N.D.
O-coumaric acid	N.D.	3.25	N.D.
Quercetin	247.40	N.D.	N.D.
T-cinnamic acid	22.26	27.08	586.40
Apigenin	N.D.	N.D.	N.D.
Kaempferol	67.10	179.53	623.50

Table 1. Phenolic compounds of some bee products.

Bee products include many biochemical components that can be found on almost all functional foods. Proteins, saccharides, fatty acids, prebiotics, probiotics, fibre, phytochemicals, bioactive peptides, mineral matters, vitamins and organic acids can be shown as examples to these components. Additionally, phenolic acids, flavonoids and carotenoids that are found in the structure of bee products affect diseases such as cancer, arteriosclerosis, cardiovascular diseases, weakening of the immune system, Parkinson's disease, Alzheimer, arthritis and photoageing dramatically as preventive and curative matters [21]. Especially flavonoids inhibit the efficacy of the enzyme systems that include lipid peroxidation, platelet aggregation, capillary permeability and lipoxygenase [22].

Bee products' functional food features do not remain limited with these contents. Phenolic components in its structure make positive impact on body's immune system in a crucial level [23]. Even in bee products which are procured by monofloral production, it is seen in **Table 1** that the phenolic component amount shows an alteration [22, 24] (**Table 1**).

The functional food feature and biological effect of bee products are fairly high [25, 26] (**Table 2**).

Product	Biological effect	Functional effects
Honey, pollen, bee bread, royal jelly, propolis	Antibacterial, fungicide, antiviral, antioxidative, immunomodulating and immunoactivating, anti-inflammatory, analgesic	Growth inhibition of pathogen bacteria, fungi, and viruses, anticancer, stimulate immune difference against inflammation
Pollen, royal jelly, propolis	Radioprotective, anti-arteriosclerotic, enhances Ca absorption	Protect against radiation, artherosclerosis and osteoporosis
Honey	Prebiotic (oligosaccharides), probiotic (contains probiotic bacteria)	Stimulates healthy digestion by promoting the growth of good intestine bacteria (*Bifidus*, etc.)
Royal jelly	Antihypertensive, vasodilative, increases reproduction and oxygen uptake of cells and has an effect on central and peripheral nervous system	Cardioprotective, stimulating and energizing against stress and fatigue, protection of the central nerve system

Table 2. Biological and functional effects of bee products [25, 26].

2.1. Honey

Honey is a bee prodomponents in its structure. Honey mainly consists of water and sugar that is formed from fructose and glucose. It includes ascorbic acid, pantothenic acid, niacin vitamins and additionally mineral substances such as manganese, phosphor, potassium and zinc [27] (see **Figure 2**).

Honey is a functional food that contains inulin and fructo-oligosaccharide (FOS) and it has prebiotic features that have beneficial effects to our gastrointestinal system. As widely known, prebiotics are undigested carbohydrates which increase the activity and number of both column bacteria and probiotics (living organisms that regulate the intestinal microbial balance). According to the identified data, prebiotics stimulate the immune system and help the inhibition of carcinogenesis inside the column [28, 29]. As a result of prebiotics' fermentation in large intestine, substances such as lactate, short-chain fatty acids (acetic, butyric and propionic acids), hydrogen gas, carbon dioxide and methane are produced. The intestine's pH drops significantly. The drop in the intestinal has many benefits such as inhibition of microorganisms that can potentially have detrimental effects, degradation of secondary bile acids, increment of the solubility and absorption of minerals such as Ca, Mg, Fe and Zn [30]. Undigested carbohydrates that are known as prebiotics include lactulose, lactitol, oligosaccharides and inulin. Undigested oligosaccharides are found naturally in fruits, vegetables, milk and honey [31].

Figure 2. Honey (*photographed by E. Topal*).

Inulin is a term that is used for heterogeneous mixture of fructose polymers that are found commonly in the form of stored carbohydrates in nature and has a polymerization degree that alters between 2 and 60°. The units that have lower polymerization degree (2–20) are called fructo-oligosaccharide or oligofructose [32]. The consumption of 4–10 g of fructo-oligosaccharides in daily basis creates bifidogenic effect, thus increasing the number of beneficial bacteria [33].

Honey is not only a functional food but also has many positive impacts on human health. Due to these impacts, it is used countless times in folkloric medicine and its usage as a drug put on the records centuries ago for us to read the crucial role of the honey in our life [34]. In addition to that, honey comes forward as an important antioxidant source due to its vitamin C, flavonoids and phenolic content [35, 36]. Honey's strong biological activity originates from its rich phenolic content. Flavonoids stimulate antibacterial, antiviral, anti-inflammatory and vasodilator effect [21, 22]. Flavonoids and phenolic acids form the most common group of plant phenolics [37]. Phenolic acids and flavonoids act as free radical scavengers due to their reducing agents besides their antioxidant and anticarcinogenic effects [38–40]. Flavonoids increase the mucosal content of prostaglandins and have an important inhibitory effect on the gastric mucosa, thus preventing ulceration. Mucosal content of prostaglandins has an important inhibitory effect on acid secretions, preventing the formation of peptic ulcers [22].

Antibacterial feature of honey that comes from its osmotic effect is an outcome of the sugar molecules in the setting which drain the water around and leave so little to the microorganisms that cause the bacteria in the honey to die from dehydration. Alongside with this osmotic effect, hydrogen peroxide that is constituted from glucose oxidase in honey creates the same level of antibacterial activity. By the time that hydrogen peroxide decomposes, it produces highly effective free radicals and consequently these situation cause bacteria to die. Due to this feature, it can be understood that honey defends the body successfully against the destructive effect caused by oxidative stress [34, 41].

In a research aimed at studying the long-time storage of the honey as a functional food, honey and pollen are fermented by probiotic lactobacillus and conserved at 40°C for 7 days to examine the prebiotic cells' liveliness. In consequence of the examination, it is found out that prebiotic stains preserve their liveliness at a significant level [42].

Honey's role as a protector of the overall health and creator of an antioxidant effect makes it useful in the food sector for long ages. The sugar known as 'Maillard Reaction' forms a composition of aldehyde-ketone and amine interaction that has an antioxidant feature. In other words, honey diminishes the microbial activity that starts after cutting the meat which gets harmful for the health [43]. Furthermore, the usage of the honey in chicken meat makes its water-holding capacity to increase, prevents boiling of loss, balances pH, gives a light colour to the meat by Maillard reaction and provides richness in flavour [44–46]. Correlative to its fructose content, honey enhances the flavour of the meat and increases its nutritional value due to its 'functional food' feature [47].

The antioxidant effect of honey can be evaluated with the lipid peroxidation system. Honey stops the bacterial growth in dry-aged and wet-aged meats and even acts like a bactericide when it gets together with propolis. The strong antioxidative and antibacterial activities of honey clean the active oxygen that cause meat to go bad [48]. Additionally, it is found out that in chicken meats that have been marinated with honey, the formation of heterocyclic amine that occurs as a result of Maillard reaction after cooking drops at a rate of 92–99%. As a reason behind this drop, it is submitted with the acidic structure of honey that prevents the occurrence of amine [49]. Phenolic and flavonoid compounds inside the honey's composition increase the capacity of the antioxidants and prevent the activation of hydrogen peroxide [48]. According to a conducted research, chicken meats that have been marinated with different portions of honey significantly prevent bacteria growth by comparison with the group that has not been treated with the honey whatsoever [46].

2.2. Pollen

Pollen is a bee product, which serves as an important food supplement thanks to its content that not only includes rich food substances as a male reproductive unit of the flower but also consists of salivary juices of honeybee. It is accepted as a 'natural drug concentration' due to its content that consists of enzymes, coenzymes, steroids, vitamins, antibiotics, microelements, carotenoids and flavonoids. Pollen's strong antioxidant feature originates from high level of polyphenols in its content. Due to this feature, it is considered as a functional food/food supplement [50]. Generally obtained from bees from various plants, pollen is accepted as 'the most distinguished food in the World' [51]. The composition and chemistry of pollen is not standardized. The type and characteristic of pollen alters in accordance with the kind of the plant that it is obtained from, the season, climate, environmental aspects and the age of the plant. The alterations, which depend on the structure of the plant that the pollen is acquired from, change the compounds of pollen and the level of impact that it has on health. For instance, meanwhile a willow's pollen is rich in vitamin C, a clover's and willow's pollens are rich in flavone matter. This content protects the body from sclerosis, spasmolysis and radioactive effect. Due to willow's and cherry's pollens' chlorogenic acid content, they protect vascular health, act as an anti-inflammatory and regulate thyroid and pituitary glands' secreted hormones. In addition to the effects of the pollens acquired from one plant type only, pollens collected from thousands of various types of flowers in the nature have exceptionally beneficial effects. For that reason, it is stated that the consumption of mixed pollens has positive impact on health [41] (see **Figure 3**).

Figure 3. Pollen (*photographed by M. Kosoglu*).

Pollen's important antioxidant level is mostly connected with the phenol compounds [52]. In the conducted research made in Brazil, in 25 examples which are collected on certain days for 9 months, isoquercetin, myricetin, quercetin, luteolin, selagin, kaempferol and isorhamnetin substances are identified. Even though the flora that the bee worked on has changed, it is noted that there were not excessive differences in phenolic composition level [53].

Pollen is accepted as a dense nutraceutical substance based on its phenolic compounds and high antioxidant activity. It regulates the blood flow in lecithin vessel. Pollens stimulate the production of polyphenols and biocides in liver and clean the liver by decreasing the oil level. At the same time, pollen is a food supplement that is extremely rich in selenium which plays a crucial role in protecting the cardiovascular health. The consumption of the pollen, which is considered as both 'functional and super food', is recommended for 15–20 g a day, two times a year in weekly cures for 3 weeks [54].

According to the results of the research made to examine the effects of physico-chemical and microbiological features of yogurts produced by honey, pollen and probiotic yeast, it is identified that there are significant increases in the amounts of oil and protein in yogurt due to the pollen addition [5]. Another study revealed that the probiotic yogurt produced by adding pollen has a positive impact on both physico-chemical and sensorial features [55]. Unfortunately, the positive impact of the addition of pollen in the yogurt on senses of taste and smell is revealed to be not the same in the case of adding the pollen in milk. It is reported that even though the addition of pollen in milk does increase the probiotic feature, sensorial features are effected negatively [56].

A conducted research presented that in order to extend the shelf life of meats, pollen can be added to the mincemeat as an antioxidant and antimicrobial agent. It is revealed that the lipid oxidation and microbial growth are prevented by pollen supplementation [57].

Adding pollens inside the whole-grain biscuits is a common way of consuming pollens. It is stated that adding an amount of 5% of pollen as a food supplement inside the biscuits does not affect the oil amount but does affect the amounts of sugar and protein at a significant level [58].

2.3. Bee bread

Bees mix pollen with honey and their digestive enzymes and then the whole composition will be fermented by the lactic acid. The final state of this mixture is called 'bee bread' which is an extremely valuable product. In apiculture, this valuable product can be put onto the market as a nutraceutical human food/food supplement. Bee bread includes a significant amount of proteins, vitamins and one of the natural antioxidants, the phenolic compounds. The flora that is found at the region where colonies are present has a significant impact on the compounds of the bee bread. Honey bees ferment hidden pollens inside the comb with various enzymes and the addition of honey then stores it as bee bread. This implementation provides a more advantageous storage against the risk of nutrition loss of the dried and frozen pollen [20] (see **Figure 4**).

The positive effect it has on health and its content that is full of rich polyphenols make bee bread medically the focus of interest [59]. In a research, which has examined the bioactive features of cherry pollens and bee bread, it is found out that these bee products have high antioxidant and anti-inflammatory capacities with their content, including 1371 mg/100 and 1428 mg/100 g high phenolic compounds [24]. A converted research made in Colombia to analyse the bee bread, flavonoid and phenolic matters' content are found 3.2 ± 1.0 mg (quercetin/g) and 8.9 ± 3.1 mg (galia acid/g.), respectively. Additionally, ferric-reducing antioxidant power (FRAP) and 2,2′-azino-bis(3-ethylbenzothiazoline-6-sulphonic acid) (ABTS) antioxidant activities are reported at a level of 46.1 ± 13.0 versus 61.5 ± 10.2 mmol (trolox/g), respectively, and according to the data bee bread can be digested with more ease and biologically used at a higher level than pollen. This case shows the fact that bee bread has a higher nutritional level and more amount of usable bioactive compounds than pollen. According to the obtained results, as a functional food supplement bee bread is a product that has a high level of utilization potential [60].

Figure 4. Bee bread (*photographed by M. Kosoglu*).

2.4. Royal jelly

It has different colours changing from cream to dark yellow, an acidic structure and an astringent taste. Royal jelly includes all the food substances necessary for the larva's growth. It is quiet rich in proteins, oils, sugars, hormones, vitamins and mineral matters. Chemically, royal jelly comprises water (50–60%), proteins (18%), carbohydrates (15%), lipids (3–6%), mineral salts (1.5%) and vitamins together with a large number of bioactive substances such as 10-hydroxyl-2-decenoic acid with immunomodulating properties, antibacterial protein, fatty acids and peptides. The royal jelly also demonstrated significantly improved recovery from 5-fluorouracil-induced damage [61, 22]. The quality of the royal jelly is evaluated by 10-hydroxy-2-decenoic acid (10-HDA) level in its content and the level of the matter is expected to be found among 1.4 and 1.8% levels. This value can differ due to the vegetation of the region where royal jelly is obtained from and the implementation techniques in harvesting royal jelly. Unlike honey, the mineral matter composition in royal jelly is not affected by the topography nor the flora [62] (see **Figure 5**).

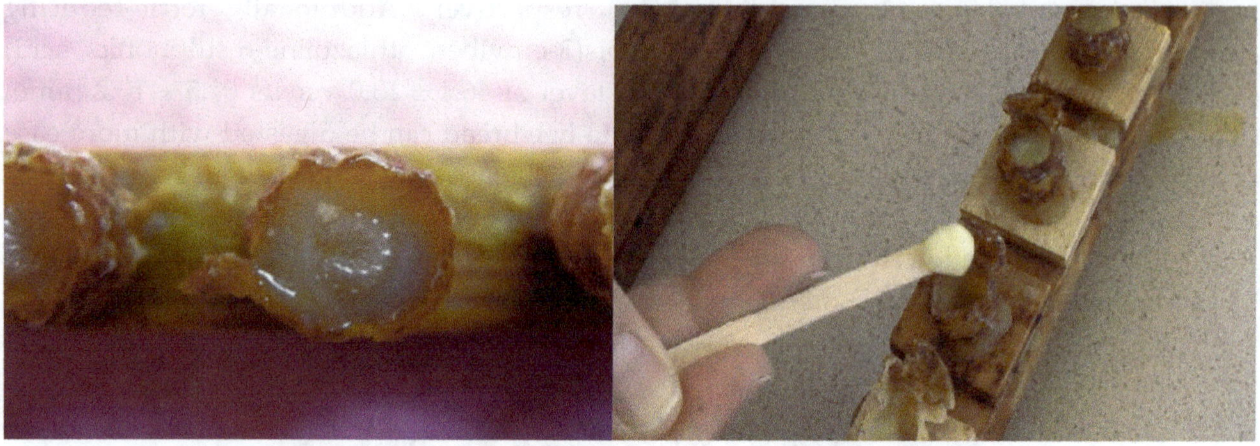

Figure 5. Royal jelly in cells and harvesting royal jelly (*photographed by M. Kosoglu*).

Royal jelly is a very complex bee product in terms of the content of compounds. Components in its compound target various biological functions both known and unknown, and therefore they play a crucial role on the biomedical effect of the royal jelly [63].

Royal jelly stimulates cell renewal, production and metabolism and due to these effects it creates liveliness, health, energy and high immunity and vigour in all the tissues of the organism. Its content is rich in natural hormones, vitamins, essential fatty acids, amino acids, sterols, phosphor compounds and acetylcholines. Acetylcholine is effected in the transmission of the neurons' messages and regulation of the work of endocrine glands. Being rich in building blocks of life, royal jelly's rich content of the gelatinous amino acid, which is the fundamental component of collagen and therefore the royal jelly, has an anti-ageing effect. The gammaglobulin in the content of the royal jelly effects the body's capacity to fight with the infections and acts as a strengthening factor to the immune system. With its decanoic acid content, royal jelly has a strong antibiotic effect to many bacteria and fungus. Due to its fight with tumours and metastasis, it is reported that royal jelly is used in oncology, psychiatry and neurology, geriatrics, bone and cartilage tissue regeneration, vessel stiffness, growth and development. It is actively used to stop physical exhaustion. Royal jelly not only lowers the cholesterol, total lipid, phospholipid, triglyceride and b-lipoprotein levels in blood but also lowers the overall cholesterol and acts as a vasodilator. Because of the fact that it consists of inulin-like peptides, royal jelly shows hypoglycaemic and immunological effects. In the case of feeding premature and undernourished babies with 8–100-mg royal jelly, it is reported that there were significant improvement on their overall health condition, weight gain, increase on the levels of red blood cells and haemoglobin [41–65]. In addition to all these features, the significant impact of the consumption of royal jelly has on sperms and its mobility and the positive results obtained from the infertility treatment when consumed regularly are reported [66].

2.5. Propolis

Propolis is a resinous substance that bees collect from the exudates of plants and which they use to seal holes in the bee hive [67]. Propolis, too, forms part of traditional medicine, and chemical analysis has pointed to the presence of at least 300 compounds in its composition. It is mainly composed of resin (50%), wax (30%), essential oils (10%), pollen (5%) and other organic compounds (5%). Among these organic compounds, we may find phenolic compounds and esters, flavonoids in all their forms (flavonoles, flavones, flavonones, dihydroflavonoles and chalcones), terpenes, beta-steroids, aromatic aldehydes and alcohols, sesquiterpenes and stilbene terpenes. Caffeic acid phenethyl ester (CAPE) is a biologically active ingredient of propolis with several interesting biological properties, including apoptosis, metastasis and radiation sensitivity of cancer cells [68–71] (see **Figure 6**).

Figure 6. Raw propolis and honey bee (*photographed by M. Kosoglu*).

The impact that propolis has on cancer bases on especially its regulative and strengthening role on the immune system. Based on this strong effect of propolis, whether propolis shows cytotoxic and apoptotic features in acute lymphoblastic leukaemia cell lines or not is searched, and found out that due to its caffeic acid and phenethyl ester content, propolis shows a significant level of cytotoxic effect and stops tumour growth [72]. According to the results of the research which searches especially for the inhibitory effects of propolis on osteogenic sarcoma cell growth, due to the anticarcinogenic effect of the apoptosis mechanism induced with the propolis extract, it is reported that propolis can be beneficial in cancer treatment [73]. Along with many polyphenols and antimetastatic effects propolis has, it is been observed that the compounds CAPE obtained from populus' propolis and Artepillin C obtained from *Baccharis propolis* show antitumoural effects [74].

The usage of propolis is recommended by patients who take radiotherapy during their cancer treatment. Propolis acts almost like a protective barrier and prevents the radioactive ray from affecting the healthy cells and makes it effective only to the distorted tissues. By this way, the level of radiation entering inside the patient's body decreases. Additionally, due to the strengthening feature of propolis on the body's immune system, it supports patient's body functions and therefore increases body's resistance. Propolis should be implemented as a complementary supportive care that is accompanied by conventional surgery, chemotherapy and radiotherapy. It is crucial to use it with an ethical and responsible approach and in right doses and for enough time [75].

Propolis and its derivates have the capacity to inhibit virus propagation. Several *in vitro* studies have shown the effect of propolis on the DNA and RNA of different viruses, among them Herpes simplex type 1, Herpes simplex type 2, adenovirus type 2, vesicular stomatitis virus and poliovirus type 2. The effects observed involve a reduction in viral multiplication and even a virucidal action [22].

The antiseptic, antibiotic, antibacterial, antifungal and antiviral features of propolis come from its galangin, caffeic and ferulic acid content. In converted researches, propolis is used in branches and diseases such as dentistry, oto-rhinolaryngology, ophthalmology, gynaecology, dermatology, digestive system diseases and pulmonary diseases besides its cancer and radiation treatment [76].

The antioxidant, antimicrobial and antifungus effects of propolis are benefited by food technology. In a converted study, 0.02 and 0.4% of ethanolic propolis extract (EEP) and 0.28% of potassium sorbate are added in meat products that are mixed with stored fats and it is stated that meat products treated with propolis have a longer shelf life than the ones mixed with potassium sorbate [77]. Similar studies determine the fact that propolis has a significant level of antibacterial effect and propolis extracted in 0.3% ethyl alcohol extract and in water can be used successfully in meat products as a natural coating compound [78, 79]. In these studies, propolis forms a surface such as a film badge during the marination and its prevention of the microbial activity is remarked. Additionally, the utilization of propolis decreases the level of the total volatile nitrogen content (TVB-N) and thiobarbituric acid (TBA) that affects the processed meats badly. Due to these features of propolis, it is recommended for use as a decent preservative in processed and unprocessed meats [77].

In a research that has specified the physico-chemical and microbiologic alterations in meat's structure, as a result of treating the beef hamburger meats with propolis ethanol extract (PEE) and freezing them for 8 months at −18°C, it is observed that propolis extends the shelf life without affecting the quality of meat negatively [80].

The impact of different propolis concentrations (250, 500 and 1000 ppm, respectively) on the microbial load on the surface of the Roumy cheese is searched and it is found that propolis inside the 1000 pm concentration completely prevents the production of mould and stigmatosis in the cheese; therefore, the utility of propolis in foods as an economic and natural preservative is indicated as a result of the research [81].

In a research examining the bactericidal and bacteriostatic activity against the *Escherichia coli bacteria in vitro*, 20% EEP is recommended to be evaluated as a natural food preservative due to its neutralization of *E. coli bacteria* [82]. In a similar way, the antibacterial effect of the propolis on some of the fermentative bacteria in yogurt is examined and the inhibitory impact of the propolis even at low concentration on the normal bacteria growth is stated [83].

In a research that has identified the antifungal effect of the propolis extract on the six types of yeasts which cause unpasteurized fruit juices to go bad, it is revealed that propolis can be used as an alternative to the chemical preservatives in fruit juices due to its antibacterial effect [84].

3. Conclusion

The improving technology and changing living conditions alter human life and nutritional habits, and as a conclusion the thought of preventing diseases by being nourished with functional and super foods enriched with natural preservatives and substances that protect the health become extremely popular. More studies are needed on the contribution of the magnificent bee products have on our health and nourishment in order to create more awareness and influence the production of qualified bee products and increase the conscious selection and consumption of the bee products made by consumers. Making multidisciplinary scientific researches about the protective effects of the bee products as functional foods that meet with the common trend of 'not getting sick' will significantly contribute to spreading the results of the researches all around the world and improving the sector.

Acknowledgements

The authors are thankful to Ms. Irem Balcioglu for her assistance of translation during the preparation of the chapter.

Author details

Banu Yucel[1*], Erkan Topal[2] and Mustafa Kosoglu[2]

*Address all correspondence to: banu.yucel@ege.edu.tr

1 Ege University, Faculty of Agriculture, Department of Animal Science, Bornova, Izmir, Turkey

2 Aegean Agricultural Research Institute, Beekeeping Unit, Menemen, Izmir, Turkey

References

[1] Ozdemir PO, Fettahlioglu S, Topoyan, M: A Study on Determining the Consumer Attitudes Towards Functional Food Products. Ege Academic Review, 2009; 9(4):26 1079–1099.

[2] Boyacioglu D: Functional Food interview. Available from: http://www.google.com.tr/url?sa=t&rct=j&q=&esrc=s&source=web&cd=1&ved=0ahUKE-wip56_F4Z_PAhWC1xQKHfLdA24QFggbMAA&url=http%3A%2F%2Fwww.dilek-boyacioglu.com %2FFonksiyonel_Gidalar_Roportaj.pdf&usg=AFQjCNE4ENohk9uHtRsTpEDXkyK-DY00yJQ&sig2=wftbo3RUzhh0noaDcwFtgg.

[3] Giannetti V, Testani E, Recchia, L: Food consumption and innovation: Functional foods. *Journal of Commodity Sciences, Technology and Quality*, 2009; 48: 213–225.

[4] Kıyak SN, Dağlı Y, Zeren Ü, Arıburnu M, Gülbandılar A, Dönmez M, Okur M: A functional food: "Şifalı Top". *Turkish Journal of Agriculture-Food Science and Technology*, 2014; *2*(6): 277–279.

[5] Chinelate GCB, Bernardino AB, Barros HV, Dos Santos EJN, Pontes DF: Probiotical yogurt honey flavor added bee pollen as a technological innovation. Third International Conference and Exhibition on Probiotics, Functional & Baby Foods. September. Naples, Italy: 2014; 2–2. pp. 23–25.

[6] Betz JM: Government Perspective on Nutraceuticals and Functional Foods, Texas AXM University, TX. 1999.

[7] ADA (Reports): Position of the American Dietetic Association: functional foods. *Journal of the American Dietetic Association*, 2004; *104*(5): 814–826.

[8] İşleroglu H, Yıldırım Z, Yıldırım M: Flaxseed as A Functional Food. Journal of Gaziosmanpasa University Agriculture Faculty, 2005; 22(2), 23–30.

[9] Sevilmis G: Upward Trend: Functional Foods. AR&GE BÜLTEN. HAZİRAN – SEKTÖREL. 2013; pp. 39–46. Available from:http://www.izto.org.tr/portals/0/argebul-19 ten/gozde_fonksiyonelgida.pdf

[10] Korhonen H: Technology options for new nutritional concept. *International Journal of Dairy Technology*, 2002; *55*(2): 79–87.

[11] Marriott BM: Functional foods: an ecologic perspective. *American Journal of Clinical Nutrition*, 2000; *71*(suppl): 1728–1734.

[12] Açıkgöz Z, Önenç SS: Functional egg production. *Hayvansal Üreti m*, 2016; *47*(1): 36–46.

[13] Misir GB: Some Sources of Marine Functional Foods and Food Ingredients. Yunus Research Bulletin, 2012; (1): 1–7.

[14] Dolekoglu CO, Sahin A, Giray FH: Factor Influencing the Consumption of Functional Food in Women: A Study in the Mediterranean Region, 2015; 21(4): 572–584.

[15] Hacioglu G, Kurt G: Consumers' Awareness, Acceptance and Attitudes Towards Functional Foods: A Research in Izmır City. Business and Economics Research Journal, 2012; 3(1): 161–171.

[16] Farr, DR: Functional foods. *Cancer Letters*, 1997; *114*: 59–63.

[17] Kwak NS, Jukes DJ: Functional foods. Part 1: the development of regulatory concept. *Food Control*, 2001; *12*: 99–107.

[18] Coskun, T: Health Benefits of Functional Foods, Journal of Hacettepe University Pediatric Health & Diseases, 2005; 48: 69–84.

[19] Guven A, Gulmez M: Functional Foods and Their Relationship with Health. Journal of Kafkas University Veterinary Faculty. 2006; 12(1): 91–96

[20] Bobis O, Marghitas lA, Dezmirean D, Morar O, Bonta V, Chirila F: Quality Parameters and Nutritional Value of Different Commercial Bee Products. *Bulletin of University of Agricultural Sciences and Veterinary Medicine Cluj-Napoca. Animal Science and Biotechnologies*, 2010; *67*:2p.

[21] Dezmirean GI, Mărghitaş LA, Dezmirean DS: Honey like component of functional food. *Scientific Papers Animal Science and Biotechnologies*, 2011; *44*(2): 406–411.

[22] Viuda-Martos M, Ruiz-Navajas Y, Fernández-López J, Pérez-Álvarez JA: Functional properties of honey, propolis, and royal jelly. *Journal of Food Science*, 2008; *73*(9): R117–R124.

[23] Küçük M, Kolaylı S, Karaoglu S, Ulusoy E, Baltaci C, Candan F: Biological activities and chemical composition of three honeys of different types from Anatolia. *Food Chemistry*, 2007; *100*: 526–534.

[24] Doğaroğlu M, Kolaylı S, Yücel B, Yıldız O, Hotaman H, Topal E, Kösoğlu M. Bioactive qualifications and characteristics of cherry honey, pollen and bee bread. 44th International Apicultural Congress, September 15–20, Daejeon-Korea: 2015; p. 456.

[25] Bogdanov, S: Functional and biological properties of the bee products. 2009. Available from: http://www.apimondia.com/congresses/2009/Apitherapy/Symposia/Functional%20and%20Biological%20Properties%20of%20Bee%20Products%20%20BOGDANOV%20Stefan.pdf. [Accessed: 24.05.2016].

[26] Bogdanov, S: Functional and biological properties of the bee products: a review. *Bee Product Science*, 2011;12p.

[27] Vallianou NG, Gounari P, Skourtis A, Panagos J, Kazazis C: Honey and its anti-inflammatory, anti-bacterial and anti-oxidant properties. *General Medicine (Los Angel)*, 2014; *2*: 132. doi: 10.4172/2327-5146.1000132

[28] Roberfroid, M. Fructooligosaccharides. *Critical Reviews in Food Science and Nutrition*, 1999; *39*(3): 267–274.

[29] Inanc N, Sahin H, Cicek B: The Impact of Probiotics and Prebiotics on Health. Erciyes Medical Journal, 2005; 27(3): 122–127.

[30] Lidestri M, Agosti M, Marini A, Boehm G: Oligosaccharides might stimulate calcium absorption in formula-fed preterm infants. *Acta Paediatrica Supplement*, 2003; *91*: 91–92.

[31] Tannock GW: Probiotics: A Critical Review. Norfolk, England. Horizon scientific press. 1999; pp. 1–3.

[32] Ninesse K: Inulin and oligofructose: what are they? *Journal of Nutrition,* 1999; *129*(suppl): 1402–1406.

[33] Green C: Fibre in Enteral Nutrition, Nutrica Research Communications, Netherlands, June 1997; pp. 28.

[34] Yucel B, Akçiçek E: Using Honey in Modern Medicine. Hasad(Food), 2005; 244: 22–25

[35] Kolankaya D:Antioxidant Effect and Honey. Mellifera, 2001; 1: 13–17

[36] Zamudio F, Kujawska M, Hilgert NI: Honey as medicinal and food resource. comparison between Polish and multiethnic settlements of the Atlantic forest, Misiones, Argentina. *Open Complementary Medicine Journal,* 2010; *2,* 58–73.

[37] Manthey J, Grohmann K: Phenols in citrus peel by-products. Concentrations of hydroxycinnamates and polymethoxylates flavones in citrus peel molasses. *Journal of Agricultural and Food Chemistry,* 2001; *49*: 3268–3273.

[38] Frankel S, Robinson GE, Berenbaum MR: Antioxidant capacity and correlated characteristics of 14 unifloral honeys. *Apicultural Research,* 1998; *37*(1): 27–31.

[39] Gheldof N, Engeseth NJ: Antioxidant capacity of honeys from various floral sources based on the determination of oxygen radical absorbance capacity and inhibition of in vitro lipoprotein oxidation in human serum samples. *Journal of Agricultural and Food Chemistry,* 2002; *50*(10), 3050–3055.

[40] Ghasemzadeh A, Ghasemzadeh N: Flavonoids and phenolic acids: role and biochemical activity in plants and human. *Journal of Medicinal Plants Research,* 2011; *5*(31): 6697–6703.

[41] Doğaroğlu M: Story of Honey from Flower to Table. Yapı Kredi Yayınları, Mas Matbaacılık, İstanbul. ISBN: 978-975-08-1323-8. 2007. pp. 207.

[42] Barbu V, Ploscutanu G: Researches regarding stability of the functional product honeylact during storage. *Annals of Food Science and Technology,* 2014; *15*(2): 346–352.

[43] Dawson PL, Mathew S: Honey ' s antioxidative properties, maillard antioxidative effects and flavour compounds produced by the addition of honey in cooked poultry products. Honey Hotline Fact Sheet. 1994;4p.

[44] Yang CC, Chen TC: Effects of refrigerated storage, pH adjustment and marinade on color of raw and microwave cooked chicken meat. *Poultry Science,* 1993; *72*: 355–362.

[45] Honikel KO, Hamm R: Measurement of water-holding capacity and juiciness. Chapter: Quality Attributes and tehir Measurement in Meat, Poultry and Fish Products. Advances in Meat Research, 1994;9:125–161.

[46] Yücel B, Önenç A, Bayraktar H, Açıkgöz Z, Altan Ö: Effect of honey treatment on some quality characteristics of broiler breast meat. *Journal of Applied Animal Research*, 2005; *28*: 53–56.

[47] Hashim IB, McWatters KH, Hung YC: Marination method and honey level affect physical and sensory characteristics of roasted chicken. *Journal of Food Science*, 1994; *64*(1): 163–166.

[48] Nagai T, Inoe R, Kanamori N, Suzuki N, Nagashima T: Characterization of honey from different floral sources. Its functional properties and effects of honey species on storage of meat. *Food Chemistry*, 2006; *97*(2): 256–262.

[49] Salmon CP, Knize MG, Felton JS: Effects of marinating on heterocyclic amine carcinogen formation in grilled chicken. *Food and Chemical Toxicology*, 1997; *35*: 433–441.

[50] Kroyer G, Hegedus N: Evaluation of bioactive properties of pollen extracts as functional dietary food supplement. *Innovative Food Science & Emerging Technologies*, 2001; *2*(3): 171–174.

[51] Mărghitas LA: The Bees and Their Products, Ed. Ceres. Bucuresti. 2005; pp. 344–355.

[52] Aličić D, Šubarić D, Jašić M, Pašalić H, Ačkar Đ: Antioxidant properties of pollen. *Hrana U Zdravlju I Bolesti*, 2014; *3*(1): 6–12.

[53] Freire KR, Lins A, Dórea MC, Santos FA, Camara CA, Silva, T: Palynological origin, phenolic content, and antioxidant properties of honeybee-collected pollen from Bahia, Brazil. *Molecules*, 2012; *17*(2): 1652–1664.

[54] Mateescu C: Personal Interview. Apimondia, Bucuresti, Romania. 2016

[55] Glušac JR, Stijepić MJ, Milanović SD, Đurđević-Milošević DM: Physicochemical properties of honeybee pollen enriched acidophilus milk and probiotic yoghurt. *Acta Periodica Technologica*, 2015; (46): 45–54.

[56] Yerlikaya O: Effect of bee pollen supplement on antimicrobial, chemical, rheological, sensorial properties and probiotic viability of fermented milk beverages. *Mljekarstvo*, 2014; *64*(4): 268–279.

[57] Turhan S, Saricaoglu FT, Mortas M, Yazici F, Genccelep, H: Evaluation of Color, Lipid Oxidation and Microbial Quality in Meatballs Formulated with Bee Pollen During Frozen Storage. Journal of Food Processing and Preservation. Korean J Food Sci Anim Resour. 2014; 34(4): 423–433.

[58] Krystyjan M, Gumul D, Ziobro R, Korus A: The fortification of biscuits with bee pollen and its effect on physicochemical and antioxidant properties in biscuits. *LWT-Food Science and Technology*, 2015; *63*(1): 640–646.

[59] Graefe EU, Wittig J, Mueller S, Riethling AK, Uehleke B, Drewelow B, Pforte H, Jacobasch G, Derendorf H, Veit M: Pharmacokinetics and bioavailability of quercetin glycosides in humans. *The Journal of Clinical Pharmacology*, 2001; *41*(5): 492–499.

[60] Zuluaga C, Serrato JC, Quicazan M: Chemical, nutritional and bioactive characterization of Colombian bee bread. *Chemical Engineering Transactions*, 2015; *43*: 175–180. DOI: 10.3303/CET1543030.

[61] Šımúth J, Bílıková K, Kováčová E: Royal Jelly proteins as a tool for development of functional ingredients for health. 2003; Available from: http://apimondiafoundation.org/foundation/files/312.pdf. [Accessed: 30.06.2016].

[62] Kosoglu M, Yucel B, Gokbulut C, Konak R, Bircan C: The Effect of Harvesting Time on Some Biochemical and Trace Element Compositions of Royal Jelly. Journal of Kafkas University Veterinary Faculty, 2013; 19(2): 233–237.

[63] Stocker A, Schramel P, Kettrup A, Bengsch E: Trace and mineral elements in royal jelly and homeostatic effects. *Journal of Trace Elements in Medicine and Biology*, 2005; *19*(2): 183–189.

[64] Filipič B, Potokar J: Effect of Royal Jelly (rj) on Human Interferon-alpha (huifn-α) Inhibition of Human Colon Cancer Cells (caco-2) Proliferation in Vitro. 2013; Available from: http://www.apimondia.com/congresses/2013/Apitherapy/PlenarySession/Effect%20Of%20Royal%20Jelly(Rj)%20On%20Human%20Interferon-Alpha%20%20Bratko%20Filipic,%20Jana%20Potokar.pdf. [Accessed: 17.07.2016].

[65] Sahinler, N: Medical Properties of Royal Jelly. Bee Products and Health(Apitherapy), Sidas Media Ltd. Sti, İzmir. 2015; p. 120–124. ISBN: 978-605-5267-26-1.

[66] Ahmadnia H, Sharifi N, Alizadeh S, Roohani Z, Kamalati A, Marjan SS: Wonderful effects of royal jelly on treatment of male-factor related infertility. *Austin Journal of Reproductive Medicine & Infertility*, 2015; *2*(6): 1031.

[67] Marcucci MC, Ferreres F, Garca-Viguera C, Bankova VS, De Castro SL, Dantas AP, Valente PHM, Paulino N: Phenolic compounds from Brazilian propolis with pharmacological activities. *Journal of Ethnopharmacology*, 2001; *74*: 105–112.

[68] Russo A, Longo R, Vanella A: Antioxidant activity of propolis: role of caffeic acid phenethyl ester and galangin. *Fitoterapia*, 2002; *73*(1): 21–29.

[69] Liao HF, Chen YY, Liu JJ, HsuML, ShiehHJ, Liao HJ, Shieh CJ, Shiao MS, Chen YJ: Inhibitory effect of caffeic acid phenethyl ester on angiogenesis, tumor invasion, and metastasis. *Journal of Agricultural and Food Chemistry*, 2003; *51*: 7907–7912.

[70] Chen YJ, Liao HF, Tsai TH, Wang SY, Ming-Shi Shiao MS: Caffeic acid phenethyl ester preferentially sensitizes CT26 colorectal adenocarcinoma to ionizing radiation without affecting bone marrow radioresponse. *International Journal of Radiation Oncology, Biology, Physics*, 2005; *63*(4): 1252–1261.

[71] Gomez-Caravaca AM, Gomez-Romero M, Arraez-Roman D, Segura-Carretero A, Fernandez-Gutierrez A: Advances in the analysis of phenolic compounds products derived from bees. *Journal of Pharmaceutical and Biomedical Analysis*, 2006; *41*: 1220–1234.

[72] Biray C, Gunduz C, Yilmaz B, Sahin F, Togcuoglu N. The evaluation of cytotoxic and apoptotic effect of propolis and its extracts caffeic acid phenethyl ester and cinnamic acid in human acute t-cell lymphoblastic leukemia cell line (CCRF-CEM). Ege Medical Journal. 2006; 45(2) : 83–92.

[73] Kurt FÖ, Vatansever SH, Sorkun K, Gurhan SID, Turkoz E, Gencay Ö, Salih B. Inhibitory effects of propolis on human osteogenic sarcoma cell proliferation mediated by caspase pathway. *Kafkas Üniversitesi Veteriner Fakültesi Dergisi*, 2010; *16*(3): 397–404.

[74] Erdogan PA, Sezgin C. Propolis and Antitumoral effect of Propolis. Bee Products and Health(Apitherapy), Sidas Media Ltd. Sti, İzmir. 2015; p: 132–136. ISBN: 978-605-5267-26-1

[75] Yucel B, Topal E, Akcicek E, Kosoglu M: Effects of Propolis on Human Health. Anadolu Journal of AARI, 2014; 24(2): 41–49.

[76] Stangaciu, S: Apitherapy Course Notes, 1999; 126 p.

[77] Ali FH, Kassem GM, Atta-Alla OA: Propolis as a natural decontaminant and antioxidant in fresh oriental sausage. *Journal Veterinaria Italiana*, 2010; *46*(2): 167–172.

[78] Han SK, Yamauchi K, Park HK: Effect of nitrite and propolis preservative on volatile basic nitrogen changes in meat products. *Microbios*, 2001; *105*(411): 71–75.

[79] Han SK, Park HK: Accumulation of thiobarbituric acid-reactive substances in cured pork sausages treated with propolis extracts. *Journal of the Science of Food and Agriculture*, 2002; *82*(13): 1487–1489.

[80] Moghazy EA, El-Shaarawy MOA: Quality attributes of beef burger as affected by using propolis and frozen storage. *Egyptian Journal of Agricultural Research*, 2001; *79*(4): 1499–1512.

[81] Aly SA, Elewa NA: The effect of Egyptian honeybee propolis on the growth of Aspergillus versicolor and sterigmatocystin biosynthesis in Ras cheese. *Journal of Dairy Research*, 2007; *74*(1): 74–78.

[82] Tosi EA, Ré E, Ortega ME, Cazzoli AF: Food preservative based on propolis: Bacteriostatic activity of propolis polyphenols and flavonoids upon *Escherichia coli*. *Food Chemistry*, 2007; *104*(3): 1025–1029.

[83] Jian-xin G, Hai-ying C, Zhao-yun L: He Influence of Propolis on Bifidobacteria and Lactobacillus in Yoghurt. 2011; http://en.cnki.com.cn/Article_en/CJFDTOTAL-ZGXD201102021.htm.

[84] Koç AN, Silici S, Mutlu-Sarıgüzel F, Sağdıç O: Antifungal activity of propolis in four different fruit juices. *Food Technology and Biotechnology*, 2007; *45*(1): 57–61.

3

Functional Properties of Brazilian Propolis: From Chemical Composition Until the Market

Andresa A. Berretta, Caroline Arruda,
Felipe Galeti Miguel, Nathalia Baptista,
Andresa Piacezzi Nascimento,
Franciane Marquele-Oliveira, Juliana Issa Hori,
Hernane da Silva Barud, Bianca Damaso,
César Ramos, Raul Ferreira and Jairo Kenupp Bastos

Abstract

Propolis is a product obtained from resins and exudates of different plants from different regions in order to protect the comb, with peculiar organoleptic, chemicals and biological properties. Considering this, this chapter presents the types of Brazilian propolis as the types available nowadays, their chemical compositions, as well as, some of their important biological properties enabling employing them as important health food, such as antimicrobial, antioxidant, and immunomodulation action. Various "*in vivo*" and clinical trial studies, conducted in different regions, on the safety and dosage of propolis, technologies used to obtain propolis extract, and several innovative presentations of this promising bee product are also presented in this chapter. Finally, this chapter aims to present the regulatory affairs, potential market for propolis around the world, and perspectives for a near future.

Keywords: Brazilian propolis, green, red, brown, chemical composition, antimicrobial, immunomodulatory, antioxidant, extraction process, innovation, regulatory affairs, potential market

1. Introduction and a brief history

The antique civilizations always used bee products as valuable therapeutic resources in their medicinal practices. The history of medicine of the Assyrian civilizations, Chinese, Tibetan, Inca, Egyptian, and also the Greco-Roman is very rich and possess records of centenary formulations, including propolis to treat or prevent diseases. Old Egyptians, Greeks, and Romans used propolis to treat wound, cutaneous lesions, ulcers, and chirurgical interventions [1].

In Egypt, propolis was used as one of the main ingredients used in the formulations to embalm cadavers. It was also used by Aristotle, Dioscorides, Pliny, and Galneo as an antiseptic and wound-healing. The Greeks, including Aristotle and Hippocrates, adopted it as an internal and external healing. Pliny Roman historian refers to propolis as a medicine to reduce swelling and relieve pain [2].

The term "propolis", "pro" in favor of and "polis", "city of bees", which means in defense of the honey comb, was described in the sixteenth century in France [3], and in the seventeenth century, propolis was considered an official drug by London Pharmacopoeia [4]. In the subsequence centuries, propolis has attracted growing interest due to its medical properties, especially in Eastern Europe. In 1908, the first scientific article about propolis chemical properties and composition [5] was published, indexed on Chemical Abstracts. In 1968, the abstract of the first patent was published on Chemical Abstracts [6].

In South Africa, during the war in the end of the nineteenth century, propolis was largely used because of its healing properties [3] and in the Second World War it was used by several Soviet clinicians [2].

In the last decades, propolis has gained wide acceptance as traditional medicine in several parts of the world. This disseminated interest in propolis in several countries encouraged a large number of studies considering chemical and biological properties of propolis [7].

Nowadays, in the Brazilian market and in several other countries, it is possible to find propolis in different presentations, such as liquid or powder extract: in bottles, capsules, tablets, vaporizers, syrups, creams, and among others, aiming to act as an antimicrobial [8–10], antioxidant [11], immunoregulatory [12–14], anti-inflammatory [12, 15, 16], antiviral [17] agent, besides several other functionss. A very large number of publications endorsing these biological benefits in *"in vitro"*, *"in vivo,"* and in some clinical trials, is be discussed further in this chapter.

Thus, this chapter presents recent studies about Brazilian types of propolis such as green, red, and brown, considering their chemical composition and some biological properties such as antimicrobial, antioxidant, and immunoregulatory, safety aspects, extraction process, and technology associated, regulatory aspects, potential market and challenges, as well as tendencies for a near future.

2. Chemical composition of Brazilian propolis

Propolis is formed by a complex set of components collected by *Apis mellifera* from different parts of plant resins (twigs, flowers, pollen, buttons, and exudates of trees) which are deposited in the hive with saliva and enzymes of the insect to seal the cracks and maintain the temperature (**Figure 1A**) [18].

Figure 1. Presentation of different types of Brazilian propolis. (A) *Apis mellifera* collecting Green propolis from *Baccharis dracunculifolia* species, (B) green propolis on the intelligent collector, (C) brown propolis and (D) red propolis. Figures (A) and (B) were gently donated by César Ramos, Natucentro Company. Figures (C) and (D) were gently donated by Felipe Galeti Miguel, Apis Flora Company. Both companies are associated with ABEMEL.

It consists of resin (50% of the mixture is composed by flavonoids and phenolic acids), wax (30%), essential oils (10%), pollen (5%), and other organic substances (5%). Among the present compounds, it can be consisted of hydrocarbons, alcohols, aliphatic and aromatic acids, esters and its derivatives, aldehydes, ketones, flavonoids, fatty acids, terpenoids, amino acids, sugars, lignans, vitamins, minerals, etc. [19].

The chemical composition of propolis differs significantly according to the geographic region where resins were collected due to the flora of each region, allowing the selection of different plants as source of resin [20].

When this product is derived from Europe or China, for example, the main plant metabolites found are flavonoids and phenolic acids, unlike the stemmed ones from southeastern Brazil, which, besides phenolic compounds, contain high amounts of terpenoids and prenylated derivatives of p-coumaric acid [21]. This difference in composition reveals the collection of resinous material, in temperate zones, from poplar, especially species of *Populus* and in southeastern Brazil, especially from *Baccharis dracunculifolia* DC (Compositae), popularly known as "vassourinha do campo" [22].

Regarding Brazilian propolis, it was further divided into 12 classes according to Park et al. [23]: the first five ones originate from the south and have the colors yellow, light brown, dark brown, light brown, and greenish brown, respectively. Regarding propolis found in the Northeast, it was divided into six groups such as reddish brown, greenish brown, dark brown, yellow, dark yellow, and yellow. Finally, the last of these classes regards the kind of propolis that comes from the Southeast and is known to have a greenish brown or green color and so-called green propolis (**Figure 1B**) [23]. After 2007, the 13 types of propolis was added: this new kind comes from the mangroves of the Brazilian states of Sergipe, Alagoas, Paraiba, Pernambuco, and Bahia. Among the Brazilian propolis, the green, the brown (**Figure 1C**), and the red (**Figure 1D**) ones are the most studied and relevant to the Brazilian economy due to their biological activities and exports to other countries, such as Japan [24].

2.1. Green propolis

Green propolis is composed of large amounts of phenolic compounds such as artepillin C, baccharin, kaempferide, isosakuranetin, dihydrokaempferide, drupanin, *p*-couma ric acid, caffeic acid, aromadendrin, caffeoylquinic acid derivatives, and other compounds, such as the triterpene lupeol-3-(3'R-hydroxy)-hexadecanoate. The key source of these compounds is *B. dracunculifolia* [12, 25–30] (**Figure 2**).

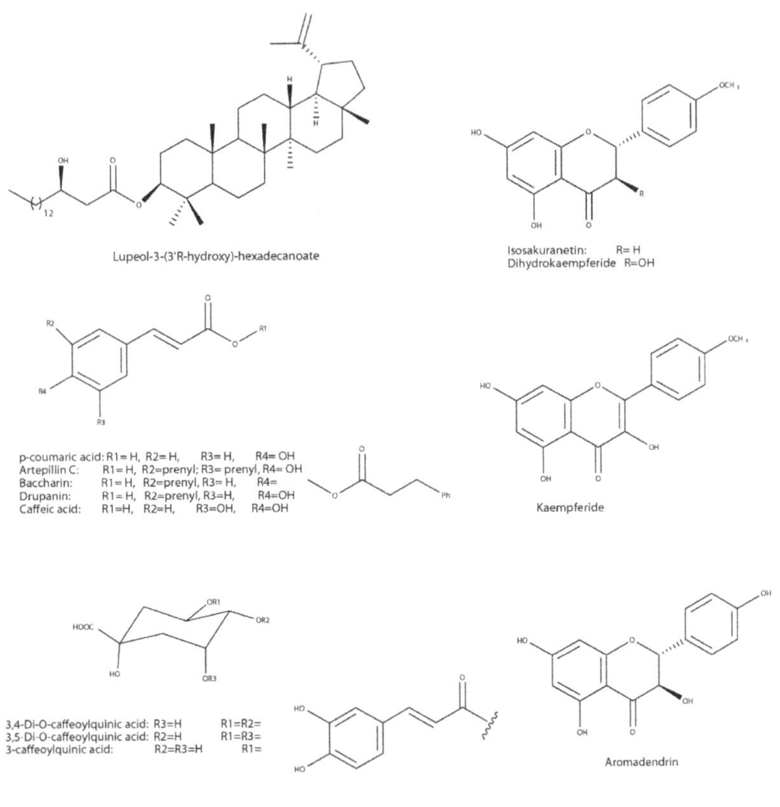

Figure 2. Chemical structures of compounds found in Brazilian green propolis.

Regarding the volatile compounds found in Brazilian green propolis, the major ones are sesquiterpenes, such as (E)-nerolidol, β-caryophyllene, spathulenol, and δ-cadinene. Further more, other compounds such as selina-3,7(11)diene, benzenepropanoic acid and longipinene were also identified. These compounds are responsible for the pleasant aroma in this bee product and are also responsible for many biological activities reported in the scientific literature. [31, 32]. (**Figure 3**).

Considering the large amount of artepillin C presents in green propolis (**Figure 4**), in addition to other biologically active compounds, the commercial value of Brazilian green propolis in the international market is high [16].

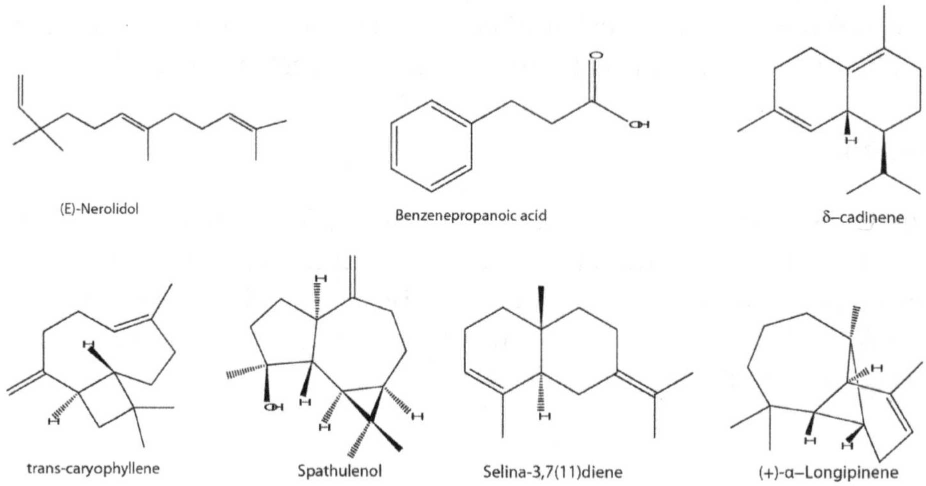

Figure 3. Volatile compounds from Brazilian green propolis.

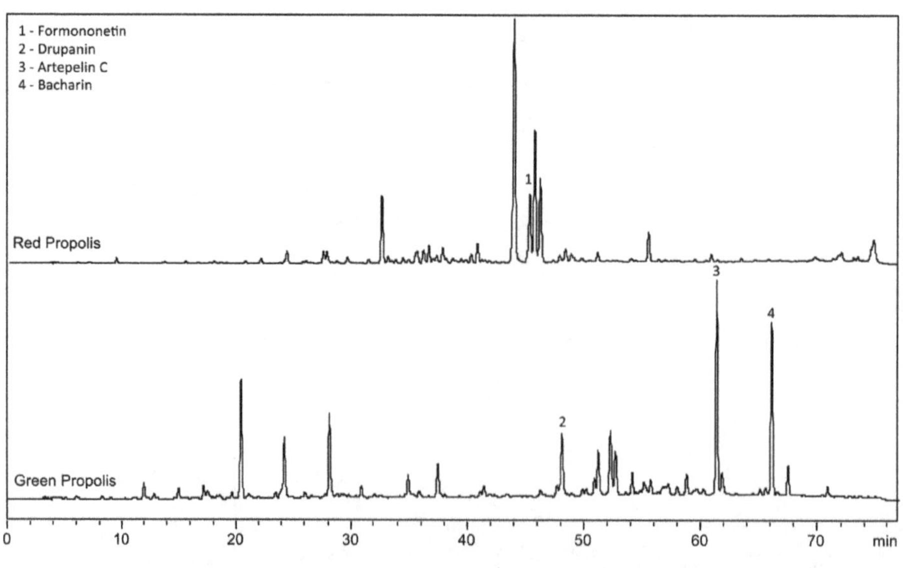

Figure 4. Chemical fingerprint for red and green propolis. Chromatographic profile was obtained using HPLC-DAD, Shimadzu Shim-pack CLC-ODS column (4.6 mm · 250 mm, particle diameter of 5 μm, pore diameter of 100). Green propolis conditions followed according Berretta et al. [8] and red propolis, Cavendish et al. 2015.

2.2. Red propolis

Many of chemical compounds of red propolis, as well as green propolis, have been determined. Some of them are elemicin, isoelemicin, methyl isoeugenol, formononetin, biochanin A, isoliquiritigenin, liquiritigenin, medicarpin, homopterocarpan, quercetin, and vestitol. In the lipophilic extract, the majority of the compounds found are polyprenylated benzophenones—guttiferone E, xanthochymol, and oblongifolin A (**Figure 5**). Because the isoflavones— 7 formononetin, biochanin A, pinocembrin, and medicarpin—are abundant in red propolis, they are used as chemical markers for identifying red propolis (**Figure 4**) [33, 34].

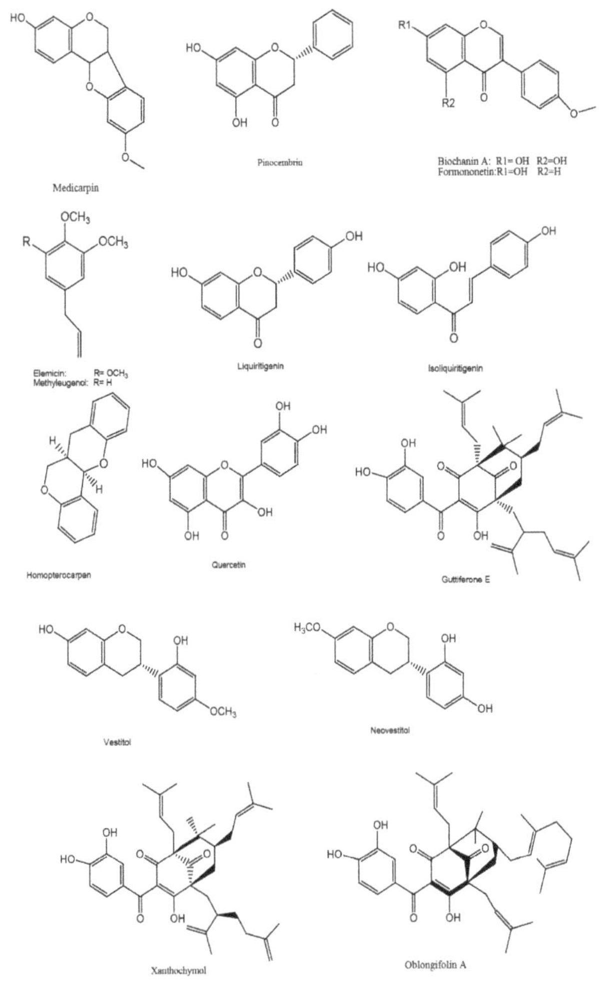

Figure 5. Chemical structures of some compounds found in Brazilian red propolis.

In addition to these compounds, De Mendonça et al. [33] identified caffeic acid, ferulic acid, umbellic acid, p-coumaric acid, genistein, kaempferol, cathechin, dalbergioidin, epicatechin, daidzein, 2′-hydroxyformononetin, evernic acid, naringenin, calycosin, (7S)-dalbergiphenol, thevetiaflavone, cycloartenol, guttiferone C, and other compounds, using LC-Orbitrap-FTMS, a powerful tool to detect compounds because it does not require chromophores such as ultra-violet detector and it can detect very low amounts.

Red propolis has these chemical constituents, mainly due to the collection of resin from *Dalbergia ecastophyllum* by the bees. Its red color is due to the presence of cationic C_{30} isoflavans, retusapurpurins A and B (**Figure 6**). This is characteristic of the propolis found in Brazilian Northeast, found especially in hives nearby mangroves in the states of Sergipe, Bahia, Alagoas, Paraiba, and Pernambuco. This bee product has attracted wide interest because of its numerous biological activities, such as cytotoxic against several cancer cell lines, antibacterial, antifungal, anti-cariogenic, antioxidant, antiproliferative, anti-inflammatory, and others [24, 33, 34]. In addition to its extracts, numerous biological activities of the isolated compounds have been described [24].

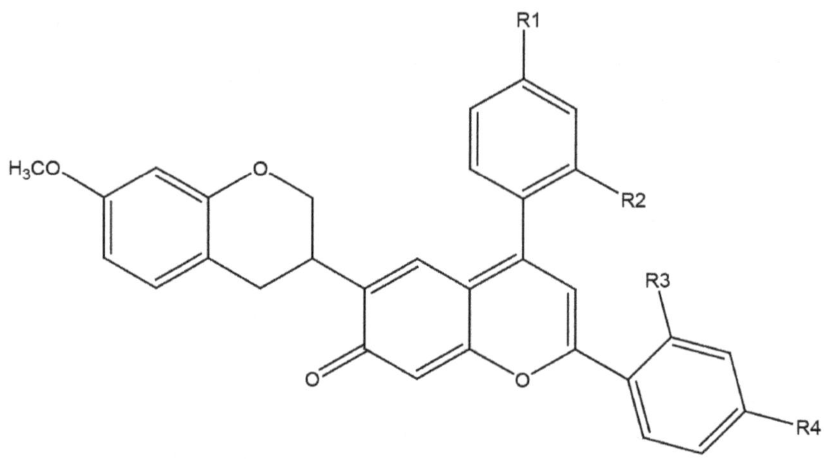

Retusapurpurin A: R1= OH R2=H R3=OCH_3 R4=OH
Returapurpurin B: R1=OCH_3 R2=OH R3=H R4=OH

Figure 6. Chemical structures of isoflavans, retusapurpurins A and B.

Righi et al. [35], besides the compounds already described, found alkanes such as *n*-tricosane, *n*-pentacosane, *n*-heptacosane, *n*-nonacosane, *n*-hentriacontane, and *n*-tritiacontane in the apolar red propolis extract (hexane extract). They also identified other compounds as β-amirin, α-amirin, lupeol, methylguaiacol, trans-anethole, resorcinol, anisylacetone, *cis*-asarone, farnesol, and among others.

Nunes et al. [36] determined 34 volatile (**Table 1**) compounds in Brazilian red propolis and found that the major ones are trans-anethole, α-copaene, and methyl-*cis*-isoeugenol (**Figure 7**). They found that the chemical composition remains relatively constant during the year, considering that 17 out of the 34 compounds were detected every season of the year. But, the compounds δ-cardinol, ß-gurjunene, isocaryophyllene, and δ-cadinene were found only in the sample collected in October and the alkanes, the 1,8 cineol, and α-selinene in the sample collected in July. This can be explained by the visitation of bees, in rainy seasons of shrubs and in dry seasons of woody plants: when the apiculture pasture changes, the chemical composition of propolis also changes. Similarly, as trans-anethole, the other red propolis compounds show biological activities such as analgesic, anesthesic, antigeno-toxic, and antioxidant, making the volatile fraction pharmacologically interesting.

Class of the compound	Compound	Class of the compound	Compound
Monoterpene	*p*-Cymene	Sesquiterpene	α-Cubebene
	Limonene		α-Copaene
	1,8-Cineole		β-Gurjunene
	Linalool		β-Caryophyllene
Aromatic compound	Naphthalene		α-Bergamotene
Alcohol, aldehyde or ketone	4-Hydroxy-4-methyl-heptan-2-one		Farnesene
	6-Methyl-5-hepten-2-ona		€-β-Farnesene
Phenylpropanoid	Octanal		D-Germacrene
	Nonanal		α-Selinene
	n-Decanal		Isocaryophyllene
	Anisaldehyde		β-Bisabolene
	n-Dodecanal		δ-Cadinene
	trans-metil isoeugenol		Cadinene
	Estragole		δ-Cadinol
	Trans-anethole	Aliphatic hydrocarbon	Tetradecano
	Methyl-*cis*-isoeugenol		Pentadecano
	Elemicin		Hexadecano

Table 1. Compounds found in Brazilian red propolis.

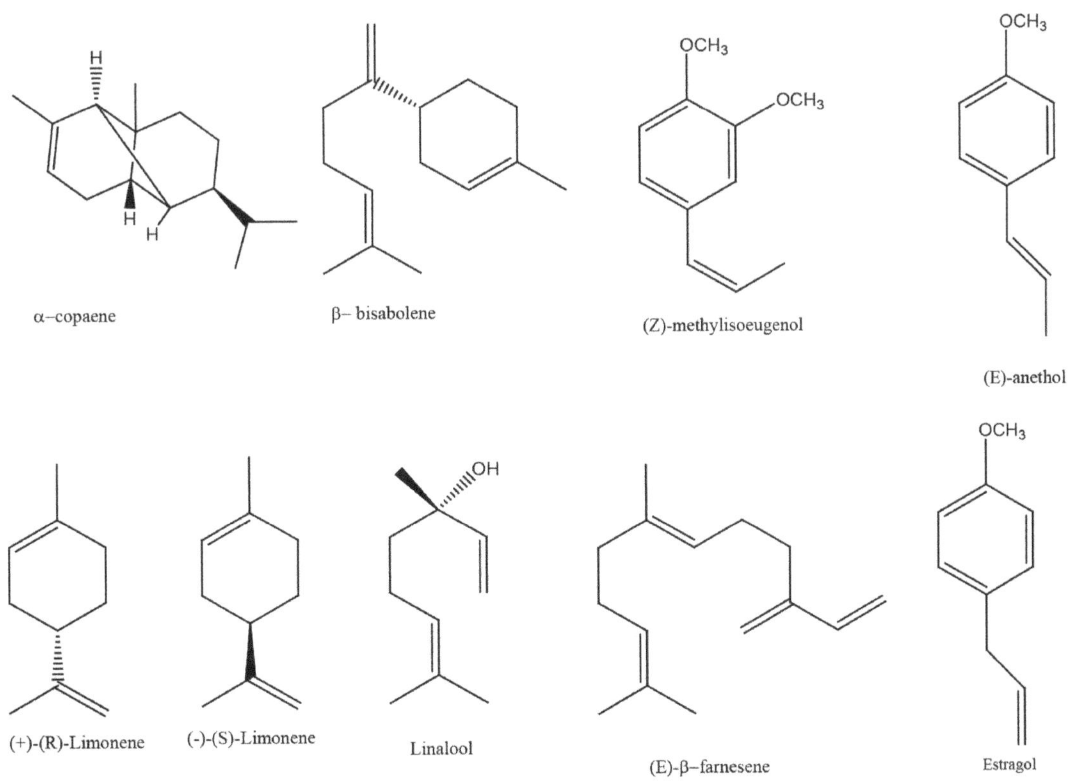

Figure 7. Some volatile compounds from Brazilian red propolis.

2.3. Brown propolis

The brown color is characteristic of propolis from different areas, but regarding Brazilian brown propolis, it is usually referred as the one that comes from the south of the country. Although many chemical compounds present in this product and their biological effects have already been identified, this type of propolis has not been well studied as red and green Brazilian propolis and many scientific reports on brown propolis are relatively old. Its botanical source seems to be mostly *Araucaria,* although some compounds found on it are also present in *B. dracunculifolia* [37].

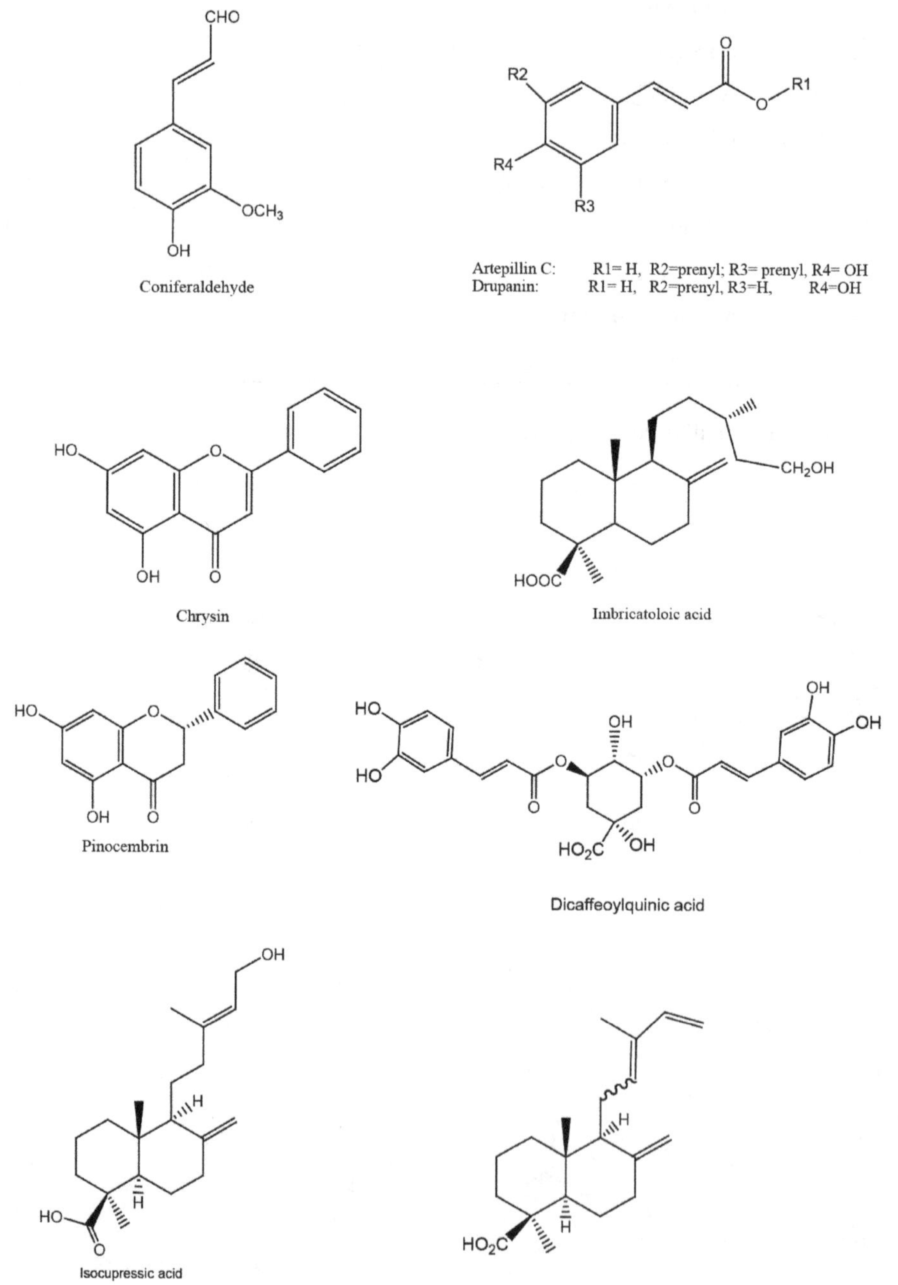

Figure 8. Chemical compounds of brown propolis.

The main compounds identified were: coniferaldehyde, 2,2-dimethyl-6-carboxyethenyl-2h-1-benzopyran, drupanin, pinocembrin, dicaffeoylquinic acid, and artepillin C, isocupressic acid, acetylisocupressic acid, imbricatoloic acid and a mixture of *cis* and *trans* isomers of communic acid [38–40] (**Figure 8**).

Among the numerous biological effects reported for brown propolis and its isolated compounds, it has been observed that both brown propolis and some of its isolated compounds have antimicrobial effect. In addition, it was possible to determine which compounds are responsible for such activity, highlighting the importance of chemically know product widely used by population [38, 41]. Moreover, brown propolis, as well as green propolis, has a significant preventive effect against oxidative stress in skin [42].

Brown propolis collected in Mato Grosso do Sul due to the significant amount of phenolic compounds in ethanol extract shows high antioxidant and antigenotoxic activities. Its volatile fraction is composed mainly of the sesquiterpenes spathulenol and (E)-nerolidol (**Figure 3**), which show an antimicrobial effect against *Cryptococcus neoformans, Enterococcus faecalis*, and *Staphylococcus aureus*. They were not mutagenic, considering that the antimicrobial activity is not because of DNA damage induction [43, 44]. The brown propolis collected from Mato Grosso also showed antimicrobial activity [45].

Therefore, considering that Brazil has a unique flora, among all types of Brazilian propolis three types of propolis are highly noticeable: green, red, and brown propolis due to their singular chemical composition, leading to their biological effects, culminating in the high value in the international market of Brazilian bee products.

3. Biological properties

3.1. Antimicrobial activity

Antimicrobial properties of Brazilian propolis are well-documented, including the antibacterial, antifungal, and antiviral activities. The biological activities of propolis are related to its chemical composition that varies with the collection period of the resin and the flora of the region visited by bees [46]. Therefore, in Brazil there are different types of propolis, since the different geographical regions of the country have a diversity of plant species. The most popular types of Brazilian propolis are green and red propolis.

Brazilian green propolis, whose most important plant source is *B. dracunculifolia*, has been extensively studied. Several studies have shown the activity of green propolis against several pathogenic bacteria, including Gram-positive bacteria (*S. aureus, Staphylococcus epidermidis, Streptococcus pneumoniae*, and *Kocuria rhizophila*) and Gram-negative bacteria (*Haemophilus influenzae, Porphyromonas gingivalis, Porphyromonas endodontalis*, and *Prevotella denticola*) [9, 10, 46–48]. The last three bacteria cause periodontal diseases, which affect the periodontal tissues (tooth supporting tissues). Furthermore, green propolis is active against cariogenic bacteria, such as *Streptococcus mutans, Streptococcus sobrinus, Streptococcus salivarius, Streptococcus sanguinis*, and *Lactobacillus casei* [48, 49]. However, some Gram-negative bacteria are not susceptible to green propolis, such as *Escherichia coli* and *Pseudomonas aeruginosa* [9, 10, 46]. *E. coli* can cause urinary tract infections and gastroenteritis, among others, while *P. aeruginosa* is associated with nosocomial infections, since it is an opportunistic bacterium.

Antifungal activity of green propolis has been reported against all three morphotypes of *Candida albicans* (yeast, pseudohyphae, and hyphae) [50]. At the cellular level, green propolis is able to induce apoptosis and secondary necrosis in yeasts, as showed in a study using *Saccharomyces cerevisiae* as a model organism [51]. Green propolis is also active against filamentous fungi (molds), such as *Trichophyton rubrum, Trichophyton tonsurans*, and *Trichophyton mentagrophytes* [52], which cause dermatophytosis. Ngatu et al. [53] reported the antimycotic effect of green propolis in patients with tinea pedis interdigitalis and tinea corporis caused by *T. rubrum*.

Green propolis also has the capacity to inhibit virus propagation. Shimizu et al. [54] reported that the ethanol extract of green propolis exhibited moderate efficacy in limiting herpetic skin lesions in mice infected with herpes simplex virus type 1 (HSV-1). Urushisaki et al. [17] showed the anti-influenza effect (H1N1 influenza virus) of the water extract of green propolis and its caffeoylquinic acids, which may have a cytoprotective action by affecting the internal cellular process. Takemura et al. [55] also reported the anti-influenza effect of the water and ethanol extracts of green propolis and their 3,4-dicaffeoylquinic acid, which enhance viral clearance by increasing tumor necrosis factor-related apoptosis-inducing ligand (TRAIL) in the lungs of mice infected with H1N1 influenza virus.

Batch-to-batch variability is a common problem in the manufacture of propolis extracts. Since medicinal use of these extracts must rely on appropriate quality requisites, batch-to-batch reproducibility is essential to ensure consistent quality. Therefore, Berretta et al. [8] developed the propolis standardized extract (EPP-AF®), an ethanolic extract which contains green propolis and has batch-to-batch chemical reproducibility. Furthermore, it has several biological activities, including antibacterial and wound-healing activities [8]. **Figure 9** shows some results obtained by our research group, showing the antibacterial activity of EPP-AF® and extracts of brown, red and green propolis.

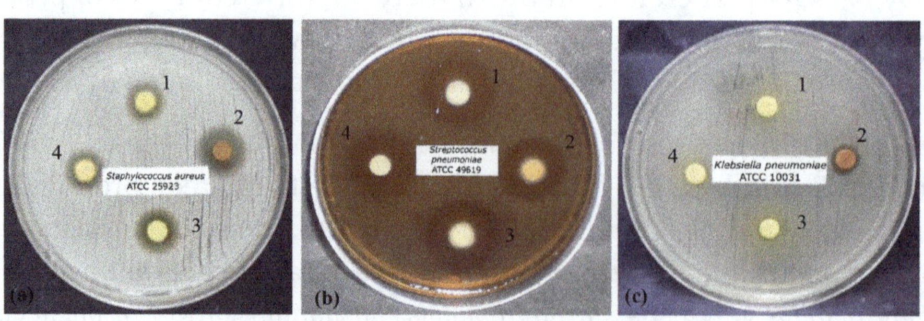

Figure 9. Zones of inhibition (disk diffusion method) provided by 1: Extract of brown propolis from the south of Brazil; 2: Extract of red propolis from the northeast of Brazil; 3: Extract of green propolis from the southeast of Brazil; 4: Propolis standardized extract (EPP-AF®); (a): Staphylococcus aureus ATCC 25923; (b): Streptococcus neumonia ATCC 49619; (c): Klebsiella neumonia ATCC 10031.

Our research group also has developed and evaluated different pharmaceutical forms of green propolis extracts, including propolis ethanolic extract (PEE), propolis water extract (PWE), propolis soluble dry extract (PSDE), and propolis matricial microparticles (PMM). With respect to antifungal activity (*S. cerevisiae* and *C. albicans*), PEE was the most potent followed by PWE, PMM, and PSDE [50]. The same results were obtained against *Lactobacillus* species (Gram-positive bacteria) (data not published yet).

Brazilian red propolis, in its turn, is produced from resinous exudates of *D. ecastophyllum*, found mainly in Northeastern Brazil (states of Alagoas, Bahia, Paraíba, Pernambuco, and Sergipe) [56]. Ethanolic extracts of red propolis showed activity against Gram-positive bacteria (*S. aureus* and *Bacillus subtilis*) and Gram-negative bacteria (*E. coli* and *P. aeruginosa*) [56–58]. These results are very interesting, since green propolis is not active against *E. coli* and *P. aeruginosa* [9, 10, 46]. Red propolis also has antifungal activity. Siqueira et al. [52] reported its activity against some dermatophytes (*T. rubrum*, *T. tonsurans*, and *T. mentagrophytes*).

Isoflavone formononetin is one of main chemical compounds in red propolis. Das Neves et al. [59] evaluated the activity of this compound against some bacteria (*S. aureus*, *S. epidermidis*, and *P. aeruginosa*) and yeasts (*C. albicans*, *Candida tropicalis*, and *C. neoformans*). The MIC value was 200 μg/ml for all bacteria and 25 μg/ml for the yeasts [59]. (6aS,11aS)-Medicarpin is the other chemical compound in red propolis, which exhibits a strong antibacterial activity, since MIC values of 16, 32 and 32 μg/ml were obtained against *S. aureus*, *B. subtilis*, and *P. aeruginosa*, respectively [57].

Kamuyama et al. [60] evaluated the use of green propolis to control microorganisms in minimally processed carrot. The study involved the comparison between: (i) carrot sanitation with 200 mg/l of total available chlorine, (ii) chlorinated solution "A" together with edible film with 0.4% propolis solution, and (iii) carrot sanitation with 0.4% propolis solution, prepared from 25% propolis alcoholic extract. Mesophilic and psychrotrophic aerobic bacteria, mold, and yeast were counted during the storage of samples of processed carrots at 10°C. The results demonstrated that the results for all treatments were similar to mesophilic and psychrotrophic bacteria. For mold and yeast count, the application of treatments (ii) and (iii), in the end of study, was similar to T0, suggesting that the use of propolis as a food preservative is viable and promising.

Borges et al. [61] evaluated the antibacterial and antifungal properties of different concentrations of a propolis hydroalcoholic extract in fresh pork sausage. This product is target of microbiological contamination, with consequent commitment of "shelf-life" and ability to cause diseases, factors that stimulate food companies to use synthetic preservatives as sodium nitrate, which possess high toxicity.

Interestingly, the results demonstrated that propolis extract (0.03 g/100 g of food) used showed greater antibacterial and antifungal results when compared to sodium nitrate.

3.2. Antioxidant activity

The propolis antioxidant property is one of the most studied biological activities worldwide. This biological property presents outstanding importance in the general benefits that propolis may bring to human health as the free-radical scavenging capacity of propolis compounds may be closely related to the anti-inflammatory, antimicrobial, anticancer activities, as well as, prevention of atherosclerosis, skin damages, ageing, and among others.

Several antioxidant methods are available to study propolis, i.e., the DPPH assay, scavenging of hydroxyl radical by the deoxyribose assay, inhibition of lipid peroxidation, inhibition of chemiluminescence produced in the H_2O_2/luminol/horseradish peroxide (HRP) system and inhibition of chemiluminescence produced in the xanthine/luminol/xanthine oxidase (XOD) system, and among others.

Propolis origin	Antioxidant activity/method	Reference
Brazil (marketed-standardized extract—PI 0405483-0)	0.016 μl/ml (IC_{50}) inhibition of lipid peroxidation 0.22 μl/ml (IC_{50}) inhibition of chemiluminescence produced in the H_2O_2/luminol/horseradish peroxide (HRP) system 0.005 μl/ml (IC_{50}) inhibition of chemiluminescence produced in the xanthine/luminol/xanthine oxidase (XOD) system 0.024 μl/ml (IC_{50}) scavenging of hydroxyl radical by the deoxyribose assay	[11]
Campo grande Brazil (raw-material*propolis from stingless bees)	3 μg/ml (IC_{50}) scavenging of DPPH No hemolyis—oxidative hemolysis inhibition assay 50–125 μg/ml—efficiency in inhibiting of AAPH-induced lipid peroxidation	[70]
Nan Province innorthern Thailand (raw-material)	Not indicated - scavenging of DPPH Observation: the authors inform the higher the ethanol amount in the ethanol aqueous solution, the higher the antioxidant activity	[71]
Alagoas, Brazil (brown propolis—raw-material)	8.01 μg/ml (IC_{50}) scavenging of DPPH—ethanolic extract	[33]
Mediterranean propolis	Most prepared extracts inhibited lipid oxidation—oxidation of sunflower oil method	[72]
Brazil aqueus extract	0.62 μg/ml (IC_{50})—inhibition of lipid peroxidation employing brain homogenate	[27]

Table 2. Antioxidant activity of propolis from several regions in the world.

Some of these methods mimic the physiological conditions found in human body, this is the case of lipid peroxidation assay, in which membrane fractions (from mitochondria or brain) are used. In this method, the addition of iron salts triggers the decomposition of lipid peroxides into peroxyl (LOO•) and alkoxyl (LO•) radicals that can abstract hydrogen from polyunsaturated acyl chains and propagate lipid peroxidation. Any antioxidant capable of scavenging LOO• and LO• will decrease peroxidation. However, other methods can be considered more accessible, such as DPPH, which can easily demonstrate, in large scale, the antioxidant capacity of propolis samples from different sources or different batches. According to Marquele-

Oliveira et al. [62], this method could even be employed as an alternative for worldwide characterization and standardization of natural products. A good correlation of the DPPH method was observed against lipid peroxidation assay. This assay is based on the ability of DPPH, a stable free radical, to be quenched and thereby decolorize in the presence of antioxidants resulting in a reduction in absorbance values. In the DPPH test, the antioxidants reduce the DPPH radical to a yellow-colored compound, diphenylpicrylhydrazine. The extension of the reaction will depend on the hydrogen-donating ability of the antioxidants [63].

Propolis antioxidant properties have been fully investigated and both propolis raw material and propolis commercial extracts have been studied. **Table 2** shows examples of the antioxidant profile and the method employed for each sample, focusing on their collecting origin. Phenolic compounds have been reported as the main propolis compounds responsible for the antioxidant property. The antioxidant role of polyphenols results from the donation of hydrogen atoms from an aromatic hydroxyl group to the free radical, leading to stabilization of the radical [64]. During the evaluation of propolis fractions (from Brazil), Wang et al. [65] observed a strong inhibition of lipid peroxidation using rat liver homogenate at a concentration of 2 mg/ml, and this activity was related to the presence of flavonoids. However, it is known that other than phenolic compounds, flavonoids are involved in the antioxidant activity of propolis. So a series of phenolic compounds, including flavonoids, were assessed against the peroxidation of linoleic acid in a micellar solution. The results demonstrated that polyphenols in general present higher activity than BHT (butylated hydroxytoluene), a well-known antioxidant [66]. In a study using cell culture, artepillin C has been proposed as a strong candidate to be responsible for the antilipoperoxidative activity of Brazilian propolis [67].

Santos et al. [68] assessed the antioxidant activity of flavonoids and reported that the presence of structural groups, i.e., the B ring dihydroxyl, double bond in C2 and C3 in conjunction with the 4-oxo function, and the additional presence of hydroxyl groups in C3 and C5 (except for quercetin and 3'-O-methyl-quercetin), were the most potent inhibitors of lipid peroxidation using mitochondria. This antioxidant activity was also due to Fe chelation, which may explain the activity of flavonoids and polyphenols which do not have the above described structural groups [69].

After screening the antioxidant properties of propolis around the world, not only in the presented references, but also in the vast literature about this topic, one can observe a wide variation of responses. On the one hand, the antioxidant ability of each extract is related to the type and amount of phenolic compounds present in each extract, closely dependent on the propolis origin. But, on the other hand, no standardization regarding the solid soluble amount in each sample is presented, making comparisons among them not adequate. However, the presence of antioxidant activity in every propolis source studied is clearly observed and this activity has special importance to propolis biological properties.

Tian et al. [73] have shown that ethanolic propolis extract (EPE) protects endothelial cells from oxidized low-density lipoprotein (ox-LDL)-induced apoptosis and inhibits atherosclerotic lesion development. This research group has also demonstrated the effect of propolis extract on endoplasmic reticulum stress-C/EBP homologous protein pathway-mediated apoptosis. Apoptosis, especially in macrophages present in atherosclerotic lesions, is considered as a prominent feature of advanced atherosclerotic plaques, suggesting that macrophage apoptosis is closely related to the atherosclerotic development and subsequent plaque rupture, which is the prominent event that results in the majority of clinical manifestations of acute coronary syndrome such as acute myocardial infarction and sudden coronary death [74]. Thus, protecting macrophages from apoptosis is believed as an effective approach to attenuate plaque instability and combat acute vascular events.

Additional studies investigated the potential use of topically and orally administered propolis extracts to prevent UV irradiation-induced oxidative stress in skin. Brazilian propolis extracts both green and brown successfully prevent UV-induced GSH (endogenous antioxidant) depletion *in vivo* and are both promising antioxidant systems against oxidative stress in skin [75].

Propolis also due to its antioxidant properties was tested against acute lung inflammation (ALI) caused by cigarette smoke (CS) *in vivo*. The researchers observed that propolis (P) treatment (200 mg/kg) normalized all biochemical parameters in the CS+P group compared with the CS group, including nitrite, myeloperoxidase level, antioxidant enzyme activities (superoxide dismutase, catalase and glutathione peroxidase), reduced glutathione/oxidized glutathione ratio, and malondialdehyde. Additionally, TNF-α expression reduced in the CS+P group when compared with the CS group. They suggested, therefore, the potential antioxidant and anti-inflammatory role for propolis with regard to ALI caused by CS in mice [76].

Regarding the influence of the propolis antioxidant activity in food preservation, when combined with heat treatment in apple juice, propolis (0.1 mg/ml) reduced the thermal treatment time and temperature needed to inactivate 5 log10 cycles of *E. coli*. No influence on organoleptic properties of the apple juice, which implies the possibility of obtaining a sensorially appealing, low-pasteurized apple juice with the functional properties provided by propolis was reached [77]. In another study, Costa et al. [78] studied the bifunctional biobased packing containing red propolis. In addition to the antimicrobial effect on coagulase-positive *Staphylococci* in cheese curds, the authors observed the reduced oxidation of butter during storage due to the antioxidant properties of propolis.

3.3. Immunoregulator

One of the biological effects of propolis is its immunomodulatory effect—by either enhancing or suppressing the immune system. This contradictory effect is probably due to its complex chemical variety, the presence in different geographic regions, and the different forms of extraction.

Little was known about the biological role of propolis until the 1990s, but recently numerous studies have been published, providing an important contribution to this research field.

Immunomodulatory as well as anti-inflammatory effects of propolis have been widely demonstrated both *in vitro* and *in vivo* [15, 79–82].

These effects are mainly related to its constituents, especially the phenolic compounds, including flavonoids as major components. Among the main types of flavonoids contained in propolis are: pinocebrin, chrisin, and caffeic acid phenethyl ester (CAPE). In addition to flavonoids, propolis can also contain cinnamic acid derivatives such as caffeic acid and its esters, besides sesquiterpenes, quinones, and coumarins [83–85]. The typical constituents of Brazilian propolis, especially the Brazilian green propolis, are: caffeoylquinic acid and prenylated derivatives of cinnamic acid, such as artepillin C, p-coumaric acid, baccharin, and drupanin [23, 86, 87].

Despite the intensive search for the main constituent of propolis responsible for its immunomodulatory role, its effect seems to be associated with a combination of its different components [88].

Bachiega et al. [89] evaluated the propolis extract and its phenolic compounds, such as cinnamic and coumaric acids on cytokine production (IL-1b, IL-6, and IL-10) before or after macrophage challenge with LPS, to assess a possible immunomodulatory action. They observed a significant reduction in IL-6 and IL-10 in macrophages treated with the compounds only when the LPS was added before the stimulus, whereas the propolis extract was capable to inhibit the cytokine production both before and after the LPS addition. Thus, concluding that this efficiency could have occurred due to the synergistic effect of all compounds present in the extract [89]. On the other hand, the effect of polyphenolic compounds isolated from propolis and propolis extract was investigated on the growth and metastatic potential of a transplantable mammary carcinoma of CBA mouse. The results indicated that water-soluble extract of propolis (WSDP), caffeic acid (CA), quercetin (QU), and CAPE could be useful tools in the control of tumor growth in experimental tumor models [13].

The immunomodulatory activity of propolis extract was also investigated *in vivo* using the ovalbumin (OVA)-induced asthma model. Sy et al. [90] demonstrated that propolis extracts can suppress the serum levels of OVA-specific antibody IgE and IgG1 and attenuate the airway inflammation in treated mice, probably by the ability of propolis to modulate cytokine production. These findings suggest that propolis extracts may be a potential novel therapeutic agent for asthma [90].

Park et al. [91] evaluated another ethanolic extract of propolis (EEP) from Korea in an inflammatory animal model of hind paw edema induced by carrageenan. They observed a significant inhibition of the development of paw edema and increased vascular permeability coupled with an excellent analgesic effect in treated animals. They also showed a significant inhibitory effect on granuloma and exudate formation. The authors suggested that the anti-inflammatory effects of propolis observed might be due to its inhibitory effect on prostaglandin production [91].

In fact, Mirzoeva et al. [92] demonstrated the effect of another ethanolic extract of propolis in suppressing the prostaglandin and leukotriene generation by murine peritoneal macrophages *in vitro* and during zymosan-induced acute peritoneal inflammation *in vivo*. Furthermore, the authors described the caffeic acid phenethyl ester (CAPE) as being the most potent modulator of the arachidonic acid cascade among the propolis components examined [92].

Similarly, Borrelli et al. [93] investigated two ethanolic propolis extracts (EPE): with and without the caffeic acid phenethyl ester (CAPE) for their anti-inflammatory activity in rats using carrageenan foot edema and carrageenan pleurisy models. They observed that only EPE with CAPE and CAPE alone significantly inhibited the carrageenan edema in the rat paw and the number of leukocytes in the pleural exudate in rats, suggesting that the anti-inflammatory activity of propolis is due to CAPE [93].

It is important to say that, despite Brazilian green propolis does not present CAPE in its composition, it presents a wide range of studies describing its beneficial properties such as antiulcerogenic, anti-inflammatory, antimutagenic, antifungal, angiogenesis, antioxidant, and immunomodulatory [14, 94–98]. Different from most European propolis extracts, which present flavonoids as the major component responsible for their effects, the biological activities of Brazilian green propolis are due to its high levels of phenolic acids such as artepillin C [99].

Studies with Brazilian green propolis have showed its role in inhibiting the development of pulmonary cancers [100], an antiviral activity *in vivo* [101], anticancer [102], an anti-inflammatory activity *in vivo* and *in vitro* [12, 16, 87], an antioxidant function in patients with type 2 diabetes mellitus [101, 102], antiherpetic activity [103], and among others [104, 105].

Despite several and growing studies involving the biological effects of Brazilian propolis, the detailed molecular and cellular basis of the action of propolis on immune cells is still unknown.

The administration of green propolis in animals subjected to chronic stress increased the generation of hydrogen peroxide, suggesting a modulation in the macrophage activation [106]. Machado et al. [12] verified an immunomodulatory effect of Brazilian green propolis extracts in acute and chronic inflammation models *in vivo* where the treated animals showed a decrease production of proinflammatory cytokines such as TNF-α and IL-6 and an increase in the IL-10 and TGF-b anti-inflammatory cytokines [12].

Most of the studies reported to date are associated with the immunomodulatory effect presented by propolis extracts with the modulation of the transcription factor NFkB [107–110].

Recently, it has been demonstrated that Brazilian green propolis can also act in a new inflammatory pathway named inflammasome. The inflammasomes are a large molecular platform formed in the cell cytosol in response to stress signals, toxins, and microbial infections. Once activated, the inflammasome induces the molecule caspase-1, which in turn provokes the processing of inflammatory cytokines IL-1β and IL-18. The Brazilian green propolis analyzed in this study (EPP-AF®) was capable of inhibiting the NLRP3 inflammasome and hence significantly reduces the IL-1β secretion in mouse macrophages. Thus, indicating that Brazilian green propolis EPP-AF® extract has a significant role in regulating the inflammasomes [15].

In conclusion, the immunomodulation caused by propolis has been amply demonstrated in recent years, both in the stimulation and suppression of the immune system, making it potentially applicable as an alternative adjuvant therapy or even in the treatment of various diseases.

Table 3 presents a summary of the activities presented in Section 3.

Kind of propolis	Biological properties	References
Brown	Antigenotoxicity	[43, 44]
	Antimicrobial	[38, 41, 44, 45]
	Antioxidant	[42, 75]
Green	Antibacterial	[8–10, 46, 47, 49]
	Anticancer	[102]
	Antifungal	[51–53]
	Anti-inflammatory	[12, 15, 16, 87]
	Antimutagenic,	[96]
	Antioxidant	[27, 42, 75, 76, 103]
	Antiulcerogenic	[94]
	Antiviral	[17, 54, 55]
	Immunomodulatory	[12, 88]
	Inhibition of angiogenesis	[98]
	Preservative	[60]
Red	Antibacterial	[24, 56–59]
	Anticancer	[33]
	Anticariogenic	[24]
	Antifungal	[24, 52, 59]
	Anti-inflammatory	[24, 80]
	Antioxidant	[24, 33]
	Antiproliferative	[24]
	Immunomodulatory	[24]

Table 3. Biological activities presented in Section 3—summary.

4. Safety aspects

4.1. Nonclinical studies

Although propolis has been used for centuries around the world demonstrating to be safe, several scientific studies have been done in order to evaluate propolis safety by oral or topical route. Here we intend to present Brazilian propolis studies done in animals and, the studies found in humans, as clinical trials are not so numerous, despite the clinical trials did not focus on safety, we are presenting them in order to compare the dosages previously used and documented in humans.

Sforcin et al. [111] evaluated some biochemical parameters of animals treated with several different types of propolis aiming to study propolis safety and the differences in the propolis source could interfere in the results. The authors determined total proteins, glucose, urea, creatinine, triglycerides, cholesterol, cholesterol-HDL, aminotransferases, and lactic dehydrogenase (LDH). The results demonstrated that all parameters were under standard values for the species studied and the propolis sources did not affect the results.

Reis et al. [95] evaluated the safety of propolis standardized extract (EPP-AF®), a Brazilian propolis composition that presents more than 50% of green propolis, by oral route in mice, in an acute model. DL50 was determined to be 3000 mg/kg after 24 h of treatment, and dosages under this value did not demonstrate intoxication signs in the animals. In the subchronic protocol (30 days) done in Wistar rats, there were no differences in the food and water intake, animals' weight and diuresis. Hematological and biochemical analysis did not show statistical differences between the treated (propolis 650 mg/kg) and placebo group. All parameters were in accordance with reference standards for the species studied. Microscopic analysis of all tissues did not show any differences with the placebo group, and it was not possible to detect any lesions, hemorrhages or cells infiltration, demonstrating the safety of oral administration of Brazilian propolis up to 650 mg/kg during 30 days of ingestion.

Mani et al. [112] evaluated the safety of Brazilian propolis in distinct treatments: (i) rats treated with 1, 3, and 6 mg/kg/day during 30 days; (ii) rats treated with 1 mg/kg/day of propolis alcoholic or aqueous during 30 days, and (iii) rats treated with 1 mg/kg/day during 90 and 150 days, demonstrating that all levels of seric cholesterol, HDL-cholesterol, total lipids, triglycerides, aminotransferases (AST), and lactic dehydrogenase (LDH) of propolis treated group were similar to the control group. The authors suggested that Brazilian propolis in the dosages used during the period of treatment were safe (**Table 4**).

Type of study (oral route)	Dosage	Propolis source	Species	Dosage converted to human according FDA guideline (mg/day)*
Biochemical parameters, 150 days [112]	1, 3 and 6 mg/kg	Brazilian	Rats	67.74
Biochemical parameters, 60 days [113]	2000 mg/kg	Iranian	Rats	22,580
Acute safety study [114]	2000 mg/kg	Polish	Rats	22,580
DL50 determination [95]	3000 mg/kg	Brazilian	Mice	17,073
Biochemical parameters—30 days [95]	650 mg/kg	Brazilian	Rats	7338
Acute toxicity parameters**	2500 mg/kg	Brazilian	Wistar rats	28,225
Subchronic study (28 days)**	1000 mg/kg	Brazilian	Wistar rats	11,290
Subchronic study (28 days)**	100, 300 and 1000 mg/kg	Brazilian	Rabbits	22,580

* Conversion considering adult weight around 70 kg.
** Results of our group and not published yet.

Table 4. Safety non-clinical results for propolis administration for oral route.

According to Dobrowolski et al. [114], LD_{50} for different sources of propolis varied from 2 to 7.3 g/kg in mice, suggesting a safe dose for humans of 1.4 and 70 mg/day (when using safety factor of 1000). In conclusion, considering Brazilian propolis LD_{50} as 17,073 mg/day for humans [95], the application of a safety factor of 10 suggested by FDA guidelines, we would have a safe dose of 1700 mg or 1.7 g/day of propolis for an adult.

4.2. Clinical studies

Khayyal et al. [115] evaluated propolis extract activity in asthmatic patients with oral administration of 260 mg of propolis/day, for 2 months. The results demonstrated reduced night attacks (2.5 attacks/week for 1/week) and improved ventilatory functions, as a consequence of a decrease of TNF-, ICAM-1, IL-6, and IL-8, and an increase of 3· of IL-10, besides a decrease of prostaglandins E2, F2, and leukotriene D4.

Cohen et al. [116] evaluated 430 children aged 1–5 year old. Treated group (n=215) received a mixture of echinacea (50 mg/ml), propolis (50 mg/ml), and vitamin C (10 mg/ml), during 12 weeks, and compared to the placebo group. Children aged 1–3 year old received 5.0 ml, 2·/day, orally while children aged 4–5 year old received 7.5 ml. They were benefits in the incidence and severity of respiratory tract infections, with a decrease of 55% in the number of sick children, 50% in the incidence of respiratory diseases, and 60% decrease in the number of days with fever.

Type of study (oral route)	Dosage	Dosage converted to human adult (mg/day)	Reference
Double-blind study with children—prevention to respiratory infections	50 mg/ml—10 ml/day (children 1–3 year old)	2482.27	[116]
Double-blind study with children—prevention to respiratory infections	50 mg/ml—15 ml/day (children 4–5 year old)	2876.71	[116]
Pilot clinical trial with asthmatic volunteers	2–3 tablets/day with 88.4 mg propolis each	265.20	[116]
Pilot clinical trial with healthy volunteers—prophylactic study	500 mg propolis (2 capsules/day)	500.0	[117]
Pilot clinical trial—recurrent stomatitis	500 mg propolis/day	500.0	[118]
Propolis for wound healing	500 mg propolis/day	500.0	[119]
Clinical trial with asthmatic patients	1 sachet with 260 mg propolis/day	260.0	[115]
Pilot clinical trial with Brazilian propolis for treatment of Helicobacter pylori	20 drops, 3× /day	˜350.0	[120]

Table 5. Clinical trials done with propolis administrated for oral route.

Brätter et al. [117] evaluated the oral administration of 500 mg of propolis for 13 days in healthy volunteers focusing on the evaluation of the immune response (TNF-α, IL-6, and IL-8). There was an increased ability in the cytokines secretion, however, without plasmatic levels. Then, prophylactic administration of propolis depends on the immune system reactivity and time, with no adverse effects.

Samet et al. [118] tested the oral administration of 500 mg of propolis in a randomized, placebo-controlled double-blind study, in which it was possible to demonstrate the benefits of propolis treatment in the repeated stomatitis, especially important in cases of resistance to treatment.

Finally, Zedan et al. [119] evaluated the administration of 500 mg of propolis/day in 45 patients aiming to offer an alternative treatment to cutaneous healings. The study compared propolis with echinacea and placebo, and propolis demonstrated to be more efficient than the other groups, especially in usual and superficial healings (**Table 5**).

Jasprica et al. [121] studied the antioxidant effects of propolis (propolis soluble in water and maltodextrin, 0.65 g of propolis, presenting 2.5% of flavonoids, equivalent to 16.25 mg expressed as galangin, Specchiasol, Italy) when administered in healthy volunteers (n=47, women and men), 3 doses/day (total daily dose of 48.75 mg of flavonoids) for 15 and 30 days, with the following parameters under investigation: superoxide dismutase, glutathione peroxidase, and catalase, malondialdehyde, total cholesterol, low- and high-density lipoprotein cholesterol, triglycerides, glucose, uric acid, ferritin and transferrin, and all routine red blood cell parameters. Interestingly, only men with 30 days of treatment presented differences in malondialdehyde (decrease), superoxide dismutase activity (increase), and a few changes in some parameters of red blood cell were detected.

Considering all the previously presented studies, it is possible to suggest that propolis dosages ranging from 260.0 mg to 2.87 g, which have already been used in humans can be considered safe. The biological results observed also varied much. Considering that propolis around the world is largely used as a supplement or functional food, it is reasonable to assume that dosages within this range will probably be safe, since none of the articles suggested any damage or complications for the volunteers. Regarding the antioxidant evaluation proposed by Jasprica et al. [121] in humans and the literature available until now, it is likely that the propolis dosage used was very high for this purpose. Some data previously published suggested that propolis can have a "pro" or "anti" action and, because several *in vitro* and *in vivo* studies demonstrated antioxidant actions at certain does, it is possible that the best result may be achieved using dosages around 500 mg/day successfully, but further investigation is needed.

5. Extraction technologies and innovation products

5.1. Extraction technology

Propolis is a very complex material depending on the vegetation present in the area visited by bees. Besides the propolis source, the extraction process associated with solvents used will definitely provide a completely different extract [20]. It is well established that chemical compounds possess several particularities such as solubility, volatility, partition coefficient oil/water, pKa, and therefore, different solvents will probably extract different compounds [122]. Depending on the temperature and equipment necessary, volatile compounds can be lost and then, it is common to use cold procedures such as maceration or percolation, and when it is necessary to use higher temperatures, it is important to be careful in order to preserve every compound of interest.

Park and Ikegaki [122] studied propolis extracts obtained from water with 96% ethanol solution as solvents. The results demonstrated that propolis extract obtained with 80% ethanol grade showed higher absorption at 290 nm. Using ethanol solution at 60%, higher quantities of isosakuranetin, quercetin, and kaempferol were extracted, while pinocembrin and sakuranetin were better extracted with ethanol solution at 70% and kaempferide, acacetin, and isorhamnetin were most extracted with ethanol solution 80%. More expressive antimicrobial activities were found in propolis extracts from 60 to 80% of ethanol solution extraction and higher antioxidant results came from propolis extract obtained with ethanol 70–80%.

The most common extraction process for propolis to oral administration is the alcoholic (70%) extraction using maceration, percolation and/or turboextraction. The ratio of propolis raw material:extract used is completely variable and usually, in Brazil, 1:3–4, i.e., 1 part of propolis raw material may offer 3 or 4 parts of extract, considering the production of a liquid extract with at least 11%w/v of propolis dry matter. Of course, this ratio may vary and it is completely dependent on the quality of propolis raw material used.

Jorge et al. [46] studied green propolis extracts obtained from four different locations in São Paulo and Minas Gerais States, using the same extraction process, i.e., hydroalcoholic solution 70% with maceration for 30 days. Although all samples evaluated were Brazilian green propolis using the same extract, different results were found for drupanin, baccharin, and artepillin C during the same month, in spite of seasonal differences. Therefore, it is possible to conclude that, using the same floral source, solvent and process extraction, different regions, and seasonal variations also offer a different chemical composition. Interestingly, these differences did not affect the safe of propolis ingestion [111].

Besides maceration, percolation or turboextraction, Trusheva et al. [123] compared ultrasound extraction and microwave-assisted extraction with the maceration process. The careful analysis of the results obtained with each process demonstrated that statistical differences were found for total phenolics and propolis total extractable matter for ultrasound process (30 min) and microwave (2 · 10 s) when compared to maceration extraction. For ultrasound, higher total phenolics were found (52 3%) in comparison to maceration (43 2%) while the reduced total extractable matter (53 3% versus 55%). In turn, microwave offered reduced values for total phenolics (40.4 0.6% versus 43 2%) and expressively higher amounts of the total extractable matter (75% versus 55%). Flavonoids analysis did not show important differences among the procedures evaluated. It is important to consider the time of extraction in each process, since maceration takes around 72 h, and ultrasound was effective with 30 minutes and microwave 2 · 10 s. Another important thing is to define the objective of the extraction: for analytical purposes, it is more practical and cheaper to use ultrasound or microwave, however, for industrial scale, these latter may not be easily implemented.

Propolis water extract was also obtained by some authors [12, 17, 98, 124] using completely different procedures, demonstrating some interesting biological activities that had been previously studied for propolis alcoholic extract, such as anti-inflammatory [12], inhibition of inflammatory angiogenesis [98], and antiviral [17]. Although the demonstration of these interesting results, chemical characterization was poorly explored in the manuscripts, except by Urushisaki et al. [17] that presented the caffeoylquinic derivatives as the most important compounds of this extraction process; however, the manuscript does not present the extraction process used. De Moura et al. [98] performed the extraction from propolis raw material properly crushed in water maceration (500 ml) with temperature around 70°C (30–60 min), two fractions were obtained from the same propolis raw material, followed by filtration. The filtrate was then lyophilized. A similar procedure was used by others too. Nafady et al. [124] in turn, used an innovative process with -cyclodextrin as encapsulate agent. To obtain this extract, 10 g of crushed propolis was dispersed in 1 l of water containing 10 g of -cyclodextrin previously dissolved. The inconvenient of this procedure is the elevate costs involved in the acquisition of -cyclodextrin besides the limitation of the propolis concentration obtained in the final product. Machado et al. [12] proposed the extraction using hydroalcoholic solvent (70%) as usual, however, the solvent was evaporated and after a hydrolysis step the propolis soft extract was then resuspended in water, in this last case, the obtained extract demonstrated similar chemical results of the alcoholic extract in the moment of preparation.

5.2. Nanoparticles and innovation products

Nowadays, finding natural additives has increased the efforts both to obtain bioactive compounds from natural raw materials and develop stable and functional derivative products. The former mentioned properties attributed to propolis are valuable and find applications in several industries, such as pharmaceutics, agrochemical, and food. The growing interest in propolis has also promoted technological development for the suitable application of propolis.

Propolis in the powder form, for example, exhibits several advantages as increased concentration of propolis dry matter, higher chemical stability of the compounds, and longer preservation of the biological properties. Additionally, the powder form also permits the production of presentations with higher compliance in therapeutics, i.e., sachets, tablets, and capsules. The drying process may also involve the encapsulation of the product resulting in micro/nanoencapsulation systems, which can minimize sensory flavor and odor and control the release of the active compounds.

Propolis dry extract was obtained by Da Silva et al. [125] by employing arabic gum and octenyl succinic anhydride (OSA) starch as carriers by spraydrier. The process allowed obtaining propolis in the powder form with preserved antioxidant activity, stability, and low hygroscopicity. Microencapsulated propolis extract obtained by complex coacervation was reached and presented inhibitory activity against *S. aureus* [126]. Bruschi et al. [127] obtained gelatin microparticles containing propolis extractive solution by spray-drying technique. The microencapsulation by spray-drying technique maintained the activity of propolis against *S. aureus*. In another study, the effect of spray drying parameters on the chemical and biological properties of alcoholic extract of green propolis was investigated [128]. Several parameters of the process demonstrated to influence the polyphenol and flavonoid content, as well as the antioxidant activity, but under an optimized condition, the dried propolis extract showed significant antioxidant activity, with 50% lipid peroxidation inhibition at concentrations ranging from 2.5 to 5.0 mg/ml.

More recently, Marquiafável et al. [129] aimed to develop a propolis dry extract with high propolis (~40%w/w of propolis dry matter) and artepillin C contents by employing a combination of silicon dioxide with arabic or modified starch and silicon dioxide by spray-drier. They have successfully obtained a standardized propolis extract with high amount of propolis, flavonoid content, expected amounts of artepillin C, and with maintained antibacterial activity, and obtained microparticles with both excipients used. Recently, results of the same group obtained dry extracts of propolis with 70–80% of dry matter; however, the microparticles were not obtained (data not published yet), and then, the odor, color, and taste are not similarly reduced as it is possible to observe when microparticles are obtained (**Figure 10**). Although microparticles were not obtained with 70–80% of propolis dry matter, this extract is the most concentrated one found in the market until now and can be used in several products with very good results, for example soft or hard capsules or tablets.

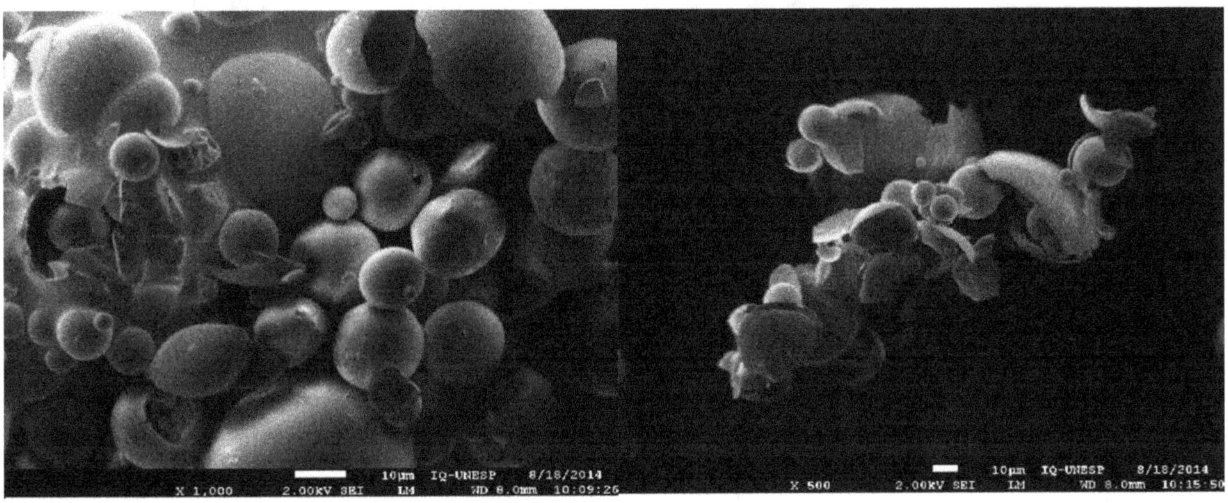

(A) (B)

Figure 10. Propolis standardized water extracts of green propolis; C: propolis standardized extract (EPP-AF®).

In general, the propolis powder extracts obtained by spray-drying technique investigated in the literature demonstrated the formation of particles at the micrometer scale, from 1 to 10–20 µm. On the other hand, as nanotechnology can offer new opportunities for propolis application, in another line of research, nanosized particles have been developed. Patil et al. [130] have obtained and characterized silver nanoparticles containing propolis [130]. Propolis nanoparticles have also been obtained employing lipid carriers. Our research group has focused on developing solid lipid nanoparticles (SLN) and nanostructured lipid carriers (NLC) encapsulating propolis. **Figure 11** shows atomic force microscopy (AFM) images of propolis-loaded NLC. Additional studies were also conducted covering NLC surface with chitosan. The chitosan-covered particles presented positive residual surface charge [≈ +40 mV], while the uncoated ones presented negative charges [≈ −30 mV]. Particles were anisometric in shape and approximately 150–200 nm in size. The images demonstrate the particle surface and confirm the nanometric size of the particle. Additionally, no roughness was observed on the particle surfaces.

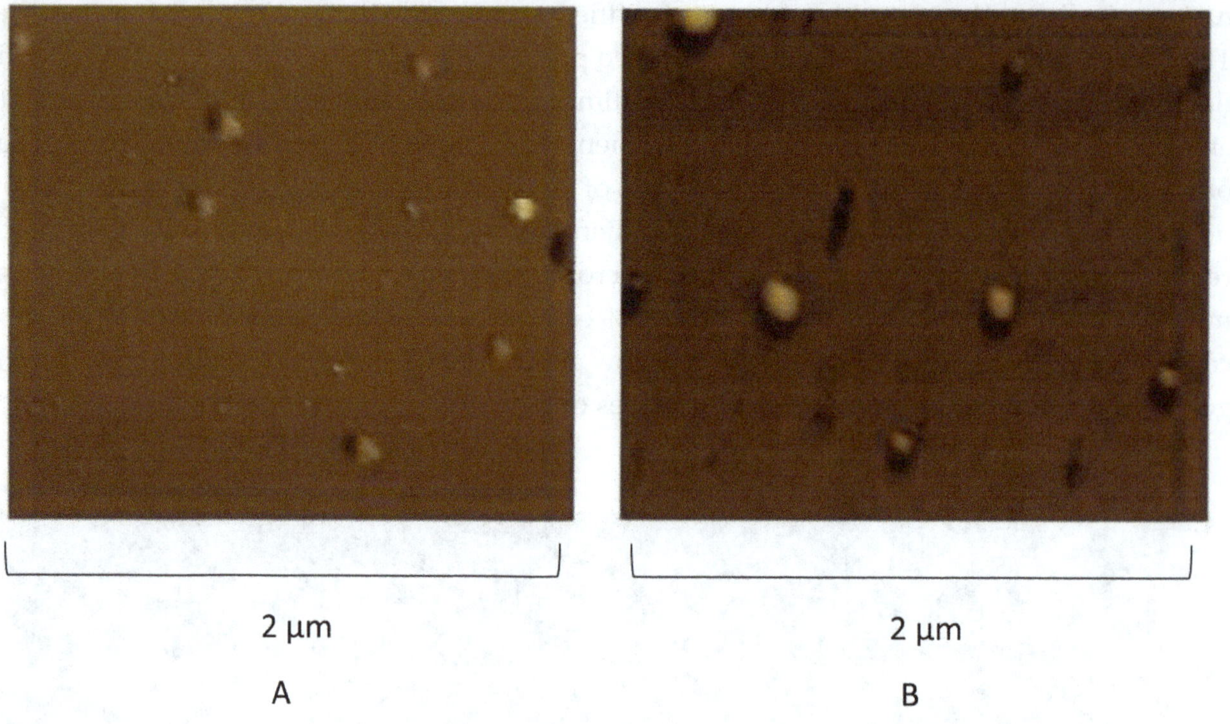

Figure 11. Atomic force microscopy (AFM) images of propolis-NLC (A) and chitosan coated propolis-NLC.

Several important applications can be carried out with innovative propolis extracts. Different presentations of propolis were previously showed as in a liquid presentation without alcohol, usually using propylene glycol or polyethylene glycol, propolis powders in different systems or concentrations, micro or nanoparticles, and others, as soft or hard capsules, with immediate or sustained release systems. Considering the applications, it is possible to formulate capsules, tablets, pills, or others with a specific amount of propolis dry matter, or with a focus on some groups of compounds (total flavonoid or polyphenols) and finally, on a biomarker or a group of these substances such as artepillin C, drupanin, or baccharin, all presentations completely applicable to functional or supplement food, or medicines. Besides oral administration, it is possible to use all of these propolis presentations in topical products, as previously published by Berretta et al. [8] who presented a propolis thermoreversible gel to treat cutaneous lesions or burns, Barud et al. [131] with a propolis biomembrane for the same application, or Berretta et al. [50] that demonstrated the benefits of the application of a propolis mucoadhesive gel in vulvovaginal candidiasis. Several other products can be found in the market and in the literature, such as mouthwashes, toothpastes dental creams, among others.

6. Regulatory affairs

Functional food and natural health products have become an important part of people's daily diet, contributing to the general health of the population and boosting the global food industry. Therefore, its importance is reflected in the interest in regulation of health claims and standards from industry stakeholders and policymakers.

In this chapter, we examine propolis product regulations and policies in many important producing and consuming countries around the world. The goal of this study is to incentive legislators to update the regulation on propolis products in order to improve information available to consumers so they can make better choices and also be provided with healthier and more innovative options.

The regulatory climate worldwide appears to be tending toward propolis classification into the health food products category, although this category also has different names, registration requirements, and allowed claims throughout the world.

Nevertheless, there are still some countries that categorize propolis as a conventional food together with the other bee products, such as honey, royal jelly, and bee pollen. That is the case of Brazil, where the product is regulated by the Ministry of Agriculture with very stringent regulation that limits the product's presentations, information to consumer and does not allow health claims. In 2005, the Brazilian "National Health Surveillance Agency" (ANVISA) published a technical note allowing the registration of propolis as a topical medicine with the claims of anti-inflammatory, antiseptic, and wound healing [132]. The publication of another regulation [133] reinforced the same rules but, due to the very strict rules for medicines, although Brazil is one of the biggest propolis markets, there are no propolis medicines registered to this date.

In the United States, propolis is encompassed together with a wide range of substances by the definition of a dietary supplement in the Dietary Supplement Health Education Act of 1994 (DSHEA) [134]. The use of function claims is also regulated by the above-mentioned regulation that established some special regulatory requirements and procedures for claims of general well-being. These claims are not preapproved by FDA, but the manufacturer must have substantiation that the claim is truthful and not misleading and must submit a notification with the text of the claim to FDA no later than 30 days after marketing the dietary supplement with the claim.

In the European Union (EU), propolis belongs to the food supplement group, regulated by the Directive 2002/46/EC [135], which defines the category as concentrated sources of nutrients or other substances with a nutritional or physiological effect. Since 2006, EU has been engaged in assessing generic health claims to surpass local regulation of member states and after this harmonization product's labels can only bear health claims authorized by the European Food Safety Authority (EFSA) [136], which evaluates scientific data on claims provided by the applicant. Up to this date, there is still no authorized health claim for propolis.

In Australia, all food supplements fall within the category of "complementary medicines" under the *Therapeutic Goods Act 1989* and the supporting Therapeutic Goods Regulations 1990 [137], in which the substances are evaluated according to their level of risk. It includes vitamin, mineral, herbal, aromatherapy, and homeopathic products. A positive list of low-risk substances that may be used has been established and propolis is one of them. It can be used as an active, excipient, or component in all listed medicine formulations. Propolis products can make indications for health maintenance and health enhancement or certain indications for nonserious, self-limiting conditions. It is the manufacturer responsibility to hold evidence to support any indications as well as any other claims made for the medicine (according to Requirements of section 26A of the Act).

Food supplements in Canada are regarded as "Natural Health Products" under the Natural Health Products Regulations (SOR/2003-196) [138–141] and may contain a wide range of substances, such as vitamins and minerals, herbal remedies, homeopathic medicines, traditional medicines, and probiotics. All products must be safe to use as over-the-counter products and not need a prescription to be sold. Propolis is positive listed to be used orally in multiple pharmaceutical dosage forms as a source of antioxidants for the maintenance of good health and to help relieve sore throat and/or other mouth and throat infections. It can also be used topically to assist in minor wound healing.

Japan is one of the first countries to move toward regulating functional foods. There are lists containing a broad range of substances that are not restricted to medicinal use and can therefore be used in food supplements. Propolis in this scenario can be used as an authorized excipient under the Food Sanitation Act 2010, as regular health food without any claims or as an active of a "Food for Specified Health Uses" (FOSHU) [142] with health claims.

The Republic of Korea defines functional food significantly differently from other countries, restricting functional food to nutraceuticals. They are regulated under the Health Functional Food Act of 2004 [143] and there is a positive list in the Health Functional Food Code with 37 categories. Propolis preparations in all forms are allowed and may include two health claims: antioxidant activity and antimicrobial activity in oral cavity.

The People's Republic of China is another example of an Asian country that uses a product-specific system of registration. The State Food and Drug Administration in China (SFDA) [144] regulates these food supplements as "health foods" and maintains positive and negative lists of substances that may be used in health foods. Propolis is in the positive list. There are 27 categories of health function claims approved by the SFDA, but the regulatory process for achieving approval of these health claims is very strict and expensive, requiring the applicant to conduct nonclinical or even clinical studies through an approved agency in addition to the regular scientific literature review.

With this brief regulatory framework on propolis products, we have presented different policies and regulations around the world and we hope that policymakers can improve the regulatory scenario in the near future in order to accelerate and foster innovation in the sector.

7. Propolis nowadays market and potential

The green propolis has gained preference in the world market since the 1980s, unveiling new horizons for the product. The growing interest of the market for green propolis in the context of international food trade follows the increasing trend in search of healthier habits, which have gained ample space in people's daily lives all over the world.

In the Japanese market, green propolis has a high commercial value: according to SEBRAE, the price of 1 kg of this product in 2010 was around $ 87 and the same amount of honey was priced at about 3 dollars. In Tokyo's market, this product is even more valued: a bottle of green propolis, in 2010, was sold for around $ 150 [145, 146]. In 2008, it was estimated that in Japan, 700 million dollars a year started to be moved by green propolis [147]. Japan's interest in green propolis is justified not only by consumption: one of the most important examples of its use is as an adjuvant in the treatment of cancer; but also by Japanese research related to the chemical composition and biological activities of this type of propolis, especially studies with artepillin C [16, 147].

The high demand for green propolis, especially from Asian countries, such as Japan, is essential for sustaining the economy that revolves around this product, which is fairly lucrative. Green propolis is produced mainly in the southeastern region, highlighting the state of Minas Gerais, where there are over 8000 beekeepers, which produces more than 35 tons of propolis per year [148]. These data show the importance of production and exports of green propolis, which is one of the pillars of Brazilian apiculture economy.

Red propolis found in the Brazilian state of Alagoas has been internationally certified by the Brazilian National Institute of Industrial Property (INPI) as the only producer of this kind of propolis in the world and most of its compounds were not found in other types of Brazilian propolis, which makes it a singular bee product [24]. Due to that, its commercial value is internationally high. It has been reported that a kilogram of this product can cost around R $ 500. Its importance to the Brazilian economy and to red propolis producer states is immeasurable. Many propolis producers are being qualified and thereby, they are improving their product quality and the production process. Like green propolis, red propolis is also highly exported to Japan due to its chemical composition and biological effects [149].

Brazil is currently the world's third largest producer of propolis, second only to Russia and China [150]. Although it represents 10–15% of world production, Brazil fulfill about 80% of Japanese demand. Minas Gerais State (Brazil) Beekeepers Federation data show that the propolis produced in the Midwest region of the state is considered the best in the world by the Japanese market, where the kilogram of product has jumped from $ 5 to $ 200 in recent years [150].

The propolis production in Brazil is estimated at around 140 tons, and the major part is destined for international market, both in raw form and as finished products. It is estimated that 100 tons are green propolis and 40 tons other types of propolis. About 80% of the green propolis produced in Brazil comes from the Midwest region of Minas Gerais State, close to the source of the São Francisco river at Serra da Canastra, region where are the highest number of producers. Despite the great Brazilian beekeeping potential, the current production is not enough to fulfill a growing global demand. The Brazilian honey bees are Africanized, presenting defensive and disease-resistant features, with no need to use chemical treatments as in other countries, which ensures Brazilian bee products excellent quality and free from contamination.

Many research fronts have been opened in the pursuit of development and adaptation of professional management techniques in the production of green propolis. In addition to the improvement actions and training of producers in beekeeping management practices, a group of green propolis producers in the Midwest region of Minas Gerais in the Source of the São Francisco River created an independent association supported by governmental agencies, universities, researchers, and local private institution.

This association aims to establish the technical and scientific cooperation between the scientific community and the beekeepers, aimed at regional development, improving the quality and increasing the amount of green propolis produced in the region. Among the main projects carried out, stand out the training of beekeepers in the professionalization of beekeeping, conservation and cultivation of *B. dracunculifolia* fields, periodical replacement of old to younger queen bees, and others projects. It is believed that the interaction of technical and practical knowledge of beekeepers in conjunction with the application of scientific knowledge by researchers and universities will contribute significantly to a comprehensive training in the professionalization of Brazilian beekeepers to fulfill the goal of maintaining the quality and increase the amount of green propolis produced in the region.

Trade promotion strategies are being constantly designed and implemented by the Brazilian Association of Honey and Propolis Exporters (ABEMEL) and the Brazilian Trade and Investment Promotion Agency (APEX-Brasil) to disseminate Brazilian bee products around the world. The result is the increasing demand from Asia, Europe, and North America countries.

In recent years, Brazil has been prominent on the international scene by winning important prizes at the World Beekeeping Awards of Apimondia, the main world beekeeping event that brings together representatives of over 130 countries and is held every 2 years. In the last editions, Brazil won gold and silver medals in the category honey and gold in the category propolis.

8. Future perspectives

Considering all information presented here, it is easy to imagine the important potential of propolis in the health of the population and in the Brazilian and international market, especially because of the important biological activities and safety demonstrated with scientific reports as *"in vitro"*, *"in vivo,"* and in some clinical trials. It is possible to generate several innovative products in different fields considering food, cosmetics, and medicines, and this choice obviously will be related to propolis dosages, formulations, and indications.

It is important to consider the investment in more clinical trials aiming to explore the benefits observed with traditional use and in animal studies, in order to refine the dosages and formulations, with regard to the development of medicines. Besides clinical studies, another important area is the improvement in the investments in productivity in field, since the Brazilian propolis available nowadays is not enough to supply all the countries that may be interested in working with this fabulous natural material produced by bees with the support of Brazilian Biodiversity. And finally, the effort of beepers and entities, such as ABEMEL, is crucial to stimulate and support this work.

Acknowledgements

We would like to thank Nivia Alcici, from Essenciale Propolis Company (Minas Gerais, Brazil), for sending regulatory documentation of propolis of several countries, and also to the Brazilian Association of Honey Exporter, ABEMEL, for financial resources for the publication of this chapter.

Author details

Andresa A. Berretta[1,2,3*], Caroline Arruda[2], Felipe Galeti Miguel[1], Nathalia Baptista[1], Andresa Piacezzi Nascimento[1], Franciane Marquele-Oliveira[1], Juliana Issa Hori[1], Hernane da Silva Barud[4], Bianca Damaso[3,5], César Ramos[3,5], Raul Ferreira[1] and Jairo Kenupp Bastos[2]

*Address all correspondence to: andresa.berretta@apisflora.com.br

1 Apis Flora Indl. Coml. Ltda., Ribeirão Preto, Brazil

2 Faculty of Pharmaceutical Sciences of Ribeirão Preto, University of São Paulo, Ribeirão Preto, Brazil

3 Brazilian Association of Honey Exporters, ABEMEL, Rio Claro, Brazil

4 Uniara, University of Araraquara, Araraquara, Brazil

5 Natucentro Própolis, Bambuí, Brazil

References

[1] Ghisalberti EL. Propolis: a review. Bee World. 1979;60:59–84.

[2] Ioirish N. As Abelhas: Bees: Pharmaceutical with Wings. Editora Mir: Moscou; 1982. 228p.

[3] Marcucci MC. Use of propolis in the cosmetic industry. Aerosol e Cosméticos. 1996; 34–36.

[4] Bankova VS, Popov SS, Marekov NI. A study on flavonoids of propolis. Journal of Natural Products. 1982;46:471–474.

[5] Helfenberg KD. The analysis of beeswax and propolis. Chemiker Zeitungum. 1908;31:987–988.

[6] Iuliu P. Patente n. RO 48101, 1965.

[7] Salatino A, Fernanes-Silva CC, Righi AA, Salatino MLF. Propolis research and the chemistry of plant products. Natural Product Reports. 2011;28:928–925.

[8] Berretta AA, Nascimento AP, Bueno PCP, Vaz MMOLL, Marchetti JM. Propolis standardized extract (EPP-AF®), an innovative chemically and biologically reproducible pharmaceutical compound for treating wounds. International Journal of Biological Sciences. 2012;8:512–521. DOI: 10.7150/ijbs.3641.

[9] Nascimento AP, Ferreira NU, Barizon EA, Rocha BA, Vaz MMOLL, Berretta AA. Methodologies for the evaluation of the antibacterial activity of propolis. African Journal of Microbiology Research. 2013;7:2344–2350. DOI: 10.5897/AJMR12.2362.

[10] Rocha BA, Bueno PCP, Vaz MMOLL, Nascimento AP, Ferreira NU, Moreno GP, Rodrigues MR, Costa-Machado ARM, Barizon EA, Campos JCL, de Oliveira PF, Acésio NO, Martins SPL, Tavares DC, Berretta AA. Evaluation of a propolis water extract using a reliable RP-HPLC methodology and *in vitro* and *in vivo* efficacy and safety characterisation. Evidence-Based Complementary and Alternative Medicine. 2013;2013 Article ID 670451. DOI: 10.1155/2013/670451.

[11] Marquele FD, Di Mambro VM, Georgetti SR, Casagrande R, Valim YML, Fonseca MJ V. Assessment of the antioxidant activities of Brazilian extracts of propolis alone and in topical pharmaceutical formulations. Journal of Pharmaceutical and Biomedical Analysis. 2005;39(3–4):455–462. 10.1016/j.jpba.2005.04.004.

[12] Machado JL, Assuncão AKM, Silva MCP, Reis AS, Costa GC, Arruda DS, Rocha BA, Vaz MMOLL, Paes AMA, Guerra RNM, Berretta AA, do Nascimento FRF. Brazilian green propolis: anti-inflammatory property by an immunomodulatory activity. Evidence-Based Complementary and Alternative Medicine. 2012;2012:157652. DOI:10.1155/2012/157652.

[13] Orsolić N, Knezević AH, Sver L, Terzić S, Basić I. Immunomodulatory and antimetastatic action of propolis and related polyphenolic compounds. Journal of Ethnopharmacology. 2004;94(2–3):307–315.

[14] Orsatti CL, Missima F, Pagliarone AC, Bachiega TF, Búfalo MC, Araújo JP Jr, Sforcin JM. Propolis immunomodulatory action in vivo on Toll-like receptors 2 and 4 expression and on pro-inflammatory cytokines production in mice. Phytotherapy Research. 2010;24(8):1141–1146. DOI: 10.1002/ptr.3086.

[15] Hori JI, Zamboni DS, Carrão DB, Goldman GH, Berretta AA. The Inhibition of inflammasome by Brazilian Propolis (EPP-AF). Evidence-Based Complementary and Alternative Medicine. 2013;2013:418508. DOI: 10.1155/2013/418508.

[16] Paulino N, Abreu SR, Uto Y, Koyama D, Nagasawa H, Hori H, Dirsch VM, Vollmar AM, Scremin A, Bretz WA. Anti-inflammatory effects of a bioavailable

compound, Artepillin C, in Brazilian propolis. European Journal of Pharmacology. 2008;587(1–3):296–301. DOI: 10.1016/j.ejphar.2008.02.067.

[17] Urushisaki T, Takemura T, Tazawa S, Fukuoka M, Hosokawa-Muto J, Araki Y, Kuwata K (2011). Caffeoylquinic acids are major constituents with potent anti-influenza effects in Brazilian green propolis water extract. Evidence-Based Complementary and Alternative Medicine. 2011;2011:254914. DOI: 10.1155/2011/254914.

[18] Daleprane JB, Freitas VS, Pacheco A, Rudnicki M, Faine LA, Dörr FA, Ikegaki M, Salazar LA, Ong TP, Abdalla DSP. Anti-atherogenic and anti-angiogenic activities of polyphenols from propolis. Journal of Nutritional Biochemistry. 2002;23:557–566. DOI:10.1016/j.jnutbio.2011.02.012.

[19] Santana LCLR, Carneiro SMP, Caland-Neto LB, Arcanjo DDR, Moita-Neto JM, Citó AMGL, Carvalho FAA. Brazilian brown propolis elicits antileishmanial effect against promastigote and amastigote forms of Leishmania amazonensis. Natural Product Research: Formerly Natural Product Letters. 2002;28:5, 340–343. DOI: 10.1080/14786419.2013.856904.

[20] Sforcin JM, Bankova V. Propolis: is there a potential for the development of new drugs? Journal of Ethnopharmacology. 2011;133:253–260. DOI: 10.1016/j.jep.2010.10.032

[21] Mello BCBS, Petrus JCC, Hubinger MD. Concentration of flavonoids and phenolic compounds in aqueous and ethanolic própolis extracts through nanofiltration. Journal of Food Engineering. 2010;96:533–539. DOI:10.1016/j.jfoodeng.2009.08.040.

[22] Ishida VFC, Negri G, Salatino A, Bandeira MFCL. A new type of Brazilian propolis: prenylated benzophenones in propolis from Amazon and effects against cariogenic bacteria. Food Chemistry. 2011;125:966–972. DOI:10.1016/j.foodchem.2010.09.089

[23] Park YK, Alencar SM, Aguiar CL. Botanical origin and chemical composition of Brazilian propolis. Journal of Agricultural and Food Chemistry. 2002;50:2502–2506. DOI: 10.1021/jf011432b

[24] Freires IA, Alencar SM, Rosalen PL. A pharmacological perspective on the use of Brazilian Red Propolis and its isolated compounds against human diseases. European Journal of Medicinal Chemistry. 2016;110:267–279. DOI: 10.1016/j.ejmech.2016.01.033

[25] Saito Y, Tsuruma K, Ichihara K, Shimazawa M, Hara H. Brazilian green propolis water extract up-regulates the early expression level of HO-1 and accelerates Nrf2 after UVA irradiation. BMC Complementary and Alternative Medicine. 2015;15:421. DOI 10.1186/s12906-015-0945-4.

[26] Hattori H, Okuda K, Murase T, Shigetsura Y, Narise K, Semenza GL, Nagasawa H. Isolation, identification and biological evaluation of HIF-1-modulating compounds from Brazilian green própolis. Bioorganic & Medicinal Chemistry 19, 2011; 5392–5401. DOI:10.1016/j.bmc.2011.07.060.

[27] Nakajima Y, Shimazawa M, Mishima S, Hara H. Water extract of propolis and its main constituents, caffeoylquinic acid derivatives, exert neuroprotective effects via antioxidant actions. Life Sciences. 2007;80: 370–377. DOI:10.1016/j.lfs.2006.09.017

[28] Figueiredo-Rinhel ASG, Kabeya LM, Bueno PCP, Tiossi RFJ, Azzolini AEC S, Bastos JK, Lucisano-Valim YM. Inhibition of the human neutrophil oxidative metabolism by *Baccharis dracunculifolia* DC (Asteraceae) is influenced by seasonality and the ratio of caffeic acid to other phenolic compounds. Journal of Ethnopharmacology. 2013;150: 655–664. DOI:10.1016/j.jep.2013.09.019

[29] Castro ML, Cury JA, Rosalen PL, Alencar SM, Ikegaki M, Duarte S, Koo H. Propolis from southeast and northeast of Brazil: influence of seasonality on antibacterial activity and phenolic composition. Quimica Nova. 2007;30:1512–1516. DOI: 10.1590/S0100-40422007000700003

[30] Lustosa SR, Galindo AB, Nunes LCC, Randau KP, Rolim Neto PJ. Propolis: updates on chemistry and pharmacology. Revista Brasileira de Farmacognosia. 2008;18:447–454, 2008. DOI: 10.1590/S0102-695X2008000300020

[31] Bankova V, Popova M, Trusheva B. Propolis volatile compounds: chemical diversity and biological activity: a review. Chemistry Central Journal. 2014;8:28. DOI: 10.1186/1752-153X-8-28

[32] SOUSA J P B, JORGE R F, LEITE M F, FURTADO N A J C, DA SILVA FILHO A A, QUEIROGA C L, MAGALHÃES P M, SOARES A E E,BASTOS J K. Seasonal Variation of the (E)-Nerolidol and Other Volatile Compounds Within Ten Different Cultivated Populations of Baccharis dracunculifolia D.C.(Asteraceae). The Journal of Essential Oil Research. 2009;21: 308–314. DOI 10.1080/10412905.2009.9700179

[33] De Mendonça ICG, Porto ICCM, Do Nascimento TG, De Souza NS, Oliveira JMS, Arruda RES, Mousinho KC, Dos Santos AF, Basílio-Júnior ID, Parolia A, Barreto FS. Brazilian red propolis: phytochemical screening, antioxidant activity and effect against cancer cells. BMC Complementary and Alternative Medicine. 2015;15:357. DOI 10.1186/s12906-015-0888-9.

[34] Fasolo D, Bergold AM, Poser GV, Teixeira HF. Determination of benzophenones in lipophilic extract of Brazilian red propolis, nanotechnology-based product and porcine skin and mucosa. Analytical and bioanalytical assays. Journal of Pharmaceutical and Biomedical Analysis. 2016. DOI:10.1016/j.jpba.2016.02.018.

[35] Righi AA, Alves TR, Negri G, Marques LM, Breyer H, Salatino, A. Brazilian red propolis: unreported substances, antioxidant and antimicrobial activities. Journal of the Science of Food and Agriculture. 2011;91:2363–2370. DOI 10.1002/jsfa.4468

[36] Nunes LC, Galindo AB, De Deus ASO, Rufino DA, Randau KP, Xavier HS, Citó AMGL, Rolim Neto PJ. Seasonal variability of the constituents of propolis and bioactivity in saline Artermia. Brazilian Journal of Pharmacognosy. 2009;19:524–529. DOI:10.1590/S0102-695X2009000400003

[37] Salatino A, Teixeira EW, Negri G, Message D. Origin and chemical variation of Brazilian propolis. Evidence-Based Complementary and Alternative Medicine. 2005;2:33–38. DOI:10.1093/ecam/neh060

[38] Bankova V, Marcuccib MC, Simova S, Nikolova N, Kujumgievc A, Popov S. Antibacterial diterpenic acids from Brazilian propolis. Zeitschrift für Naturforschung. 1996;51:277–280.

[39] Sawaya ACHF, Tomazela DM,Cunha IBS, Bankova VS, Marcucci MC, Custodio AR, Eberlin MN. Electrospray ionization mass spectrometry fingerprinting of Própolis. Analyst. 2004;129:739–744. DOI: 10.1039/b403873h

[40] Huang S, Zhang CP, Wang K, Li GQ, Hu FL. Recent advances in the chemical composition of propolis. Molecules. 2014;19:19610–19632. DOI:10.3390/molecules191219610

[41] Silva CSR, Barreto CLP, Peixoto RM, Mota RA, Ribeiro MF, Da Costa MM. Antibacterial effect of Brazilian brown propolis in different solvents against *staphylococcus* spp. Isolated from caprine mastitis. Ciência Animal Brasileira. 2012;13:247–251, 2012.

[42] Fonseca YM, Oliveira FM, Vicentini FTMC, Furtado NAJC, Sousa JPB, Valim YML, Fonseca MJV. Evaluation of the potential of Brazilian propolis against UV-induced oxidative stress. Evidence-Based Complementary and Alternative Medicine. 2011;2012:863917. DOI:10.1155/2011/863917

[43] Fernandes FH, Guterres ZR, Garcez WS, Lopes SM, Corsino J, Garcez FR. Assessment of the (anti)genotoxicity of brown propolis extracts from Brazilian Cerrado biome in a *Drosophila melanogaster* model. Food Research International. 2014;62:20–26. DOI: 10.1016/j.foodres.2014.02.029.

[44] Fernandes FH, Guterresb ZR, Violante IMP, Lopes TFS, Garcez WS, Garcez F R. Evaluation of mutagenic and antimicrobial properties of brown propolis essential oil from the Brazilian cerrado biome. Toxicology Reports. 2015;2:1482–1488. DOI: 10.1016/j.toxrep.2015.11.007.

[45] Pimenta HC, Violante IMP, Musis CR, Borges AH, Aranha AMF.*In vitro* effectiveness of Brazilian brown propolis against *Enterococcus faecalis*. Brazilian Oral Research [online]. 2015;29(1):1–6. DOI: 10.1590/1807-3107BOR-2015.vol29.0058

[46] Jorge R, Furtado NAJC, Sousa JPB, Da Silva Filho AA, Gregório Júnior LE, Martins CHG, Soares AEE, Bastos JK, Cunha WR, Silva MLA. Brazilian Propolis: seasonal variation of the prenylated p-coumaric acids and antimicrobial activity. Pharmaceutical Biology. 2008;46:889–893. DOI: 10.1080/13880200802370373

[47] Fiordalisi SA, Honorato LA, Loiko MR, Avancini CA, Veleirinho MB, Machado Filho LC, Kuhnen S. The effects of Brazilian propolis on etiological agents of mastitis and the viability of bovine mammary gland explants. Journal of Dairy Science. 2016;99:2308–2318. DOI: 10.3168/jds.2015-9777

[48] Koo H, Gomes BPFA, Rosalen PL, Ambrosano GMB, Park YK, Cury JA. *In vitro* antimicrobial activity of propolis and *Arnica montana* against oral pathogens. Archives of Oral Biology. 2000;45:141–148.

[49] De Luca MP, Franca JR, Macedo FAFF, Grenho L, Cortes ME, Faraco AAG, Moreira AN, Santos VR. Propolis varnish: antimicrobial properties against cariogenic bacteria, cytotoxicity, and sustained-release profile. Biomed Research International. 2014; Article ID 348647. DOI: 10.1155/2014/348647

[50] Berretta AA, Castro PA, Cavalheiro AH, Fortes VS, Bom VP, Nascimento AP, Marquele-Oliveira F, Pedrazzi V, Ramalho LNZ, Goldman GH. Evaluation of mucoadhesive gels with propolis (EPP-AF) in preclinical treatment of candidiasis vulvovaginal infection. Evidence-Based Complementary and Alternative Medicine. 2013;2013:641480. DOI: 10.1155/2013/641480

[51] Castro PA, Savoldi MS, Bonatto D, Barros MH, Goldman MHS, Berretta AA, Goldman GH. Molecular characterization of propolis-induced cell death in *Saccharomyces cerevisiae*. Eukaryotic Cell. 2011;10:398–411. DOI: 10.1128/EC.00256-10

[52] Siqueira ABS, Gomes BS, Cambuim I, Maia R, Abreu S, Souza-Motta CM, de Queiroz LA, Porto ALF. Trichophyton species susceptibility to green and red propolis from Brazil. Letters in Applied Microbiology. 2009;48:90–96. DOI: 10.1111/j.1472-765X.2008.02494.x

[53] Ngatu NR, Saruta T, Hirota R, Eitoku M, Luzitu NS, Muzembo BA, Matsui T, Suganuma N. Brazilian green propolis extracts improve *Tinea pedis interdigitalis* and *Tinea corporis*. Journal of Alternative and Complementary Medicine ed. 2012;18:8–9. DOI: 10.1089/acm.2011.0696

[54] Shimizu T, Takeshita Y, Takamori Y, Kai H, Sawamura R, Yoshida H, Watanabe W, Tsutsumi A, Park YK, Yasukawa K, Matsuno K, Shiraki K, Kurokawa M. Efficacy of Brazilian propolis against herpes simplex virus type 1 infection in mice and their modes of antiherpetic efficacies. Evidence-Based Complementary and Alternative Medicine. 2011;2011:976196. DOI: 10.1155/2011/976196

[55] Takemura T, Urushisaki T, Fukuoka M, Hosokawa-Muto J, Hata T, Okuda Y, Hori S, Shigemi T, Araki Y, Kuwata K. 3,4-Dicaffeoylquinic acid, a major constituent of Brazilian propolis, increases TRAIL expression and extends the lifetimes of mice infected with the influenza A virus. Evidence-Based Complementary and Alternative Medicine. 2012;2012:946867. DOI: 10.1155/2012/946867

[56] Daugsch A, Moraes CS, Fort P, Pak YK. Brazilian red propolis—chemical composition and botanical origin. Evidence-Based Complementary and Alternative Medicine. 2008;5:435–441. DOI: 10.1093/ecam/nem057

[57] Inui S, Hatano A, Yoshino M, Hosoya T, Shimamura Y, Masuda S, Ahn MR, Tazawa S, Araki Y, Kumazawa S. Identification of the phenolic compounds contributing to

antibacterial activity in ethanol extracts of Brazilian redpropolis. Natural Product Research. 2014;28:1293–1296. DOI: 10.1080/14786419.2014.898146

[58] Machado BAS, Silva RPD, Barreto GA, Costa SS, da Silva DF, Brandão HN, da Rocha JLC, Dellagostin OA, Henriques JAP, Umsza-Guez MA, Padilha FF. Chemical composition and biological activity of extracts obtained by supercritical extraction and ethanolic extraction of brown, green and red propolis derived from different geographic regions in Brazil. PLoS One. 2016;11:e0145954. DOI: 10.1371/journal.pone.0145954

[59] Das Neves MVM, da Silva TMS, Lima EO, da Cunha EVL, Oliveira EJ. Isoflavone formononetin from red propolis acts as a fungicide against *Candida* sp. Brazilian Journal of Microbiology. 2016;47:159–166. DOI: 10.1016/j.bjm.2015.11.009

[60] Kamuyama O, Abrão Júnior J, Teixeira JMA, De Andrade NJ, Minin VPR, Soares LS. Extract of propolis in the sanitation and conservation of minimally processed carrots. Revista Ceres. 2008;55(3):218–223.

[61] Borges CHF, Almeida DA, Fragiorge EJ. Antibacterial and antifungal activity of different concentrations of propolis hydroalcoholic extract (EHP) in fresh pork sausage. Food Engineering. 2009;6:53–82.

[62] Marquele-Oliveira F, Fonseca YM, Georgetti SR, Vicentini FTMC, Bronzati V, Fonseca MJ V. Evaluation of the antioxidant activity as an additional parameter to attain the functional quality of natural extracts. Latin American Journal of Pharmacy. 2008;27(3):325–332.

[63] Kumazawa S, Hamasaka T, Nakayama T. Antioxidant activity of propolis of various geographic origins. Food Chemistry. 2004;84(3):329–339. DOI 10.1016/S0308-8146(03)00216-4

[64] Duthie GG, Gardner PT, Kyle JAM. Plant polyphenols: are they the new magic bullet? The Proceedings of the Nutrition Society. 2003;62(3):599–603. DOI 10.1079/PNS2003275.

[65] Wang BJ, Lien YH, Yu ZR. Supercritical fluid extractive fractionation—study of the antioxidant activities of propolis. Food Chemistry. 2004;86(2):237–243. DOI 10.1016/j.foodchem.2003.09.031

[66] Banskota AH, Tezuka Y, Kadota S. Recent progress in pharmacological research of propolis. Phytotherapy Research. 2001;15:561–571. DOI 10.1002/ptr.1029.

[67] Shimizu K, Ashida H, Matsuura Y, Kanazawa K. Antioxidative bioavailability of artepillin C in Brazilian propolis. Archives of Biochemistry and Biophysics. 2004;424(2): 181–188. DOI 10.1016/j.abb.2004.02.021.

[68] Santos AC, Uyemura SA, Lopes JLC, Bazon JN, Mingatto FE, Curti C. Effect of naturally occuring flavonoids on lipid peroxidation and membrane permeability transition in mitochondria. Free Radical Biology & Medicine. 1998;24(9):1455–1461.

[69] Ozgová Š, Heřmánek J, Gut I. Different antioxidant effects of polyphenols on lipid peroxidation and hydroxyl radicals in the NADPH-, Fe-ascorbate- and Fe-microsomal systems. Biochemical Pharmacology. 2003;66(7):1127–1137. DOI 10.1016/S0006-2952(03)00425-8.

[70] Campos JF, dos Santos UP, Macorini LFB, de Melo AMMF, Balestieri JBP, Paredes-Gamero EJ, et al. Antimicrobial, antioxidant and cytotoxic activities of propolis from Melipona orbignyi (Hymenoptera, Apidae). Food and Chemical Toxicology. 2014;65:374–380. DOI 10.1016/j.fct.2014.01.008

[71] Siripatrawan U, Vitchayakitti W, Sanguandeekul R. Antioxidant and antimicrobial properties of Thai propolis extracted using ethanol aqueous solution. International Journal of Food Science and Technology. 2013;48(1):22–7. DOI 10.1111/j.1365-2621.2012.03152.x

[72] Graikou K, Popova M, Gortzi O, Bankova V, Chinou I. Characterization and biological evaluation of selected Mediterranean propolis samples. Is it a new type? Food Science and Technology. 2016;65:261–267. DOI 10.1016/j.lwt.2015.08.025.

[73] Tian H, Sun H, Zhang J, Zhang X, Zhao L, Guo S, et al. Ethanol extract of propolis protects macrophages from oxidized low density lipoprotein-induced apoptosis by inhibiting CD36 expression and endoplasmic reticulum stress-C/EBP homologous protein pathway. BMC Complementary and Alternative Medicine. BMC Complementary and Alternative Medicine; 2015;1–12. DOI 10.1186/s12906-015-0759-4.

[74] Thorp E, Tabas I. Mechanisms and consequences of efferocytosis in advanced atherosclerosis. Journal of Leukocyte Biology. 2009;86(5):1089–95. DOI 10.1189/jlb.0209115.

[75] Fonseca YM, Marquele-Oliveira F, Vicentini FTMC, Furtado NAJC, Sousa JPB, Lucisano-Valim YM, et al. Evaluation of the potential of Brazilian propolis against UV-induced oxidative stress. Evidence-Based Complementary and Alternative Medicine. 2011;2011. 10.1155/2011/863917.

[76] Lopes AA, Ferreira TS, Nesi RT, Lanzetti M, Pires KMP, Silva AM, et al. Antioxidant action of propolis on mouse lungs exposed to short-term cigarette smoke. Bioorganic and Medicinal Chemistry. 2013;21(24):7570–7577. DOI 10.1016/j.bmc.2013.10.044.

[77] Luis-Villaroya A, Espina L, García-Gonzalo D, Bayarri S, Pérez C, Pagán R. Bioactive properties of a propolis-based dietary supplement and its use in combination with mild heat for apple juice preservation. International Journal of Food Microbiology. 2015;205:90–7. DOI 10.1016/j.ijfoodmicro.2015.03.020.

[78] Costa SS, Druzian JI, Aparecida B, Machado S, De CO, Guimara G. Bi-Functional biobased packing of the cassava starch, glycerol, licuri nanocellulose and red propolis. PLoS One. 2014;9(11):e112554. DOI. 10.1371/journal.pone.0112554.

[79] Fitzpatrick LR, Wang J, Le T. Caffeic acid phenethyl ester, an inhibitor of nuclear factor-kappaB, attenuates bacterial peptidoglycan polysaccharide-induced colitis in rats. Journal of Pharmacology and Experimental Therapeutics. 2001;299(3): 915–920.

[80] Ledón N, Casacó A, González R, Merino N, González A, Tolón Z. Antipsoriatic, anti-inflammatory, and analgesic effects of an extract of red propolis. Zhongguo Yao Li Xue Bao. 1997;18(3):274–276.

[81] Menezes H, Alvarez JM, Almeida E. Mouse ear edema modulation by different propolis ethanol extracts. Arzneimittelforschung. 1999;49(8):705–707.

[82] Song YS, Park EH, Hur GM, Ryu YS, Kim YM, Jin C. Ethanol extract of propolis inhibits nitric oxide synthase gene expression and enzyme activity. Journal of Ethnopharmacology. 2002;80(2–3):155–161.

[83] Burdock GA. Review of the biological properties and toxicity of bee propolis (propolis). Food and Chemical Toxicology. 1998;36(4):347–363.

[84] Hegazi AG, Abd El Hady FK, Abd Allah FA. Chemical composition and antimicrobial activity of European propolis. Zeitschrift für Naturforschung C. 2000;55(1–2):70–75.

[85] Banskota AH, Tezuka Y, Prasain JK, Matsushige K, Saiki I, Kadota S. Chemical constituents of Brazilian propolis and their cytotoxic activities. Journal of Natural Products1998;61(7):896–900.

[86] Teixeira EW, Message D, Negri G, Salatino A, Stringheta PC. Seasonal variation, chemical composition and antioxidant activity of Brazilian propolis samples. Evidence-Based Complementary and Alternative Medicine. 2010;7(3):307–315.

[87] Szliszka E, Kucharska AZ, Sokół-Łętowska A, Mertas A, Czuba ZP, Król W. Chemical composition and anti-inflammatory effect of ethanolic extract of Brazilian green propolis on activated J774A.1 macrophages. Evidence-Based Complementary and Alternative Medicine. 2013;2013:976415. DOI: 10.1155/2013/976415.

[88] Fischer G, Hubner SO, Vargas GD, and T. Vidor. Immunomodulation by Propolis. Arquivos do Instituto Biológico. 2008;75:247–253.

[89] Bachiega TF, Orsatti CL, Pagliarone AC, Sforcin JM. The effects of propolis and its isolated compounds on cytokine production by murine macrophages. Phytotherapy Research. 2012;26(9):1308–1313. DOI: 10.1002/ptr.3731.

[90] Sy LB, Wu YL, Chiang BL, Wang YH, Wu WM. Propolis extracts exhibit an immunoregulatory activity in an OVA-sensitized airway inflammatory animal model. International Immunopharmacology. 2006;6(7):1053–1060.

[91] Park EH, Kim SH, Park SS. Anti-inflammatory activity of propolis. Archives of Pharmacal Research. 1999;22(6):554–558.

[92] Mirzoeva OK, Calder PC. The effect of propolis and its components on eicosanoid production during the inflammatory response. Prostaglandins, Leukotrienes and Essential Fatty Acids. 1996;55(6):441–449.

[93] Borrelli F, Maffia P, Pinto L, Ianaro A, Russo A, Capasso F, Ialenti A. Phytochemical compounds involved in the anti-inflammatory effect of propolis extract. Fitoterapia. 2002;73:53–63.

[94] de Barros MP, Sousa JP, Bastos JK, de Andrade SF. Effect of Brazilian green propolis on experimental gastric ulcers in rats. Journal of Ethnopharmacology. 2007;110(3): 567–571.

[95] Reis, CM, Carvalho, JCT, Caputo, LRG. Anti-inflammatory and antiulcer activity and subchronic toxicity of propolis ethanolic extract. Revista Brasileira de Farmacognosia. 2000; (10):43–49.

[96] Tavares DC, Mazzaron Barcelos GR, Silva LF, Chacon Tonin CC, Bastos JK. Propolis-induced genotoxicity and antigenotoxicity in Chinese hamster ovary cells. Toxicology In Vitro. 2006;20(7):1154–1158.

[97] Sforcin JM, Fernandes A Jr, Lopes CA, Bankova V, Funari SR. Seasonal effect on Brazilian propolis antibacterial activity. Journal of Ethnopharmacol. 2000;73(1–2):243–249.

[98] de Moura SA, Ferreira MA, Andrade SP, Reis ML, Noviello Mde L, Cara DC. Brazilian green propolis inhibits inflammatory angiogenesis in a murine sponge model. Evidence-Based Complementary and Alternative Medicine. 2011;2011:182703. DOI: 10.1093/ecam/nep197.

[99] Bankova VS, de Castro SL, Marcucci MC. Propolis: recent advances in chemistry and plant origin. Apidologie. 2000; (31):3–15.

[100] Kimoto T, Koya-Miyata S, Hino K, Micallef MJ, Hanaya T, Arai S, Ikeda M, Kurimoto M. Pulmonary carcinogenesis induced by ferric nitrilotriacetate in mice and protection from it by Brazilian propolis and artepillin C. Virchows Archiv. 2001;438(3):259–270.

[101] Gekker G, Hu S, Spivak M, Lokensgard JR, Peterson PK. Anti-HIV-1 activity of propolis in CD4(+) lymphocyte and microglial cell cultures. Journal of Ethnopharmacol. 2005;102(2):158–163.

[102] Messerli SM, Ahn MR, Kunimasa K, Yanagihara M, Tatefuji T, Hashimoto K, Mautner V, Uto Y, Hori H, Kumazawa S, Kaji K, Ohta T, Maruta H. Artepillin C (ARC) in Brazilian green propolis selectively blocks oncogenic PAK1 signaling and suppresses the growth of NF tumors in mice. Phytotherapy Research. 2009;23(3):423–427. DOI: 10.1002/ptr.2658.

[103] Zhao L, Pu L, Wei J, Li J, Wu J, Xin Z, Gao W, Guo C. Brazilian green propolis improves antioxidant function in patients with type 2 diabetes mellitus. International Journal of Environmental Research and Public Health. 2016;13(5). DOI: 10.3390/ijerph13050498.

[104] Fukuda T, Fukui M, Tanaka M, Senmaru T, Iwase H, Yamazaki M, Aoi W, Inui T, Nakamura N, Marunaka Y. Effect of Brazilian green propolis in patients with type 2 diabetes: a double-blind randomized placebo-controlled study. Biomedical Reports. 2015;3(3):355–360.

[105] Mazia RS, de Araújo Pereira RR, de Francisco LM, Natali MR, Dias Filho BP, Nakamura CV, Bruschi ML, Ueda-Nakamura T. Formulation and evaluation of a mucoadhesive thermoresponsive system containing Brazilian green propolis for the treatment of lesions caused by herpes simplex type I. Journal of Pharmaceutical Sciences. 2016;105(1):113–121. DOI: 10.1016/j.xphs.2015.11.016.

[106] Missima F, Sforcin JM. Green Brazilian propolis action on macrophages and lymphoid organs of chronically stressed mice. Evidence-Based Complementary and Alternative Medicine. 2008;5(1):71–75. DOI: 10.1093/ecam/nel112.

[107] Washio K, Kobayashi M, Saito N, Amagasa M, Kitamura H. Propolis ethanol extract stimulates cytokine and chemokine production through NF-κB activation in C2C12 myoblasts. Evidence-Based Complementary and Alternative Medicine. 2015;2015:349751. DOI: 10.1155/2015/349751.

[108] Wu Z, Zhu A, Takayama F, Okada R, Liu Y, Harada Y, Wu S, Nakanishi H. Brazilian green propolis suppresses the hypoxia-induced neuroinflammatory responses by inhibiting NF-κb activation in microglia. Oxidative Medicine and Cellular Longevity. 2013;2013:906726. DOI: 10.1155/2013/906726.

[109] Cho MS, Park WS, Jung WK, Qian ZJ, Lee DS, Choi JS, Lee DY, Park SG, Seo SK, Kim HJ, Won JY, Yu BC, Choi IW. Caffeic acid phenethyl ester promotes anti-inflammatory effects by inhibiting MAPK and NF-κb signaling in activated HMC-1 human mast cells. Pharmaceutical Biology. 2014;52(7):926–932. DOI: 10.3109/13880209.2013.865243.

[110] Park MH, Kang DW, Jung Y, Choi KY, Min do S. Caffeic acid phenethyl ester downregulates phospholipase D1 via direct binding and inhibition of NFκB transactivation. Biochemical and Biophysical Research Communications. 2013;442(1–2):1–7. DOI: 10.1016/j.bbrc.2013.09.105.

[111] Sforcin JM, Kaneno R, Funari SRC. Absence of seasonal effect on the immunomodulatory action of Brazilian própolis on natural killer activity. Journal of Venomous Animals and Toxins. 2002;8(1). DOI.10.1590/S0104-79302002000200005.

[112] Mani F, Damasceno HCR, Novelli ELB, Martins EAM, Sforcin JM. Propolis: effect of different concentrations, extracts and intake period on seric biochemical variables. Journal of Ethnopharmacology. 2006;105(1–2):95–98.

[113] Mohammadzadeh S, Shariatpanahi M, Hamedi M, Ahmadkhaniha R, Samadi N, Ostad SN. Chemical composition, oral toxicity and antimicrobial activity of Iranian propolis. Food Chemistry. 2007;103:1097–1103.

[114] Dobrowolski JW, Vohora SB, Sharma K, Shah SA, Naqvi SA, Dandiya PC. Antibacterial, antifungal, antiamoebic, antiinflammatory and antipyretic studies on propolis bee products. Journal of Ethnopharmacology. 1991;35(1):77–82.

[115] Khayyal MT, El-ghazaly MA, El-khatib AS, Hatem AM, de Vries PJF, El-shafei S, Khattab MM. A clinical pharmacological study of the potential beneficial effects of a propolis food product as an adjuvant in asthmatic patients. Fundamental & Clinical Pharmacology. 2003;17:93–102.

[116] Cohen HA, Varsano I, Kahan E, Sarrel EM, Uziel Y. Effectiveness of an herbal preparation containing echinacea, propolis, and vitamin C in preventing respiratory tract infections in children a randomized, double-blind, placebo-controlled, multicenter study. Archives of Pediatrics and Adolescent Medicine. 2004;158:217–221.

[117] Brätter C, Tregel M, Liebenthal C, Volk HD. Prophylactic effectiveness of propolis for immunostimulation: a clinical pilot study. Forsch Komplementmed. 1999;6(5):256–260.

[118] Samet N, Laurent C, Susarla SM, Samet-Rubinstteen N. The effect of bee propolis on recurrent aphthous stomatitis: a pilot study. Clinical Oral Investigations. 2007;11(2): 143–147.

[119] Zedan H, Hofny ERM, Ismail SA. Propolis as an alternative treatment for cutaneous warts. International Journal of Dermatology. 2009;48:1246–1249.

[120] Coelho LGV, Bastos EMAF, Resende CC, e Silva CMP, Sanches BSF, de Castro FJ, Moretzsohn LD, Vieira WLS, Trindade OR. Brazilian green propolis on Helicobacter pylori infection. A pilot clinic study. Helicobacter. 2007;12:572–574.

[121] Jasprica I, Mornar A, Debeljak Z, Smolcic-Bubalo A, Medic-Saric M, Mayer L, Romic Z, Bucan K, Balog T, Sobocanec S, Sverko V. In vivo study of própolis supplementation effects on antioxidative status and red blood cells. Journal of Ethnopharmacology. 2007;110:548–554.

[122] Park YK, Ikegaki M. Preparation of water and ethanolic extracts of propolis and evaluation of the preparations. Bioscience Biotechnology and Biochemistry. 1998;62(11):2230–2232.

[123] Trusheva B, Trunkova D, Bankova V. Different extraction methods of biologically active components from propolis: a preliminary study. Chemistry Central Journal. 2007;1(13): 1–4. Doi.10.1186/1752-153X-I-13.

[124] Nafady AM, El-Shanawany MA, Mohamed MH, Hassanean HA, Nohara T, Yoshimitsu H, Ono M, Sugimoto H, Doi S, Sasaki K, Kuroda H. Cyclodextrin-enclosed substances of Brazilian própolis. Chemical Pharmaceutical Bulletin. 2003;51(8):984–985.

[125] Da Silva FC, Da Fonseca CR, De Alencar SM, Thomazini M, Balieiro JCDC, Pittia P, et al. Assessment of production efficiency, physicochemical properties and storage stability of spray-dried propolis, a natural food additive, using gum Arabic and OSA

starch-based carrier systems. Food and Bioproducts Processing. 2013;91(1):28–36. DOI 10.1016/j.fbp.2012.08.006.

[126] Nori MP, Favaro-Trindade CS, Matias de Alencar S, Thomazini M, de Camargo Balieiro JC, Contreras Castillo CJ. Microencapsulation of propolis extract by complex coacervation. LWT: Food Science and Technology. 2011;44(2):429–35. DOI 10.1016/j.lwt.2010.09.010.

[127] Bruschi ML, Cardoso MLC, Lucchesi MB, Gremião MPD. Gelatin microparticles containing propolis obtained by spray-drying technique: preparation and characterization. International Journal of Pharmaceutics. 2003;264(1–2):45–55. DOI.org/10.1016/S0378-5173(03)00386-7.

[128] Marquele FD, Stracieri KM, Fonseca MJ V, Freitas LAP. Spray-dried propolis extract. I: physicochemical and antioxidant properties. Pharmazie. 2006;61(4):325–330.

[129] Marquiafável FS, Nascimento AP, Barud H da S, Marquele-Oliveira F, de-Freitas LAP, Bastos JK, et al. Development and characterization of a novel standardized propolis dry extract obtained by factorial design with high artepillin C content. Journal of Pharmaceutical Technology and Drug Research. 2015;4(1):1. DOI 10.7243/2050-120x-4-1

[130] Patil S, Desai N, Mahadik K, Paradkar A. Can green synthesized propolis loaded silver nanoparticulate gel enhance wound healing caused by burns? European Journal of Integrative Medicine. 2015;7(3):243–250. DOI 10.1016/j.eujim.2015.03.002

[131] Barud HS, Araújo Júnior AM, Saska S, Mestieri LB, Campos JADB, de Freitas RM, Ferreira NU, Nascimento AP, Miguel FG, Vaz Mmoll, Barizon EA, Marquele-Oliveira F, Gaspar AMM, Ribeiro SJL, Berretta AA. Antimicrobial Brazilian Propolis (EPP-AF) containing biocellulose membranes as promising biomaterial for skin wound healing. Evidence-Based Complementary and Alternative Medicine. 2013. Available from: http://dx.doi.org/10.1155/2013/703024.

[132] Brazil, ANVISA. Technical Note on the Registration of Products Containing Propolis. [Internet]. 2005. Available from: http://www.anvisa.gov.br/medicamentos/catef/propolis.htm. Accessed August 22, 2016.

[133] Brazil, ANVISA. Resolution – RDC n. 24. De 14 De Junho De 2010 [Internet]. 2010. Available from: http://portal.anvisa.gov.br/documents/10181/2921710/RDC_24_2011.pdf/9a13f0fe-2e81-4fd1-9858-98eed4096813. Accessed August 22, 2016.

[134] United States, FDA. Dietary Supplement Health Education Act of 1994 (DSHEA) [Internet]. 1994. Available from: http://www.fda.gov/Food/Dietarysupplements/default.htm. Accessed August 22, 2016.

[135] European Union, Directive 2002/46/EC of the European Parliament and of the council of 10 June 2002 on the approximation of the laws of the Member States relating to food supplements [Internet]. 2002. Available from: http://eur-lex.europa.eu/LexUriServ/LexUriServ.do?uri=OJ:L:2002:183:0051:0057:EN:PDF.Accessed August 22, 2016.

[136] European Union, Regulation EC No. 1924/2006 of the Parliament and of the Council of 20 December 2006 on Nutrition and Health Claims Made on Foods [Internet]. 2006. Available from: http://eur-lex.europa.eu/LexUriServ/LexUriServ.do?uri=OJ:L:2007:012:0003:0018:EN:PDF.Accessed August 22, 2016.

[137] Australia. Therapeutic Goods Order No. 69 [Internet]. 2009. Available from: https://www.legislation.gov.au/Details/F2009C00264. Accessed August 22, 2016.

[138] Canada. Natural Health Products Regulations (SOR/2003-196) [Internet]. 2003. Available from: http://laws-lois.justice.gc.ca/eng/regulations/SOR-2003-196/. Accessed August 22, 2016.

[139] Canada. Monograph: Propolis—Oral [Internet]. 2014. Available from: http://webprod.hc-sc.gc.ca/nhpid-bdipsn/monoReq.do?id=147. Accessed August 22, 2016.

[140] Canada. Monograph: Propolis—Buccal [Internet]. 2014. Available from: http://webprod.hc-sc.gc.ca/nhpid-bdipsn/atReq.do?atid=propolis_buc&lang=eng. Accessed August 22, 2016.

[141] Canada. Monograph: Propolis—Topical [Internet]. 2014. Available from: http://webprod.hc-sc.gc.ca/nhpid-bdipsn/atReq.do?atid=propolis_top&lang=eng. Accessed August 22, 2016.

[142] Japan. Specifications and Standards for Foods, Food Additives, etc. Under the Food Sanitation Act (Abstract) [Internet]. 2010. Available from: https://www.jetro.go.jp/ext_images/en/reports/regulations/pdf/foodext2010e.pdf. Accessed August 22, 2016.

[143] South Korea. Functional Health Foods Act. [Internet]. 2014. Available from: http://www.mfds.go.kr/files/upload/eng/4_Health%20Functional%20Food%20Act.pdf. Accessed August 22, 2016.

[144] South Korea. Health Functional Food Code. [Internet]. 2014. Available from: http://www.mfds.go.kr/files/upload/eng/7_Health%20Functioanl%20Food%20Codepdf. Accessed August 22, 2016.

[145] Belmiro MS, Oki YY, Fernandes GW. 2011. Available from: http://www.apacame.org.br/mensagemdoce/112/artigo2.htm. Accessed June 12, 2015.

[146] SEBRAE. Brazilian Exportation of Propolis. Available from: http://www.sebrae.com.br/setor/apicultura/sobreapicultura/mercado/exportacoes/Exportacao%20Mel%20-%20Janeiro%202011.pdf. 2011. Accessed June 12, 2015.

[147] Nascimento EA, Chang R, Morais SAL, Piló-Veloso D, Reis DC. An easily detectable chemical marker for Propolis from Alecrim-do-campo (Baccharis dracunculifolia). Revista Brasileira de Farmacognosia. 2008;18:379–386. DOI:/10.1590/S0102-695X2008000300012

[148] FAEMG. High price of the propolis in the market. 2014. Available from: http://www.faemg.org.br/Noticia.aspx?Code=5620&ContentVersion=C&Show=all. Accessed June 12, 2015.

[149] BRASKEM. Red Propolis is recognized as an exclusive product of Brazil. 2012. Available from: https://www.braskem.com.br/detalhe-noticia/Propolis-Vermelha-e-reconhecida-como-produto-exclusivo-do-Brasil. Accessed August 04, 2016.

[150] Pereira AS, Seixas FRMS, Aquino Neto FR. Propolis: 100 years of research and its future. Química Nova. 2002;25(2): http://dx.doi.org/10.1590/S0100-40422002000200021.

4

Native Chilean Fruits and the Effects of their Functional Compounds on Human Health

Elizabeth M. Ulloa-Inostroza, Eric G. Ulloa-Inostroza, Miren Alberdi, Daniela Peña-Sanhueza, Jorge González-Villagra, Laura Jaakola and Marjorie Reyes-Díaz

Abstract

In recent years, there has been great interest in the nutraceutical compounds of fruits from native Chilean plant species. In this context, fruits of *Amomyrtus meli* (Meli), *Aristotelia chilensis* (Maqui), *Berberis microphylla* (Calafate), *Luma apiculata* (Arrayán), *Luma chequén* (Chequén), and *Ugni molinae* (Murtilla) growing predominantly in Chilean forests have been studied. This chapter has compiled the existing information about antioxidant activity and antioxidant compound contents of the above mentioned fruit species and their association with the prevention of pathophysiological disorders in humans, such as inflammation, diabetes, and cardiovascular diseases. Results show that the antioxidant compounds of these species, particularly anthocyanins, decrease inflammation as well as the risk of diabetes and cardiovascular diseases. Therefore, consumption of these fruits is a good alternative for preventing cardiovascular and age-related diseases and pathophysiological disorders.

Keywords: *Amomyrtus meli, Aristotelia chilensis*, arrayán, antioxidant, *Berberis microphylla*, calafate, cardiovascular, carcinogenesis, chequén, diabetes, inflammatory, *Luma apiculata, Luma chequén*, maqui, meli, murtilla, phenolic compounds, *Ugni molinae*

1. Introduction

Chile is a South American country bordered by Perú to the north, Bolivia to the northeast, and Argentina to the east. The continental Chilean territory is a long, narrow strip of land between the Andes in the east and the Pacific Ocean in the west. Its island territories include the Pacific islands of Juan Fernández, Salas y Gómez, Desventuradas, and Pascua and Easter Island in Oceania [1, 2]. Due to its long surface and geographical location, Chile has a large range of different climates from tropical to polar, creating great diversity in its ecosystems and richness in its biodiversity [1–4]. This promotes better use of natural resources for improving the bioeconomy of the country [1, 3]. Traditionally, native Chilean fruits have been used as medicine by Chilean ethnic groups [5]. Currently, studies into native Chilean fruits have focused on identifying and defining their key compounds with medicinal effects [6]. Native Chilean species with edible fruits include woody or shrub forest species belonging to the Elaecarpaceae, Berberidaceae, and particularly Myrtaceae families (**Table 1**). These families are distributed from Illapel to Tierra del Fuego (**Figure 1**) and are species-rich in antioxidant and nutraceutical compounds with benefits to human health, including *Amomyrtus meli* (Meli) (**Figure 2**), *Aristotelia chilensis* (Maqui) (**Figure 3**), *Berberis microphylla* (Calafate) (**Figure 4**), *Luma apiculata* (Arrayán) (**Figure 5**), *Luma chequén* (Chequén) (**Figure 6**), and *Ugni molinae* (Murtilla) (**Figure 7**) [3, 7–11]. These species inhabit the Valdivian Evergreen Forest (Bosque siempreverde Valdiviano) and the Evergreen Patagonian Forest (Bosque siempreverde Patagónico) together with other Chilean forest species that have not been domesticated. Interestingly, Murtilla is the most domesticated among them [10, 12–16].

Scientific name of plant species	Common name	Family	References
Amomyrtus meli (Phil.) D. Legrand & Kausel.	Meli	Myrtaceae	[39, 54]
Aristotelia chilensis (Mol.) Stuntz.	Maqui	Elaeocarpaceae	
Berberis microphylla G. Forst.	Calafate	Berberidaceae	
Luma apiculata (DC.) Burret.	Arrayán, Arrayán rojo, Luma, Palo colorado, Temu	Myrtaceae	
Luma chequen (Mol.) A. Gray.	Arrayán Blanco, Chequén	Myrtaceae	
Ugni molinae Turcz.	Murtilla	Myrtaceae	

Table 1. Scientific and common name as well as families of the native Chilean fruits.

Scalbert et al. [17] and Lila et al. [18] reported a relationship between consumption of fruits with high antioxidant contents and a reduction in oxidative stress in humans. The richness and abundance of enzymatic and nonenzymatic antioxidants in plant species decrease the oxidative damage, helping to inhibit the formation of free radicals by several mechanisms: (1) inhibiting the initiation of the peroxidation, (2) preventing formation of reactive oxygen species (ROS), and (3) breaking the autoxidation chain reaction in humans [19–22]. ROS are normally produced in humans but are exacerbated under exogenous stresses (ozone, cigarette smoking, air pollutants, etc.) [23–27]. Chemical effectiveness of antioxidants against ROS is due to the protection of biomolecules such as lipids, proteins, and carbohydrates to prevent damage by oxidative stress in biomembranes. For this reason, it is very important that people consume antioxidant-rich fruits or foods to decrease ROS [28]. The human diseases associated with oxidative stress are the inflammatory process, cardiovascular disorders, and carcinogenesis (gastric and colorectal) [29–32]. Therefore, fruits with a high antioxidant power are of great benefit in disease prevention related with the inflammatory process and in particular associated with oxidative stress [6, 33–35].

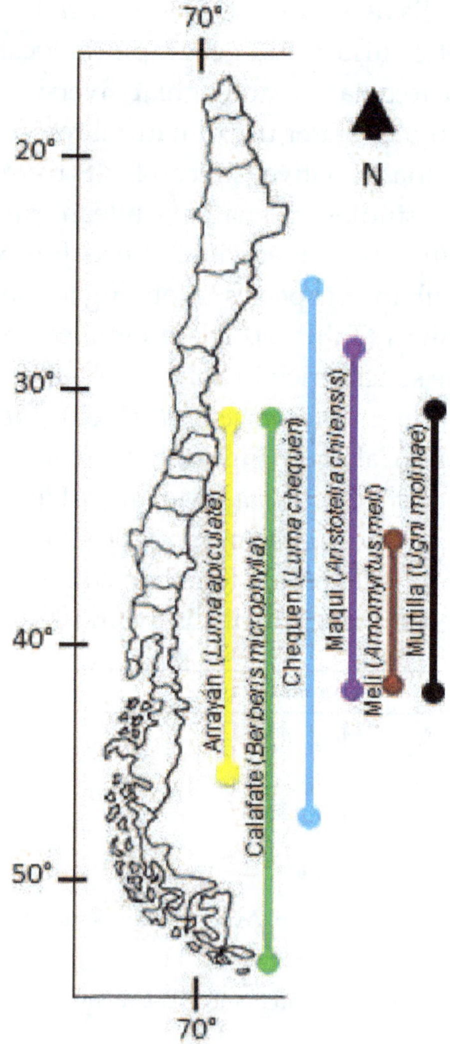

Figure 1. Geographical distribution of native species in the forests of continental Chile.

Figure 2. Pictures of *Amomyrtus meli* (Meli): tree (A), leaves (B), and flowers (C). Extracted from Donoso [40].

Figure 3. Pictures of *Aristotelia chilensis* (Maqui): tree (A), leaves (B), and fruits (C).

Figure 4. Pictures of *Berberis microphylla* (Calafate): flowers (A), fruits (B) and composed pictures of fruits, leaves, and thorns (C).

Figure 5. Pictures of *Luma apiculata* (Arrayán): tree (A), leaves (B), and fruits (C). The picture was replaced by an our picture.

Figure 6. Pictures of *Luma chequen* (Chequén): tree (A), leaves (B), and immature fruits (C).

Figure 7. Pictures of *Ugni molinae* (Murtilla): tree (A), flowers (B), and fruits (C).

Regarding native fruits rich in antioxidants and other bioactive compounds, it is important to detect the quality and quantity of the compounds responsible for the health beneficial properties [36–38]. In this context, the aim of this review is to associate the main bioactive compounds, antioxidant activity, and content of individual phenolic compounds of some native Chilean fruits such as Arrayán (**Figure 5**), Calafate (**Figure 4**), Chequén (**Figure 6**), Maqui (**Figure 3**), Meli (**Figure 2**), and Murtilla (**Figure 7**) with the prevention of diseases and pathophysiological disorders in humans.

2. Morphological characterization, geographical distribution, and ethnobotany of native Chilean fruits

2.1. *A. meli* (Meli) (Phil.) D. Legrand & Kausel

A. meli is a deciduous tree endemic to Chile, reaching a height up to 20 m (**Figure 2A**). It has a very distinctive white trunk, which makes it unmistakable among the species of Chilean trees [39]. It grows mostly on moist and shaded sites from the Arauco to Chiloé Regions (**Figure 1**) in the coastal rainforest, growing on the banks of waterways or under the canopy of other larger species [40]. No conservation problems have been reported for this species, although it is present in a few protected areas such as Chiloé National Park, the Valdivian Coast Reserve, and Oncol Park [39]. Despite its palatability, its fruits are not consumed because they have been reported to cause headaches. The infusion of its leaves is recommended as medicinal use for hypertension (**Figure 2B**) [14, 39]. It has great ornamental potential due to its abundant and very fragrant flowering (**Figure 2C**).

2.2. *A. chilensis* (Maqui) (Mol.) Stuntz

A. chilensis is an evergreen species endemic to Chile (**Figure 3A**), and it is distributed from Illapel (Coquimbo Region) to Chiloé (Los Lagos Region) (**Figure 1**) [37]. It grows naturally, forming wild populations named "macales" and can protect against erosion, since it grows rapidly in abandoned, burned, or overexploited soils [41]. The bright green Maqui leaves are oval-lanceolate with a serrated edge (**Figure 3B**) [42]. The fruits are small, purple bright berries about 5 mm of diameter (**Figure 3C**), with two to five small seeds inside and are eaten fresh, in juice or in jam, and also used as food coloring and in wine-making [42, 43]. Maqui fruits have been of great interest to consumers for their high antioxidant activity, mainly due to phenolic compounds (**Figure 3C**) [44–46], which means this species has been considered a "superfruit" [47].

2.3. *B. microphylla* (Calafate) G. Forst

B. microphylla, commonly called Calafate, is a native Chilean plant species belonging to the Berberidaceae family (**Table 1**) [10, 15]. It is an evergreen shrub 2–3 m high with thorns and single yellow flowers (**Figure 4A**). Its fruit (barberry) is a dark blue-purple berry 7–11 mm in size (**Figure 4B** and **C**) [15, 48, 49]. Berberis genus incorporate among 20–60 species in South America, and the majority is commonly known as Michay or Calafate [50, 51]. In Chile, Calafate is distributed from Curicó to Tierra del Fuego at different altitudes and grows under a wide range of ecological conditions (**Figure 1**) [15, 48, 52]. This species is not cultivated and grows in the wild or in small gardens. Its fruits are used to prepare jam, juice, and wine [10, 15, 48, 50].

2.4. *Luma apiculata* (Arrayán) (DC) Burret

L. apiculata or Arrayán is an evergreen tree of the Valdivian Evergreen Forest (**Figure 5A**). Its bark is smooth reddish with whitish parts, reaching heights from 12 to 15 m tall with a crooked trunk around 50 cm in diameter (**Figure 5A**) [14, 40, 53]. The flowers are hermaphrodites and axillary arranged in groups of three or five, which form a berry, a small, black edible fruit

(1.3–1.5 cm in diameter) (**Figure 5C**) [39, 51]. It is distributed in Chile from the Valparaiso to Aysén Regions (**Figure 1**) on lakeshores and riverbeds and on very moist soils. Trees of this species are conserved in various National Parks, including La Campana, Radal Siete Tazas, Laguna del Laja, Huerquehue, Puyehue, and Chiloé. Arrayán roots have medicinal uses as an anti-hemorrhagic and astringent, while the bark is used to treat herpes and ulcers, and the leaves (**Figure 5B**) to heal wounds, and treat stomach disorders [55].

2.5. *Luma chequen* (Chequén) (Mol.) A. Gray

The Chequén is an endemic tree of the Andean forests in Chile (**Figure 6A**). It is distributed from Coquimbo to Capitán Prat (**Figure 1**) [39]. It grows only in areas of high humidity, on the banks of streams, and in gulches [39]. It is a medium-sized branched shrub, which reaches 9 m in height, with grayish brown bark, and the whole plant gives off a pleasant aroma (**Figure 6A**) [39]. The leaves are oval, wide, and short (**Figure 6B**). The berry fruits are edible and dark purple 1 cm in diameter, ripening in early autumn (**Figure 6C**) [14, 39].

2.6. *Ugni molinae* (Murtilla) Turcz

Natural habitats of Murtilla are forests and coastal mountains from Valparaiso to Aysén (**Figure 1**). It is an evergreen shrub from 1 to 2 m tall (**Figure 7A**) [56, 57]. The dark green leaves are lanceolate, and the flowers are hermaphrodite (**Figure 7B**) [56]. The fruits are bright red globose berries 5–15 mm in diameter with a sweet taste and a strong aroma (**Figure 7C**) [56, 58]. Murtilla is adapted to most soils and is resistant to drought, wind, and cold, but not frost [59]. The traditional uses are for jams, juices, chocolates, and liqueur production (enmurtado or enmutillado).

3. Effect of Chilean fruits on diseases and pathophysiological disorders: beneficial effects on human health

3.1. Bioactive compounds in native Chilean fruits

Native Chilean fruits are naturally rich in phenolic compounds beneficial to human health (**Tables 2** and **3**) [8, 14, 48]. Phenolic compounds are plant secondary metabolites with bioactive properties and are generally involved in the defense against stress conditions in plants [60, 61]. They can be generally characterized by astringency, color, flavor, odor, and oxidative stability [62, 63]. In recent years, the phenolic compounds of wild or domesticated Arrayán, Chequén, Maqui, Meli, and Murtilla have been studied, with the results highlighting the high antioxidant activity of their leaves and fruits [8, 14–16, 48]. The main phenolic compounds in these fruits can be divided into phenolic acids, flavonoids, flavanols, and anthocyanins [63, 64] (**Tables 2** and **3**). Ruiz et al. [48] performed a comparison using total antioxidant activity and Trolox equivalent antioxidant capacity (TEAC) methods among edible fruits of Calafate, Maqui, and Murtilla, showing that Maqui and Calafate had a higher antioxidant activity with 88.1 and 74.5 Trolox equivalent (TE) g^{-1} of FW, respectively, followed by Murtilla with 11.7 TE g^{-1} FW. Afterward, this was confirmed by Dai and Mumper [65], who determined antioxidant activity in Maqui and Murtilla fruits with one of the most commonly used methods, that is, the 2,2-diphenylpicrylhydrazyl (DPPH) assay. The antioxidant activity was higher in Maqui (399.8) than in Murtilla (82.9) in milligrams of crude extract per Liter^{-1} (mg of crude extract L^{-1}). This method is based on the measurement of the antioxidant compounds able to scavenge the stable free radical DPPH. It is a simple, rapid, and inexpensive method [66]. Usually, the results are expressed as milligrams of a sample that bleached 50% of the DPPH solution (IC_{50}) [67]. Therefore, low IC_{50} values show a high antioxidant activity. In this antioxidant assay, Brand-Williams et al. [68] reported that to reach IC_{50} only 0.0016 g L^{-1} of Maqui fruit is necessary; meanwhile, for "Blueberries" (*Vaccinium corymbosum*), "Strawberries" (*Fragaria ananassa*), and "Raspberries" (*Rubus idaeus*) an average of 0.03 g L^{-1} of fruits is needed. The result of IC_{50} for Maqui fruits was confirmed by Fredes et al. [69] and Céspedes et al. [70], who reported IC_{50} values of 0.0012 and 0.0019 g L^{-1} by DPPH assay, respectively. This means that Maqui fruits exhibited the highest antioxidant activity compared with other berries cultivated in Chile. Additionally, Dai and Mumper [65] compared phenolic compounds of leaves and fruits, showing that in Maqui leaves, concentrations were 200% and in Murtilla, 50% higher than in other fruits. Another commonly used method to determine total antioxidants is the oxygen-radical absorbing capacity (ORAC), which measures the antioxidant values as TE and includes both inhibition time and extent of oxidation inhibition [71, 72]. With this method, Prior et al. [73] reported that fruits of wild Calafate have 25 and 150% higher antioxidant activity than Maqui and Murtilla, respectively.

Phenolic compounds profile	Arrayán (*L. apiculata*)	Calafate (*B. microphylla*)	Chequén (*L. chequen*)	Maqui (*A. chilensis*)	Meli (*A. meli*)	Murtilla (*U. molinae*)
Caffeic acid	ND	+	ND	ND	ND	ND
Catechin	ND	ND	ND	+	ND	+
Chlorogenic acid	+	+	+	ND	+	+
Cinnamicbenzenepropenoic acid	ND	ND	ND	ND	ND	+
Coumaric acid	ND	+	ND	ND	ND	ND
Dimethoxy-quercetin	ND	ND	ND	+	ND	ND
Ellagic acid	ND	ND	ND	+	ND	+
Epigallocatechin gallate	+	ND	ND	ND	ND	ND
Ferulic acid	ND	+	ND	ND	ND	ND
Feruloyl-quinic acid	+	+	+	ND	+	+
Furosinin	ND	ND	+	ND	ND	ND
Gallic acid	ND	+	ND	+	ND	+
Hyperoside	+	+	+	ND	+	+
Isoquercitrin	+	+	+	ND	+	+
Isorhamnetin	ND	+	+	ND	ND	ND
Isorhamnetin-3-rutinoside-7-glucoside	ND	+	ND	ND	ND	ND

Phenolic compounds profile	Arrayán (*L. apiculata*)	Calafate (*B. microphylla*)	Chequén (*L. chequen*)	Maqui (*A. chilensis*)	Meli (*A. meli*)	Murtilla (*U. molinae*)
Isorhamnetin-3-galactoside	ND	+	ND	ND	ND	ND
Isorhamnetin-3-glucoside	ND	+	ND	ND	ND	ND
Isorhamnetin-3-rutinoside	ND	+	ND	ND	ND	ND
Isorhamnetin-3-(600-acetyl)-hexoside	ND	+	ND	ND	ND	ND
Isorhamnetin-3-malonylgalactoside	ND	+	ND	ND	ND	ND
Isorhamnetin-3-malonylglucoside	ND	+	ND	ND	ND	ND
Kaempferol	ND	+	ND	ND	ND	+
Kaempferol-deoxyhexoside	ND	+	ND	ND	ND	ND
Myricetin	+	+	+	+	+	+
Myricetin-3-rutinoside-7-glucoside	ND	+	ND	ND	ND	ND
Myricetin-3-glucoside	ND	+	ND	+	ND	ND
Myricetin-3-rutinoside	ND	+	ND	ND	ND	ND
Myricetin-3-O-rhamnose	+	ND	ND	ND	ND	ND
Myricetin-3-galactoside	ND	+	ND	ND	ND	ND
Neochlorogenic acid	+	+	+	ND	+	+
Protocatechuic	ND	ND	ND	+	ND	ND
Protocatechuic 3,4-Dihydroxybenzoic	ND	ND	ND	ND	ND	+
p-coumaric acid	ND	ND	ND	+	ND	+
Quercetin	+	+	+	+	+	+
Quercetin-3-rutinoside-7-glucoside	ND	+	ND	ND	ND	ND
Quercetin-3-galactoside	ND	+	ND	+	ND	ND
Quercetin-3-rutinoside	ND	+	ND	ND	ND	ND
Quercetin-3-glucoside	ND	+	ND	+	ND	ND
Quercetin-3-malonylgalactoside	ND	+	ND	ND	ND	ND
Quercetin-3-malonylglucoside	ND	+	ND	ND	ND	ND
Quercetin-3-rhamnoside	ND	+	ND	ND	ND	ND
Quercetin-galloyl-hexoside	ND	ND	ND	+	ND	ND
Quercetin-3-(600-acetyl)-hexoside 1	ND	+	ND	ND	ND	ND
Quercetin-3-(600-acetyl)-hexoside 2	ND	+	ND	ND	ND	ND

Phenolic compounds profile	Arrayán (*L. apiculata*)	Calafate (*B. microphylla*)	Chequén (*L. chequen*)	Maqui (*A. chilensis*)	Meli (*A. meli*)	Murtilla (*U. molinae*)
Quercetin-3-rhamnoside	ND	+	ND	ND	ND	ND
Quercetin-3-O-(6″-O-galloyl)-hexose	+	ND	ND	ND	ND	ND
Quercetin-3-O-glucose (isoquercitrin)	+	ND	ND	ND	ND	ND
Rutin	+	+	+	ND	+	+
Rutin hydrate	ND	ND	ND	+	ND	ND
Vanillic 4-Hydroxy-3-methoxybenzoic acid	ND	ND	ND	ND	ND	+
Unknown quinic acid derivative	+	ND	+	ND	ND	ND
Unknown gallotannin	ND	ND	+	ND	ND	ND
References	[8, 11, 14]	[14, 15, 48]	[8, 14]	[48, 81]	[14]	[14, 74, 79]

+, presence of compounds in fruits.
ND, not detected.

Table 2. Identification of phenolic compounds in native Chilean fruits.

Anthocyanin profile	Arrayán (*L. apiculata*)	Calafate (*B. microphylla*)	Chequén (*L. chequen*)	Maqui (*A. chilensis*)	Meli (*A. meli*)	Murtilla (*U. molinae*)
Delphinidin-3-glucoside	ND	+	ND	+	ND	+
Cyanidin-3-glucoside	ND	+	ND	+	ND	+
Petunidin-3-glucoside	ND	+	ND	ND	ND	ND
Peonidin-3-glucoside	ND	+	ND	ND	ND	+
Malvidin-3-glucoside	ND	+	ND	ND	ND	ND
Delphinidin-3-rutinoside	ND	+	ND	ND	ND	ND
Cyanidin-3-rutinoside	ND	+	ND	ND	ND	ND
Petunidin-3-rutinoside	ND	+	ND	ND	ND	ND
Peonidin-3-rutinoside	ND	+	ND	ND	ND	ND
Malvidin-3-rutinoside	ND	+	ND	ND	ND	ND

Anthocyanin profile	Arrayán (*L. apiculata*)	Calafate (*B. microphylla*)	Chequén (*L. chequen*)	Maqui (*A. chilensis*)	Meli (*A. meli*)	Murtilla (*U. molinae*)
Delphinidin-3-sambubioside	ND	ND	ND	+	ND	ND
Cyanidin-3-sambubioside	ND	ND	ND	+	ND	ND
Delphinidin-3-rutinoside-5-glucoside	ND	+	ND	ND	ND	ND
Petunidin-3-rutinoside-5-glucoside	ND	+	ND	ND	ND	ND
Malvidin-3-rutinoside-5-glucoside	ND	+	ND	ND	ND	ND
Delphinidin-3-sambubioside-5-glucoside	ND	ND	ND	+	ND	ND
Cyanidin-3-sambubioside-5-glucoside	ND	ND	ND	+	ND	ND
Delphinidin-3,5-dihexoside	ND	+	ND	ND	ND	ND
Cyanidin-3,5-dihexoside	ND	+	ND	ND	ND	ND
Petunidin-3,5-dihexoside	ND	+	ND	ND	ND	ND
Peonidin-3,5-dihexoside	ND	+	ND	ND	ND	ND
Malvidin-3,5-dihexoside	ND	+	ND	ND	ND	ND
Delphinidin-3,5-diglucoside	ND	ND	ND	+	ND	ND
Cyanidin-3,5-diglucoside	ND	ND	ND	+	ND	ND
Delphinidin-3-*O*-arabinoside	ND	+	ND	ND	ND	ND
Peonidin-3-*O*-arabinoside	ND	+	ND	ND	ND	+
Peonidin-3-*O*-di hexoside	ND	+	ND	ND	ND	+
Cyanidin-3-*O*-di-hexoside	+	ND	ND	ND	ND	ND
Cyanidin-3-*O*-galactoside	ND	ND	+	ND	+	ND

Anthocyanin profile	Arrayán (*L. apiculata*)	Calafate (*B. microphylla*)	Chequén (*L. chequen*)	Maqui (*A. chilensis*)	Meli (*A. meli*)	Murtilla (*U. molinae*)
Petunidin-3-*O*-galactoside	+	+	+	ND	+	+
Malvidin-3-*O*-galactoside	+	ND	ND	ND	ND	ND
Delphinidin-3-*O*-glucoside	ND	+	ND	ND	ND	ND
Cyanidin-3-*O*-glucoside	+	ND	+	ND	+	ND
Petunidin-3-*O*-glucoside	+	ND	+	ND	+	+
Peonidin-3-*O*-glucoside	+	+	+	ND	ND	+
Malvidin-3-*O*-glucoside	+	+	+	ND	+	ND
Cyanidin-3-*O*-rutinose	ND	ND	ND	ND	ND	+
Petunidin-3-*O*-rutinoside	ND	+	ND	ND	ND	+
Cyanidin-3-*O*-(6-succinoyl)-glucoside	ND	+	ND	ND	ND	+
Malvidin-3-*O*-(6-coumaroyl) glucoside	ND	+	ND	ND	ND	ND
Petunidin-3-*O*-(6-acetyl) glucoside	ND	+	ND	ND	ND	ND
Malvidin-3-*O*-(6-acetyl) galactoside	+	+	ND	ND	ND	ND
References	[7, 8, 10]	[7, 8, 10, 14, 48, 78]	[8, 10, 14]	[38, 46, 48, 75, 78, 80, 81]	[8, 14]	[10, 14, 48, 79]

+: presence of compounds in fruits.
ND, not detected.

Table 3. Identification of anthocyanins in native Chilean fruits.

Genotype, environmental factors, and geographical location are among the main causes for the differences in the antioxidant capacity in native Chilean fruits, since in all studies, fruits were collected in different locations. Mariangel et al. [15] analyzed Calafate fruit antioxidant capacity by DPPH method from different sites in southern Chile (Mañihuales and El Blanco; Aysén and Temuco and Lonquimay; Araucanía). They found that more southern provenances (Mañihuales and El Blanco) showed a higher antioxidant capacity—9.4 and 7.5 TE g^{-1} of DW, respectively—than northernmost provenances (Temuco and Lonquimay) with 5.2 and 3.3 TE

g^{-1} DW, respectively. This suggests that growth conditions have a direct influence on the content of nutraceutical compounds in fruits, as total phenols in Calafate also showed the same trend: 16.1 mg gallic acid equivalent (GAE) g^{-1} DW (Lonquimay) and 34.6 mg GAE g^{-1} DW (Mañihuales). Thus, fruits from the southernmost region exhibited higher levels of total phenols.

Several types of phenolic compounds have been reported in native Chilean "superfruits," including caffeic acid, ferulic acid, gallic acid, myricetin, *p*-coumaric acid, and others (**Tables 2** and **3**) [8, 14, 15, 47, 48, 74]. In native berries, total phenolics have been analyzed by Ruiz et al. [48] using the Folin-Ciocalteu method. The results showed a higher total phenol content for Maqui (97 μmol GAE g^{-1} FW) followed by Calafate (87 μmol GAE g^{-1} FW) and Murtilla (32 μmol GAE g^{-1} FW). However, no statistically significant differences were found between Maqui and Calafate. Afterward, Brito et al. [14] found higher values for Calafate (65 mg GAE g^{-1} DW) than for Arrayán (27 mg GAE g^{-1} DW), Meli (17 mg GAE g^{-1} DW), Murtilla (9 mg GAE g^{-1} DW), and Chequén (5 mg GAE g^{-1} DW). In addition, these studies have recognized anthocyanins as the most important compounds in native Chilean fruits (**Table 3**) [10, 14, 47, 48, 75–80].

Anthocyanins in Calafate fruits from the Aysén and Magallanes Regions, analyzed by HPLC-DAD, showed total anthocyanin concentrations between 14 and 26 μmol g^{-1} FW, corresponding to the highest values in fruits from the Aysén Region [48]. Comparable anthocyanin values as in Calafate were found in Maqui (16 and 20 μmol g^{-1} FW), whereas the lowest values were found in Murtilla (mean 0.2 μmol g^{-1} FW) and in Blueberry (2.0 μmol g^{-1} FW). The lowest results in Murtilla could be explained by the weaker coloration (rose) of their fruits compared with the black and blue-purple color of the other analyzed fruits [48]. In this context, Mariangel et al. [15] reported differences among Calafate fruits collected in different sites of the Araucanía (Temuco and Lonquimay) and Aysén Regions (Mañihuales and El Blanco). Higher values of cyanidin were found in El Blanco (0.6 mg g^{-1} DW), followed by Temuco (0.2 mg g^{-1} DW), Mañihuales (0.1 mg g^{-1} DW), and Lonquimay (0.06 mg g^{-1} DW). These results suggest that anthocyanin concentrations vary depending on the different agro-characteristics of the growth areas and the fruit-ripening time. Brito et al. [14] reported anthocyanin contents (in mg cyaniding 3-O-glucoside g^{-1} of DW) with higher values in Calafate fruits (51.6) than in Arrayán (15.2), Meli (13), Murtilla (6.85), and finally Chequén (1). Therefore, it is reported that there is higher anthocyanin content in fruits of Calafate and Maqui compared with other native berries [7, 14, 48, 81].

Due to the difficulty of having fresh fruits rich in antioxidants for consumption out of season, it is very important to know the effect of fruit preservation techniques on the content of bioactive compounds. Among the common techniques in use (convective hot-air, freeze drying, and direct cold), native Chilean fruits have demonstrated minor variation in the concentration of phenolic compounds and antioxidant activity compared with the fresh fruits [46, 82]. The effect of freeze-drying and direct cold on the content of bioactive compounds in native Chilean fruits are the least studied of these techniques, and therefore, they should be explored because cold storage in the postharvest period of these berries may produce fewer changes in the antioxidant levels.

3.2. Oxidative stress and antioxidant response in human pathophysiological disorders

Under normal conditions, the human body produces free radicals and other reactive oxygen species (ROS) [29, 83]. ROS are molecules characterized by an unpaired electron in an atomic orbital, being highly unstable, such as hydroxyl radical, superoxide anion radical, hydrogen peroxide, oxygen singlet, hypochlorite, nitric oxide radical, and peroxynitrite radical [83]. In the human body, the ROS molecules are produced in the mitochondria, peroxisomes, in inflammation, phagocytosis processes, and ischemia. They are exacerbated by exposure to stress conditions such as ozone, cigarette smoking, air pollutants, and industrial chemicals, among others [84–86]. The ROS have a high affinity for organic molecules (proteins, carbohydrates, and lipids), causing oxidative stress, damage in biomembranes, and altering body homeostasis [87]. The body counteracts the damage induced by ROS by activating antioxidant systems, which are stable molecules able to donate an electron to the free radical, neutralizing it and reducing ROS damage [88, 89]. The human metabolism produces enzymatic antioxidants such as superoxide dismutase, catalase, xanthine oxidase, lipogenase, and cyclooxygenase, and nonenzymatic compounds such as glutathione, ubiquinol, and uric acid [90]. However, it is also necessary to supply other antioxidants in the diet to strengthen the antioxidant capacity. Among them, vitamin E (α-tocopherol), vitamin C (ascorbic acid), and B-carotene are important. These antioxidants reduce the process of lipid peroxidation, preventing or decreasing oxidative reactions and cell damage [91, 92]. In tissues injured by infection, heat, hypertoxia trauma, and toxin-enhanced oxidative stress, processes are induced in the short-term [36]. To counteract the oxidative damage in tissues, transport of antioxidant enzymes (e.g., xanthine oxidase, lipogenase, and cyclooxygenase) and activation of phagocytes related to the release of free iron, copper ions, or a disruption of the electron transport chains of oxidative phosphorylation take place [36]. However, when the conditions of oxidative damage are higher than antioxidant defense responses, a critical imbalance between free radical generation and antioxidant defenses occurs. This imbalance in the human body induces complications such as diabetes mellitus, onset, promotion or progression of cancer, and neurodegenerative damage such as Parkinson's disease, among others [93–95]. Epidemiological evidence suggests that diets high in antioxidants reduce the incidence of heart disease, cancer, and neurological disorders, altering the inflammatory process common in these diseases [96–100]. Despite the wealth of information that relates to the consumption of fruits in general with the prevention of diseases linked to oxidative stress, studies into native Chilean fruits are limited.

3.3. Native Chilean fruits as a source of antioxidants

Antioxidant compounds in Chilean berries, such as phenolic acids, flavonoids, flavanols, anthocyanins, and procyanidins, among others, have been widely studied for their highly protective effect on human health, particularly with respect to age-related diseases and pathophysiological disorders related to oxidative stresses [6, 11, 37, 69, 82, 86, 101–104]. Antioxidants have an important role as anti-inflammatory or cancer chemopreventive compounds and against degenerative disorders, decreasing the risk of oxidative stress [6, 99, 105]. Furthermore, anthocyanins in berry fruits have a positive effect on human health, related to their capacity to act as antioxidants, and the protective effect in chronic diseases such as diabetes, cardiovascular diseases, and different type of cancers [6, 10, 14, 64, 106, 107]. Regarding the health beneficial properties of fruits, it is important to study different aspects: the species, the composition of bioactive compounds, their antioxidant capacity, and the capacity of different compounds to modulate the transcription factors of enzymes that induce inflammatory diseases [108, 109]. In this context, the traditional use of the plants is also important. Folk medicine in Chile has used leaves and fruits of Arrayán, Calafate, Chequén, Maqui, Meli, and Murtilla to treat throat pain, ulcers, inflammation, and kidney pain disorders [5, 41, 43]. Interestingly, the richness, abundance, and diversity of bioactive compounds in berries of the Chilean native species have shown effects against some pathophysiological disorders (inflammation, diabetes, and cardiovascular) as reported by Fredes et al. [69], Reyes-Farias et al. [6], Alonso [37], Wellen and Hotamisligil [101], Glass and Witztum [104], Lipfert et al. [86], Fuentes et al. [11], and Genskowsky et al. [82].

Inflammation is a defensive mechanism of the organism to specific noxa (factor producing damage), involving different cellular or humoral agents. Its purpose is to restore the body homeostasis, eliminating the noxae. When the inflammatory process becomes self-perpetuating, it ceases to be beneficial and becomes harmful [110–113]. Cell membrane damage activates the phospholipase A2, favoring the synthesis of arachidonic acid, which serves as a substrate for the formation of lipoxygenase and promotes the oxygenase cycle and leukotriene synthesis, which in turn induce the synthesis of prostaglandins and thromboxanes [114]. This promotes neutrophil chemotaxis, ultimately phagocytizing the damaged cell [115]. In addition, leukotriene favors vascular permeability, facilitating the influx of neutrophils to the injured tissue. In general, berries are important inhibitors of inflammatory processes due to the phenolic compounds like anthocyanins present in their fruits [116–121]. This has been supported by reports about moderate consumption of raspberry, strawberry, and bilberry (*Vaccinium myrtillus*) juices and green and black tea that can help prevent the development of early atherosclerosis [118]. Consumption of lingonberry (*Vaccinium vitis-idaea*) juice for 10 weeks has an anti-inflammatory effect on salt-induced hypertension in rat models, probably due to high polyphenol concentrations in the juice [119]. In wild blueberries, the fruits have provided *in vivo* evidence of the improvement or the prevention of metabolic disturbances associated with developing obesity, particularly a systemic low-grade inflammation and hypertension in mice [120]. In the case of strawberry, its tannins (as enriched extract or as pure compounds) are able to act on gastric epithelial cells, thereby inhibiting the inflammatory response [121]. More detailed information about the berries and their anti-inflammatory properties has been reported by Yang and Kortesniemi [122] and Joseph et al. [123]. Studies with native berries such as Maqui and Calafate fruits have given evidence that the compounds of these fruits can be considered a good anti-inflammatory agent [116, 124]. The anti-inflammatory activity has been shown *in vivo* in the ears of mice and *in vitro* using macrophages and guinea pigs [125–127]. Rouanet et al. [118] also showed that leaf extracts of these species have anti-inflammatory activity in mice, showing that leaf extracts containing quercetin and kaempferol can reduce inflammation.

Diabetes has been recognized as one of the most important chronic diseases in the world [128]. According to the International Diabetes Federation (IDF), 400 million people worldwide had diabetes in 2013, and it could reach 642 million by 2040 [128]. In 2013 alone, US$548 billion was spent on diabetes management [128]. Therefore, it is a great challenge to find alternatives to reduce this global epidemic. Type II diabetes comprises disease groups of diverse etiology characterized by the presence of chronic hyperglycemia, altering the secretion and action of insulin, as well as alterations in the metabolism of carbohydrates, proteins, and lipids [128, 129]. Type II diabetes represents over 90% of diabetes cases, and its etiology involves both genetic and environmental factors [130]. In the last few years, a gradual increase in incidences has been reported, inducing metabolic alterations and cardiovascular complications [129–131]. Early treatment with dietary and/or pharmacological hygienic measures in patients with prediabetic states can reduce the incidence of diabetes [132]. However, clearly there is a genetic predisposition that is favored by some factors such as obesity or a sedentary lifestyle [133–135]. In this sense, the adipose tissue sets free inflammatory mediators such as interleukins, tumor necrosis factor (TNF-alpha), or free fatty acids, which increase insulin resistance and oxidative stress [130, 136, 137]. Most of the reports about this disease comprise the effects of commercial berries [138]. Thus, preclinical and clinical studies have suggested that consumption of commercial berries has health benefits with preventive effects on diabetes, improving insulin resistance [122, 139–142]. Studies in obesity-prone rats with a diet containing 2% (wt/wt) freeze-dried powder of highbush blueberry reduced the phenotypes of metabolic syndrome, affecting the gene transcripts of the peroxisome proliferator-activated receptors in adipose and muscle tissues involved in fat and glucose metabolism [143]. This may be due to fibers and/or polyphenols present in the berry fruits. Powder from lingonberries did not change the insulin curve in humans, when consumed together with added glucose [139]. Mursu et al. [140] reported that the intake of berries in the diet may reduce risk of type 2 diabetes in Finnish men. In addition, consumption of blackcurrant (*Ribes nigrum*) extract in amounts roughly equivalent to 100 g in fruit drinks with low sugar and administered immediately before a high-carbohydrate meal reduced postprandial glycemia, insulinemia, and incretin secretion, being beneficial to human health [141]. Further information about the link between berries and their anti-diabetic properties is available in Yang and Kortesniemi [122] and Tsuda [142]. Nevertheless, there are more limited reports in native Chilean fruits in the form of *in vitro* and *in vivo* studies. In this context, Dai and Mumper [65] reported that in *in vitro* studies, Maqui fruits have hypoglycemic effects that inhibit α-amylases and α-glucosidases, enzymes involved in the carbohydrate metabolism. In line with this evidence, the inhibition of both enzymes by Maqui fruits was also confirmed by Schreckinger et al. [144]. Using *in vivo* assays, Fredes et al. [78] reported the anti-diabetic properties of Maqui fruits in mice. They showed that oral administration of a standardized anthocyanin-rich formulation from Maqui fruits decreased blood glucose in obese hyperglycemic mice. It has been suggested that Maqui fruits could act by inhibiting sodium glucose cotransporter in the small intestine [145]. However, more studies are needed to confirm this physiological mechanism involved in reducing blood glucose. The results of *in vitro* studies of the biological effects of phenolic compounds are questioned due to the limited bioavailability and absorption of these compounds in the human body [146].

Cardiovascular diseases are a heterogeneous group of pathologies, the common substrate of which is the alteration of different arteries of the body regardless of caliber of arteries [147, 148]. In particular, these supply the brain, heart, lower limbs, and aorta [149]. The cascade of events for atherosclerosis are proliferation of smooth muscle cells, recruitment of inflammatory cells, and lipid deposits within the blood vessel walls, forming plaques of atheroma [149, 150]. This formation prevents normal tissue irrigation, inducing ischemia reduction or loss of blood flow in a tissue [150]. If the atheromatous plaque ruptures, the body in an attempt to repair activated coagulation cascade induces the formation of a platelet plug, which totally or partially obstructs the flow distally [151]. Therefore, the tissues supplied by the artery suffer hypoxia or anoxia due to the induction of tissue necrosis [149, 152, 153]. Plant antioxidants have been shown to reduce cellular oxidative damage and to protect against cardiovascular diseases [154]. Berry consumption has been known to benefit human health with preventive effects on cardiovascular diseases. Thus, Erlund et al. [155] reported that the consumption of moderate amounts of berries such as bilberries, nectar of lingonberries, blackcurrant, strawberry puree, cold-pressed chokeberry (*Aronia melanocarpa*), and raspberry juice resulted in favorable changes in platelet function, high density lipoprotein (HDL) cholesterol, and blood pressure in male and female volunteers, concluding that the berries may play a role in the prevention of cardiovascular disease. Afterward, Basu et al. [156] deepened the evidence about berry-rich diet that controls the risk of chronic diseases among them the cardiovascular risk. Interestingly, Oudot et al. [157] reported that a high salt diet (8% NaCl) with 2 g/day berries can prevent the cardiac alterations independently of changes in systolic pressure in rats. This was verified by Yang and Kortesniemi [122] and Huang et al. [158], who highlighted that berries are an essential fruit group in heart-healthy diets as a supplementary option to better prevent and control cardiovascular disease in humans. More details are available in Zhu et al. [159] and Rodriguez-Mateos et al. [160]. In the case of native Chilean fruits, the preclinical and clinical studies are more limited, although a recent interest in the study of these berries as a beneficial food to protect the heart is growing. Fredes et al. [69] found that Maqui fruits can significantly reduce the cardiac injury produced by ischemia-reperfusion (I/R) in rat heart *in vivo*. The I/R injury occurs after a myocardial ischemia, and it is known to generate free radicals, heart injury, and necrosis [69, 161, 162]. Therefore, the high level of phenolic compounds and antioxidant activity of this endemic berry can scavenge free radicals produced by I/R and protect the heart [69]. Similarly, concentrated Maqui juice has a high capacity for the oxidation of low-density lipoproteins, which is considered one of the first steps in the development of atherosclerosis [163]. In this sense, Fuentes et al. [11] suggested that Arrayán fruit extracts could protect endothelium-dependent vasodilation (measure to probe endothelial function in different pathophysiological disorders), which is impaired by high glucose [11]. Consequently, they consider that the extract may have an important use in the prevention of vascular damage induced by high glucose. In addition, Falkenberg et al. [164] showed an inhibition of "platelet aggregation" (induced by adenosine diphosphate and collagen) in sheep and human blood through the application of Arrayán and Chequén extracts, which was confirmed by the inhibition of platelet surface activation markers. Afterward, research showed that Murtilla fruits demonstrated vasodilator activity in the aortic rings [165]. Another important damage parameter used as a marker for the risk of developing heart disease is the low-density lipoproteins (LDL). Maqui extracts may reduce oxidative modifications of LDL in overweight people and smokers [124].

Finally, all the properties mentioned of the native Chilean fruits place them in the category of "superfruits" due to their excellent biological effects on human health. Thus, these "superfruits" could be used as nutraceuticals and functional foods with potential use in the human health industry.

4. Conclusion

We conclude that Arrayán, Chequén, Maqui, Meli, and Murtilla fruits are rich in antioxidant compounds with high antioxidant power, such as anthocyanins. The results show that the consumption of native Chilean fruits reduces the risk of pathophysiological disorders such as inflammation, diabetes, and cardiovascular disease due to their bioactive compounds. Therefore, these berries have a potential for increased use in the functional food and nutraceutical industries.

Acknowledgements

We are very grateful to PIA-UFRO PIA16-0006, DI 16-2013 and DI 16-2011 projects from the Dirección de Investigación at the Universidad de La Frontera, Temuco, Chile. Also, PhD fellowship N°21110919 provided by the Comisión Nacional de Investigación Científica y Tecnológica (CONICYT) of the Government of Chile. We would like to thank to Professor Claudio Donoso Zegers for authorizing the use of pictures and Sra. Olga López Sandoval and Dr. Manuel Chacón Fuentes for providing pictures for this chapter.

Author details

Elizabeth M. Ulloa-Inostroza[1], Eric G. Ulloa-Inostroza[2], Miren Alberdi[1, 3], Daniela Peña-Sanhueza[1], Jorge González-Villagra[1], Laura Jaakola[4] and Marjorie Reyes-Díaz[1, 3]*

*Address all correspondence to: marjorie.reyes@ufrontera.cl

1 Center of Plant, Soil Interaction and Natural Resources Biotechnology, Scientific and Technological Bioresource Nucleus (BIOREN), Universidad de La Frontera, Temuco, Chile

2 Department of Internal Medicine, Faculty of Medicine, Universidad de La Frontera, Temuco, Chile

3 Department of Chemical Sciences and Natural Resources, Faculty of Engineering and Sciences, Universidad de La Frontera, Temuco, Chile

4 Department of Arctic and Marine Biology UiT, The Arctic University of Norway, Tromsø, Norway

References

[1] Alvarado G, Moya JC. División Político-Administrativa y Censal 2007. Gobierno de Chile, Santiago-Chile. In: Instituto Nacional de Estadísticas, editors. 2008. 357 p.

[2] Instituto Geográfico Militar. Ejército de Chile [Internet]. 2016. Available from: https://www.igm.cl/div/MAPAIGM/CHILE%20COMPLETO/CHILE%20FISICO.jpg [Accessed: 2026-08-23].

[3] Niemeyer HM. Biologically active compounds from Chilean medicinal plants. In: Arnason JT, Matta R, Romeo JT, eds., Recent Advances in Phytochemistry. Phytochemistry of Medicinal Plants; Vol. 29, Plenum Press. New York, USA. 1995. 137–159 p. DOI: 10.1007/978-1-4899-1778-2_7.

[4] Comisión Nacional del Medio Ambiente (CONAMA). Biodiversidad de Chile, Patrimonio y Desafíos. In: Ocho Libros Editores. Ed: Rovira J, Ugalde J, Stutzin M. Santiago, Chile. 2008.

[5] Mølgaard P, Holler JG, Asar B, Liberna I, Rosenbæk LB, Jebjerg CP, Jørgensen L, Lauritzen J, Guzman A, Adsersen A, Simonsen HT. Antimicrobial evaluation of Huilliche plant medicine used to treat wounds. J Ethnopharmacol. 2011;**138**:219–227. DOI: 10.1016/j.jep.2011.09.006.

[6] Reyes-Farias M, Vasquez K, Ovalle-Marin A, Fuentes F, Parra C, Quitral V, Jimenez P, Garcia-Diaz DF. Chilean native fruit extracts inhibit inflammation linked to the pathogenic interaction between adipocytes and macrophages. J Med Food. 2015;**18**:601–608. DOI: 10.1089/jmf.2014.0031.

[7] Ruiz A, Hermosín-Gutiérrez I, Vergara C, Von Baer D, Zapata M, Hitschfeld A, Obando L, Mardones C. Anthocyanin profiles in south Patagonian wild berries by HPLC-DAD-ESI-MS/MS. Food Res Int. 2013;**51**:706–713. DOI: 10.1016/j.foodres.2013.01.043.

[8] Simirgiotis MJ, Bórquez J, Schmeda-Hirschmann G. Antioxidant capacity, polyphenol content and tandem HPLCDAD- ESI/MS profiling of phenolic compounds from the South American berries *Luma apiculata* and *L. chequen*. Food Chem. 2013;**139**:289–299. DOI: 10.1016/j.foodchem.2013.01.089.

[9] Retamales HA, Scharaschkin T. Comparative leaf anatomy and micromorphology of the Chilean myrtaceae: taxonomic and ecological implications. Flora Morphol Distribution Funct Ecol Plants. 2015;**217**:138–154. DOI: 10.1016/j.flora.2015.10.005.

[10] Ramirez J, Zambrano R, Sepúlveda B, Kennelly EJ, Simirgiotis M. Anthocyanins and antioxidant capacities of six Chilean berries by HPLC–HR-ESI-ToF-MS. Food Chem. 2015;**176**:106–114. DOI: 10.1016/j.foodchem.2014.12.039.

[11] Fuentes L, Valdenegro M, Gómez MG, Ayala-Raso A, Quiroga E, Martínez JP, Vinet R, Caballero E, Figueroa CR. Characterization of fruit development and potential health benefits of arrayan (*Luma apiculata*), a native berry of South America. Food Chem. 2016;**196**:1239–1247. DOI: 10.1016/j.foodchem.2015.10.003.

[12] Landrum L. The myrtle family (Myrtaceae) in Chile. Proc Calif Acad Sci USA. 1988;**45**:277–317.

[13] Murillo-Aldana J, Ruiz E. Revalidación de *Nothomyrcia* (Myrtaceae), un género endémico del Archipiélago de Juan Fernández. Gayana Bot. 2011;**682**:129–134. DOI: 10.4067/S0717-66432011000200002.

[14] Brito A, Areche C, Sepúlveda B, Kennelly E, Simirgiotis M. Anthocyanin characterization, total phenolic quantification and antioxidant features of some Chilean edible berry extracts. Molecules. 2014;**19**:10936–10955. DOI: 10.3390/molecules190810936.

[15] Mariangel E, Reyes-Díaz M, Lobos W, Bensch, E, Schalchli H, Ibarra P. The antioxidant properties of calafate (*Berberis microphylla*) fruits from four different locations in southern Chile. Cienc Investig Agrar. 2013;**40**:161–170. DOI: 10.4067/S0718-16202013000100014.

[16] Chacón-Fuentes MA, Lizama MG, Parra LJ, Seguel IE, Quiroz AE. Insect diversity, community composition and damage index on wild and cultivated murtilla. Cienc Investig Agrar. 2016;**4**:57–67. DOI: 10.4067/S0718-16202016000100006.

[17] Scalbert A, Manach C, Morand C, Remesy C. Dietary polyphenols and the prevention of diseases. Crit Rev Food Sci Nutr. 2005; **45**:287–306. DOI: 10.1080/1040869059096.

[18] Lila MA, Ribnicky D, Rojo L, Rojas-Silva P, Oren A, Havenaar R, Janle E, Raskin I, Yousef G, Grace M. Complementary approaches to gauge the bioavailability and distribution of ingested berry polyphenolics. J Agric Food Chem. 2011;**60**:5763–5771. DOI: 10.1021/jf203526h.

[19] Zaidi SM, Banu N. Antioxidant potential of vitamins A, E and C in modulating oxidative stress in rat brain. Clin Chim Acta. 2004;**340**:229–233. DOI: 10.1016/j.cccn.2003.11.003.

[20] Zaidi SMKR, Al-Qirim TM, Banu N. Effects of antioxidant vitamins on glutathione depletion and lipid peroxidation induced by restraint stress in the rat liver. Drugs R D. 2005;**6**:157–165. DOI: 10.2165/00126839-200506030-00004.

[21] Uttara B, Singh AV, Zamboni P, Mahajan R. Oxidative stress and neurodegenerative diseases: a review of upstream and downstream antioxidant therapeutic options. Curr Neuropharmacol. 2009;**7**:65–74. DOI: 10.2174/157015909787602823.

[22] Sen GL. Remembering one's identity: the epigenetic basis of stem cell fate decisions. FASEB J. 2011;**25**:2123–2128. DOI: 10.1096/fj.11-182774.

[23] Meerson FZ, Kagan VE, Kozlov YuP, Belkina LM, Arkhipenko YuV. The role of lipid peroxidation in pathogenesis of ischemic damage and the antioxidant protection of the heart. Basic Res Cardiol. 1982;**77**:465–485. DOI: 10.1007/BF01907940.

[24] Valavanidis A, Vlachogianni T, Fiotakis K. Tobacco smoke: involvement of reactive oxygen species and stable free radicals in mechanisms of oxidative damage, carcinogenesis and synergistic effects with other respirable particles. Int J Environ Res Publ Health. 2009;**6**:445–462. DOI: 10.3390/ijerph6020445.

[25] Khanna A, Guo M, Mehra M, Royal W. Inflammation and oxidative stress induced by cigarette smoke in Lewis rat brains. J Neuroimmunol. 2013;**254**:69–75. DOI: 10.1016/j.jneuroim.2012.09.006.

[26] Adar SD, Klein R, Klein BE, Szpiro AA, Cotch MF, Wong TY, O'Neill MS, Shrager S, Barr RG, Siscovick DS, Kaufman JD. Air pollution and the microvasculature: a cross-sectional assessment of *in vivo* retinal images in the population-based multi-ethnic study of atherosclerosis (MESA). PLoS Med. 2010;**7**:1549–1277. DOI: 10.1371/journal.pmed.1000372.

[27] Hiltermann JT, Lapperre TS, van Bree L, Steerenberg PA, Brahim JJ, Sont JK, Sterk PJ, Hiemstra PS, Stolk J. Ozone-induced inflammation assessed in sputum and bronchial lavage fluid from asthmatics: a new noninvasive tool in epidemiologic studies on air pollution and asthma. Free Radic Biol Med. 1999;**27**:1448–1454. DOI: 10.1016/S0891-5849(99)00191-4.

[28] Fiedor J, Burda K. Potential role of carotenoids as antioxidants in human health and disease. Nutrients. 2014;**6**:466–488. DOI: 10.3390/nu6020466.

[29] Bandyopadhyay U, Das D, Banerjee RK. Reactive oxygen species: oxidative damage and pathogenesis. Curr Sci. 1990;**77**:658–666.

[30] Chang KC, Chung SY, Chong WS, Suh JS, Kim SH, Noh HK, Seong BW, Ko HJ, Chun KW. Possible superoxide radical-induced alteration of vascular reactivity in aortas from streptozotocin-treated rats. J Pharmacol Exp Ther. 1993;**266**:992–1000.

[31] Hertog MGL, Kromhout D, Aravanis C, Blackburn H, Buz-ina R, Fidanza F, Iampaoli S, Jansen A, Menotti A, Nedeljkovic S, Pekkarinen M, Simic BS, Oshima H, Feskens EJM, Hollman PCH, Katan MB. Flavonoid intake and long-term risk of coronary heart disease and cancer in the seven countries study. Arch Intern Med. 1995;**155**:281–286. DOI: 10.1001/archinte.1995.00430040053006.

[32] Guerra-Araiza C, Álvarez-Mejía AL, Sánchez-Torres S, Farfan-García E, Mondragón-Lozano R, Pinto-Almazán R, Salgado-Ceballos H. Effect of natural exogenous antioxidants on aging and on neurodegenerative diseases. Free Radic Res. 2013;**47**:451–462. DOI: 10.3109/10715762.2013.795649.

[33] Owen RW, Giacosa A, Hull WE, Haubner R, Spiegelhalder B, Bartsch H. The antioxidant/anticancer potential of phenolic compounds isolated from olive oil. Eur J Cancer. 2000;**36**:1235–1247. DOI: 10.1016/S0959-8049(00)00103-9.

[34] Sala A, Recio MD, Giner RM, Manez S, Tournier H, Schinella G, Rios JL. Anti-inflammatory and antioxidant properties of *Helichrysum italicum*. J Pharm Pharmacol. 2002;**54**:365–371. DOI: 10.1211/0022357021778600.

[35] Sun J, Chu YF, Wu XZ, Liu RH. Antioxidant and antiproliferative activities of common fruits. J Agric Food Chem. 2002;**50**:7449–7454. DOI: 10.1021/jf0207530.

[36] Lobo V, Patil A, Phatak A, Chandra A. Free radicals, antioxidants and functional foods: impact on human health. Pharmacogn Rev. 2010;**4**:118–126. DOI: 10.4103/0973-7847.70902.

[37] Alonso JR. Maqui (*Aristotelia chilensis*): un nutracéutico chileno de relevancia medicina. Rev Farmacol Chile. 2012;**5**:95–100.

[38] Céspedes C, Valdez-Morales M, Avila J, El-Hafidi M, Alarcón J, Paredes-López O. Phytochemical profile and the antioxidant activity of Chilean wild black-berry fruits, *Aristotelia chilensis* (Mol.) Stuntz (Elaeocarpaceae). Food Chem. 2010a;**119**:886–895. DOI: 10.1016/j.foodchem.2009.07.045.

[39] Hoffmann AE. Flora silvestre de Chile. 3rd ed. Ediciones Fundación Claudio Gay. Santiago, Chile. 1995.

[40] Donoso C. Las especies arbóreas de los bosques templados de Chile y Argentina Autoecología. Ediciones Marisa Cuneo. Valdivia, Chile. 2006. 677 p.

[41] Hoffmann A, Farga C, Lastra J, Veghazi E. Plantas Medicinales de Uso Común en Chile. Ediciones Fundación Claudio Gay. Santiago, Chile. 1992. 273 p.

[42] Hoffman, A. Flora silvestre de Chile, zona araucana. 5 ed. Ediciones Fundación Claudio Gay. Santiago, Chile. 2005. 257 p.

[43] Muñoz, O. Plantas medicinales de uso en Chile: química y farmacología. 2 ed. Ediciones Universitaria. Santiago, Chile. 2004.

[44] Gironés-Vilaplana A, Mena P, García-Viguera C, Moreno D. A novel beverage rich in antioxidant phenolics: maqui berry (*Aristotelia chilensis*) and lemon juice. LWT Food Sci Technol. 2012;**47**:279–286. DOI: 10.1016/j.lwt.2012.01.020.

[45] Araneda X, Quilamán E, Martínez M, Morales D. Elaboration and evaluation of maqui juice (*Aristotelia chilensis* (Mol.) Stuntz) by stream drag. Sci Agropecu. 2014;**5**:149–156. DOI: 10.17268/sci.agropecu.2014.03.05.

[46] Brauch JE, Buchweitz M, Schweiggert RM, Carle R. Detailed analyses of fresh and dried maqui (*Aristotelia chilensis* (Mol.) Stuntz Berrries and juice. Food Chem. 2016;**190**:308–316. DOI: 10.1016/j.foodchem.2015.05.097.

[47] Ruiz A, Pastene E, Vergara C, Von Baer D, Avello M, Mardones C. Hydroxycinnamic acid derivatives and flavonol profiles of maqui (*Aristotelia chilensis*) fruits. J Chil Chem Soc. 2016;**61**:2792–2796. DOI: 10.4067/S0717-97072016000100010.

[48] Ruiz A, Hermosín I, Mardones C, Vergara C, Herlitz C, Vega M, Dorau C, Winterhalter P, Von Baer D. Polyphenols and antioxidant activity of calafate (*Berberis microphylla*) fruits and other native berries from southern Chile. J Agric Food Chem. 2010;**58**:6081–6089. DOI: 10.1021/jf100173x.

[49] Bottini CM, Bustos C, Bran D. Arbustos de la Patagonia, Calafates y Michay. Presencia. 1993;**8**:5–9.

[50] Bottini MCJ, De Bustos A, Sanso AM, Jouve N, Poggio L. Relationships in Patagonian species of Berberis (Berberidaceae) based on the characterization of rDNA internal transcribed spacer sequences. Bot J Linnean Soc. 2007;**153**:321–328. DOI: 10.1111/j.1095-8339.2007.00586.x.

[51] Fajardo Morales V. Estudio químico de las especies chilenas del género Berberis. Rev Latinoam Quim. 1987;**18**:46–50.

[52] Landrum L. Revision of Berberis (Berberidaceae) in Chile and adjacent southern Argentina. Ann Missouri Bot Gard. 1999;**86**:793–834. DOI: http://dx.doi.org/10.2307/2666170.

[53] Hoffman A. Flora Silvestre de Chile, Zona Araucana. 2ª ed. Ediciones Fundación Claudio Gay. Santiago, Chile. 1991. 285 p.

[54] Gut B. Trees in Patagonia. In: Birkhäuser, editor. Eds:. Birkhäuser BioSciences. 2008. DOI: 10.1007/978-3-7643-8838-6.

[55] Muñoz M, Barrera E, Meza I. El uso medicinal y alimenticio de plantas nativas y naturalizadas en Chile. In: Public. Ocac Museo Hist Nat de Chile. Santiago, Chile. 1981.

[56] Massardo F, Rossi R. Valoración de la biodiversidad: usos medicnales de la flora native chilena. A y D. 1996;**3**:76–81.

[57] Seguel I, Riveros M, Lehnebach C, Torres A. Antecedentes reproductivos y fenológicos de *Ugni molinae* Turcz. (Myrtaceae). Phyton. 1999;**65**:13–21.

[58] Aguila Chacón C, Nahuelhual Muñoz L. Cultivo de murtilla (*Ugni molinae* turcz.) como alternativa de diversificación productiva para la agricultura familiar campesina de la cordillera de la costa. Agro Sur. 2008;**36**:158–167. DOI: 10.4206/agrosur.2008.v36n3-05.

[59] Andrade N, Medel F, Montealegre J. Prospección del estado fitosanitario de arbustos frutales en la Región de los Lagos. Agro Sur. 1984;**12**:59–64.

[60] Landrum L, Donoso C. *Ugni molinae* (Myrtaceae), a potential fruit crop for regions of Mediterranean, maritime and subtropical climates. Econ Bot. 1990;**44**:536–539.

[61] Diaz Napal GN, Defago, M, Valladares G, Palacios S. Response of *Epilachna paenulata* to two flavonoids, Pinocembrin and quercetin, in a comparative study. J Chem Ecol. 2010;**36**:898–904. DOI: 10.1007/s10886-010-9823-1.

[62] Ulloa-Inostroza EM, Alberdi M, Meriño-Gergichevich C, Reyes-Díaz M. Low doses of exogenous methyl jasmonate applied simultaneously with toxic aluminum improve the antioxidant performance of *Vaccinium corymbosum*. Plant Soil. DOI: 10.1007/s11104-016-2985-z.

[63] Rice-Evans C, Miller NJ, Paganga G. Structure antioxidant activity relationships of flavonoids and phenolic acids. Free Radic Biol Med. 1996;**20**:933–956. DOI: 10.1016/0891-5849(95)02227-9.

[64] Rice-Evans C, Miller N, Paganga G. Antioxidant properties of phenolic compounds. Trends Plant Sci. 1997;**2**:152–159. DOI: 10.1016/S1360-1385(97)01018-2.

[65] Dai J, Mumper RJ. Plant phenolics: extraction, analysis and their antioxidant and anticancer properties. Molecules. 2010;**15**:7313–7352. DOI: 10.3390/molecules15107313.

[66] Rubilar M, Jara C, Acevedo F, Gutierrez C, Sineiro J, Shene C. Extracts of maqui (*Aristotelia chilensis*) and Murta (*Ugni molinae* Turcz.): sources of antioxidant compounds and α-glucosidase/α-amylase inhibitors. J Agric Food Chem. 2011;**59**:1630–1637. DOI: 10.1021/jf103461k.

[67] Kedare S, Singh RP. Genesis and development of DPHH method of antioxidant assay. J Food Sci Technol. 2011;**48**:412–422. DOI: 10.1007/s13197-011-0251-1.

[68] Brand-Williams W, Cuvelier ME, Berset C. Use of a free radical method to evaluate antioxidant activity. LWT Food Sci Technol. 1995;**28**:25–30. DOI: 10.1016/S0023-6438(95)80008-5.

[69] Fredes C, Montenegro G, Zoffoli JP, Santander F, Robert P. Comparison of the total phenolic content, total anthocyanin content and antioxidant activity of polyphenol-rich fruits grown in Chile. Cienc Investig Agrar. 2014a;**41**:49–60. DOI: 10.4067/S0718-16202014000100005.

[70] Céspedes C, El-Hafidi M, Pavon N, Alarcón J. Antioxidant and cardioprotective activities of phenolic extracts from fruits of Chilean blackberry *Aristotelia chilensis* (Elaeocarpaceae), maqui. Food Chem. 2008;**107**:820–829. DOI:10.1016/j.foodchem.2007.08.092.

[71] Fredes C, Yousef G, Robert P, Grace M, Lila M, Gomez M, Gebauer M, Montenegro G. Anthocyanin profiling of wild maqui berries (*Aristotelia chilensis* (Mol.) Stuntz) from different geographical regions in Chile. J Sci Food Agric. 2014b;**94**:2639–2648. DOI: 10.1002/jsfa.6602.

[72] Dávalos A, Gómez-Cordovés C, Bartolomé B. Extending applicability of the oxygen radical absorbance capacity (ORAC—fluorescein) assay. J Agric Food Chem. 2004;**52**:48–54. DOI: 10.1021/jf0305231.

[73] Prior RL, Wu XL, Schaich K. Standardized methods for the determination of antioxidant capacity and phenolics in foods and dietary supplements. J Agric Food Chem. 2005;**53**:4290–4302. DOI: 10.1021/jf0502698.

[74] Speisky H, López Alarcón C, Gómez M, Fuentes J, Sandoval Acuña C. First web-based database on total phenolics and oxygen radical absorbance capacity (ORAC) of fruits produced and consumed within the South Andes Region of South America. J Agric Food Chem. 2012;**60**:8851–8859. DOI: 10.1021/jf205167k.

[75] Rodríguez K, Ah-Hen K, Vega-Gálvez A, López J, Quispe-Fuentes I, Lemus-Mondaca R, Gálvez-Ranilla L. Changes in bioactive compounds and antioxidant activity during convective drying of murta (*Ugni molinae* T.) berries. Int J Food Sci Technol. 2013;**49**:1–13. DOI: 10.1111/ijfs.12392.

[76] Escribano-Bailón M, Alcalde-Eon C, Muñoz O, Rivas-Gonzalo J, Santos-Buelga C. Anthocyanins in berries of maqui (*Aristotelia chilensis* (Mol.) Stuntz). Phytochem Anal. 2006;**17**:8–14. DOI: 10.1002/pca.872.

[77] Guerrero J, Ciampi L, Castilla A, Medel F, Schalchli H, Hormazabal E, Bensch E, Alberdi M. Antioxidant capacity, anthocyanins and total phenols of wild and cultivated berries in Chile. Chil J Agric Res. 2010;**70**:537–544. DOI: 10.4067/S0718-58392010000400002.

[78] Fredes C, Montenegro G, Zoffoli JP, Gomez M, Robert P. Polyphenol content and antioxidant activity of maqui (*Aristotelia chilensis* (Molina) Stuntz) during fruit development and maturation in central Chile. Chil J Agric Res. 2012;**72**:582–589. DOI: 10.4067/S0718-58392012000400019.

[79] Rojo LE, Ribnicky D, Logendra S, Poulev A, Rojas-Silva P, Kuhn P, Dorn R, Grace M, Lila MA, Raskin I. *In vitro* and *in vivo* anti-diabetic effects of anthocyanins from Maqui berry (*Aristotelia chilensis*). Food Chem. 2012;**131**:387–396. DOI: 10.1016/j.foodchem.2011.08.066.

[80] Alfaro S, Mutis A, Quiroz A, Seguel I, Scheuermann E. Effects of drying techniques on murtilla fruit polyphenols and antioxidant activity. J Food Res. 2014:**5**;73–82. DOI: 10.5539/jfr.v3n5p73.

[81] Tanaka J, Kadekaru T, Ogawa K, Hitoe S, Shimoda H, Hara H. Maqui berry (*Aristotelia chilensis*) and the constituent delphinidin glycoside inhibit photoreceptor cell death induced by visible light. Food Chem. 2013;**19**:129–137. DOI: 10.1016/j.foodchem.2013.01.036.

[82] Genskowsky E, Puente LA, Pérez-Álvarez JA, Fernández-López J, Muñoz LA, Viuda-Martos M. Determination of polyphenolic profile, antioxidant activity and antibacterial properties of maqui [*Aristotelia chilensis* (Molina) Stuntz] a Chilean blackberry. J Sci Food Agric. 2016;**96**:4235–4242. DOI 10.1002/jsfa.7628.

[83] Rodríguez K, Ah-Henb KS, Vega-Gálvez A, Vásquez V, Quispe-Fuentes I, Pilar Rojas P, Lemus-Mondaca R. Changes in bioactive components and antioxidant capacity of maqui, *Aristotelia chilensis* [Mol] Stuntz, berries during drying. LWT Food Sci Technol. 2016;**65**:537–542. DOI: 10.1016/j.lwt.2015.08.050.

[84] Ohshima H, Tazawa H, Sylla BS, Sawa T. Prevention of human cancer by modulation of chronicinflammatory processes. Mutat Res Fundam Mol Mech Mutagen. 2005;**591**:110–122. DOI: 10.1016/j.mrfmmm.2005.03.030.

[85] Joffe M. Infertility and environmental pollutants. Br Med Bull. 2003;**68**:47–70. DOI: 10.1093/bmb/ldg025.

[86] Lipfert FW, Zhang J, Wyzga RE. Infant mortality and air pollution: a comprehensive analysis of US data for 1990. J Air Waste Manag Assoc. 2000;**50**:1350–1366. DOI: 10.1080/10473289.2000.10464168.

[87] Vergara D, Ávila D, Escobar E, Carrasco-Pozo C, Sánchez A, Gotteland M. The intake of maqui (*Aristotelia chilensis*) berry extract normalizes H_2O_2 and IL-6 concentrations in exhaled breath condensate from healthy smokers—an explorative study. Nutr J. 2015. DOI: 10.1186/s12937-015-0008-1.

[88] D'Autréaux B, Toledano MB. ROS as signalling molecules: mechanisms that generate specificity in ROS homeostasis. 2007;**8**:813–824. DOI: 10.1038/nrm2256.

[89] Valko M, Leibfritz D, Moncol J, Cronin MTD, Mazur M, Telser J. Free radicals and antioxidants in normal physiological functions and human disease. Int J Biochem Cell B. 2007;**39**:44–84. DOI: 10.1016/j.biocel.2006.07.001.

[90] Cotelle N, Bernier JL, Catteau JP, Pommery J, Wallet JC, Gaydou EM. Antioxidant properties of hydroxyflavones. Free Radic Biol Med. 1996;**20**:35–43. DOI: 10.1016/0891-5849(95)02014-4.

[91] Neergheen VS, Bahorun T, Taylor EW, Jen LS, Aruoma OI. Targeting specific cell signaling transduction pathways by dietary and medicinal phytochemicals in cancer chemoprevention. Toxicology. 2010;**278**:229–241. DOI: 10.1016/j.tox.2009.10.010.

[92] Niki E, Noguchi N, Tsuchihashi H, and Gotoh N. Interaction among vitamin C, vitamin E, and beta-carotene. Am J Clin Nutr. 1995;**62**:1322–1326.

[93] Zhang Y, Dawson VL, Dawson TM. Oxidative stress and genetics in the pathogenesis of Parkinson's disease. Neurobiol Dis. 2000;**7**:240–250. DOI: 10.1006/nbdi.2000.0319.

[94] Dröge W. Free radicals in the physiological control of cell function. Physiol Rev. 2002;**82**:47–95. DOI: 10.1152/physrev.00018.2001.

[95] Atack JM, Kelly DJ. Oxidative stress in campylobacter jejuni: responses, resistance and regulation. Future Microbiol. 2009;**4**:677–690. DOI: 10.2217/fmb.09.44.

[96] Iborra M, Moret I, Rausell F, Bastida G, Aguas M, Cerrillo E, Nos P, Beltran B. Role of oxidative stress and antioxidant enzymes in Crohn's disease. Biochem Soc Trans. 2011;**39**:1102–1106. DOI: 10.1042/BST0391102.

[97] Lin Y, Berg AH, Iyengar P, Lam TK, Giacca A, Combs TP, Rajala MW, Du X, Rollman B, Li W, Hawkins M, Barzilai N, Rhodes CJ, Fantus IG, Brownlee M, Scherer PE. The hyperglycemia-induced inflammatory response in adipocytes: the role of reactive oxygen species. J Biol Chem. 2005;**280**:4617–4626. DOI: 10.1074/jbc.M411863200.

[98] Furukawa S, Fujita T, Shimabukuro M, Iwaki M, Yamada Y, Nakajima Y, Nakayama O, Makishima M, Matsuda M, Shimomura I. Increased oxidative stress in obesity and its impact on metabolic syndrome. J Clin Invest. 2004;**114**:1752–1761. DOI: 10.1172/JCI21625.

[99] Urakawa H, Katsuki A, Sumida Y, Gabazza EC, Murashima S, Morioka K, Maruyama N, Kitagawa N, Tanaka T, Hori Y, Nakatani K, Yano Y, Adachi Y. Oxidative stress is associated with adiposity and insulin resistance in men. J Clin Endocrinol Metab. 2003;**88**:4673–4676. DOI: 10.1210/jc.2003-030202.

[100] Ferrari N, Tosetti F, De Flora S, Donatelli F, Sogno I, Noonan DM, Albini A. Diet-derived phytochemicals: from cancer chemoprevention to cardio-oncological prevention. Curr Drug Targets. 2011;**13**:1909–1924. DOI: 10.2174/138945011798184227.

[101] Wellen KE, Hotamisligil GS. Inflammation, stress, and diabetes. J Clin Invest. 2005;**115**:1111–1119. DOI: 10.1172/JCI200525102DS1.

[102] Peña J. *Aristotelia chilensis*: a possible nutraceutical or functional food. Med Chem. 2015;**5**:378–382. DOI: 10.4172/2161-0444.1000289.

[103] Stocker, R. Dietary and pharmacological antioxidants in atherosclerosis. Curr Opin Lipidol. 1999;**10**:589–597.

[104] Glass CK, Witztum JL. Atherosclerosis. The road ahead. Cell. 2001;**104**:503–516. DOI: 10.1016/S0092-8674(01)00238-0.

[105] Watson RR, Schönlau F. Nutraceutical and antioxidant effects of a delphinidin-rich maqui berry extract Delphinol®: a review. Minerva Cardioangiol. 2015;**63**:1–12.

[106] Priyadarsini RV, Nagini S. Cancer chemoprevention by dietary phytochemicals: promises and pitfalls. Curr Pharm Biotechnol. 2012;**13**:125–136. DOI: 10.2174/138920112798868610.

[107] Zamora-Ros R, Fedirko V, Trichopoulou A, González CA, Bamia C, Trepo E, Nöthlings U, Duarte-Salles T, Serafini M, Bredsdorff L, Overvad K, Tjønneland A, Halkjaer J, Fagherazzi G, Perquier F, Boutron-Ruault MC, Katzke V, Lukanova A, Floegel A, Boeing H, Lagiou P, Trichopoulos D, Saieva C, Agnoli C, Mattiello A, Tumino R, Sacerdote C, Bueno-de-Mesquita HB, Peeters PH, Weiderpass E, Engeset D, Skeie G, Argüelles MV, Molina-Montes E, Dorronsoro M, Tormo MJ, Ardanaz E, Ericson U, Sonestedt E, Sund M, Landberg R, Khaw KT, Wareham NJ, Crowe FL, Riboli E, Jenab M. Dietary flavonoid, lignan and antioxidant capacity and risk of hepatocellular carcinoma in the European prospective investigation into cancer and nutrition study. Int J Cancer. 2013;**133**:2429–2443. DOI: 10.1002/ijc.28257.

[108] Pojer E, Mattivi F, Jhonson D, Stokey C. The case for anthocyanin consumption to promote human health: a review. Compr Rev Food Sci Food Saf. 2013;**12**:483–508. DOI: 10.1111/1541-4337.12024.

[109] Kang JH, Pasquale LR, Willett W, Rosner B, Egan KM, Faberowski N, Hankinson SE. Antioxidant intake and primary open-angle glaucoma: a prospective study. Am J Epidemiol. 2003;**158**:337–346. DOI:10.1093/aje/kwg167.

[110] Zhang K, Zuo Y. GC-MS determination of flavonoids and phenolic and benzoic acids in human plasma after consumption of cranberry juice. J Agric Food Chem. 2004;**52**:222–227. DOI: 10.1021/jf035073r.

[111] Balkwill F, Mantovani A. Inflammation and cancer: back to Virchow?. Lancet. 2001;**357**:539–545. DOI: 10.1016/S0140-6736(00)04046-0.

[112] Palapattu GS, Sutcliffe S, Bastian PJ, Platz EA, De Marzo AM, Isaacs WB, Nelson WG. Prostate carcinogenesis and inflammation: emerging insights. Carcinogenesis. 2005;**26**:1170–1181. DOI: 10.1093/carcin/bgh317.

[113] Schwartsburd PM. Chronic inflammation as inductor of procancer microenvironment: pathogenesis of dysregulated feedback control. Cancer Metastasis Rev. 2003;**22**:95–102. DOI: 10.1023/A:1022220219975.

[114] Szafran B, Borazjani A, Lee JH, Ross MK, Kaplan BLF. Lipopolysaccharide suppresses carboxylesterase 2 g activity and 2-arachidonoylglycerol hydrolysis: a possible mechanism to regulate inflammation. Prostaglandins Other Lipid Mediat. 2015;**121**:199–206. DOI: 10.1016/j.prostaglandins.2015.09.005.

[115] Simmons DL, Botting RM, Ha T. Cyclooxygenase isozymes: the biology of prostaglandin synthesis and inhibition. Pharmacol Rev. 2004;**56**:387–437. DOI: 10.1124/pr.56.3.3.

[116] Jiang M, Klein M, Zanger UM, Mohammad MK, Cave MC, Gaikwad NW, Dias NJ, Selcer KW, Guo Y, He J, Zhang X, Shen Q, Qin W, Li J, Li S, Xie W. Inflammatory regulation of steroid sulfatase: a novel mechanism to control estrogen homeostasis and inflammation in chronic liver disease. J Hepatol. 2016;**64**:44–52. DOI: 10.1016/j.jhep.2015.07.022.

[117] Romanucci V, D'Alonzo D, Guaragna A, Di Marino C, Davinelli S, Scapagnini G, Di Fabio G, Zarrelli A. Bioactive compounds of *Aristotelia chilensis* Stuntz and their pharmacological effects. Curr Pharm Biotechnol. 2016;**17**:1:11.

[118] Rouanet JM, Décordé K, Del Rio D, Auger C, Borges G, Cristol JP, Lean MEJ, Crozier A. Berry juices, teas, antioxidants and the prevention of atherosclerosis in hamsters. Food Chem. 2010;**118**:266–271. DOI: 10.1016/j.foodchem.2009.04.116.

[119] Kivimäki AS, Siltari A, Ehlers PI, Korpela R, Vapaatalo H. Lingonberry juice negates the effects of a high salt diet on vascular function and low-grade inflammation. J Funct Foods. 2014;**7**:238–245.

[120] Mykkänen OT, Huotari A, Herzig KH, Dunlop TW, Mykkänen H, Kirjavainen PV. Wild blueberries (*Vaccinium myrtillus*) alleviate inflammation and hypertension associated with developing obesity in mice fed with a high-fat diet. PLoS One. 2014. DOI: 10.1371/journal.pone.0114790.

[121] Fumagalli M, Sangiovanni E, Vrhovsek U, Piazza S, Colombo E, Gasperotti M, Mattivi F, De Fabiani E, Dell'Agli M. Strawberry tannins inhibit IL-8 secretion in a cell model of gastric inflammation. Pharmacol Res. 2016;**111**:703–712. DOI: 10.1016/j.phrs.2016.07.028.

[122] Yang B, Kortesniemi M. Clinical evidence on potential health benefits of berries. Curr Opin Food Sci. 2015;**2**:36–42. DOI: 10.1016/j.cofs.2015.01.002.

[123] Joseph SV, Edirisinghe I, Burton-Freeman BM. Berries: anti-inflammatory effects in humans. J Agric Food Chem. 2014;**62**:3886–3903. DOI: 10.1021/jf4044056.

[124] Davinelli S, Bertoglio JC, Zarrelli A, Pina R, Scapagnini G. A randomized clinical trial evaluating the efficacy of an anthocyanin–maqui berry extract (Delphinol®) on oxidative stress biomarkers. J Anim Physiol Anim Nutr. 2015;**34**:28–33. DOI: 10.1080/07315724.2015.1080108.

[125] Céspedes C, Alarcon J, Avila J, Nieto A. Anti-inflammatory activity of *Aristotelia chilensis* Mol. (Stuntz) (Elaeocarpaceae). BLACPMA. 2010b;**9**:91–99.

[126] Schreckinger M, Wang J, Yousef G, Lila M, Gonzalez de Mejia E. Antioxidant capacity and *in vitro* inhibition of adipogenesis and inflammation by phenolic extracts of *Vaccinium floribundum* and *Aristotelia chilensis*. J Agric Food Chem. 2010;**58**:8966–8976. DOI: 10.1021/jf100975m.

[127] Muñoz O, Christen P, Cretton S, Backhouse N, Torres V, Correa O, Costa E, Miranda H, Delporte C. Chemical study and anti-inflammatory, analgesic and antioxidant activities of the leaves of *Aristotelia chilensis* (Mol.) Stuntz. Elaeocarpaceae. 2011;**63**:849–859. DOI: 10.1111/j.2042-7158.2011.01280.x.

[128] The IDF Diabetes Atlas Seventh Edition [Internet]. 2015. Organización Mundial de la Salud. 2016. Informe mundial sobre la diabetes. Available from: http://www.idf.org/diabetesatlas. [Accessed: 2016-01-18].

[129] Muniyappa R, Montagnani M, Koh KK, Quon MJ. Cardiovascular actions of insulin. Endocr Rev. 2007;**28**:463–491. DOI: 10.1210/er.2007-0006.

[130] Murdock KW, LeRoy AS, Lacourt TA, Duke DC, Heijnen CJ, Fagunde CP. Executive functioning and diabetes: the role of anxious arousal and inflammation. Psycho Neuro Endocrinol. 2016;**71**:102–109. DOI: 10.1016/j.psyneuen.2016.05.006.

[131] Bacha F, Gidding SS, Pyle L, Katz LL, Kriska A, Nadeau KJ, Lima JAC, On behalf of the Treatment Options for Type 2 Diabetes in Adolescents and Youth (TODAY) Study Group. Relationship of cardiac structure and function to cardiorespiratory fitness and lean body mass in adolescents and young adults with type 2 diabetes. J Pediatr. 2016. DOI: 10.1016/j.jpeds.2016.06.048.

[132] Wild SH, Morling JR, McAllister DA, Kerssens J, Fischbacher C, Parkes J, Roderick PJ, Sattar N, Christopher D. Byrne CD, On behalf of the Scottish and Southampton Diabetes and Liver Disease Group, and the Scottish Diabetes Research Network Epidemiology Group. Type 2 diabetes and risk of hospital admission or death for chronic liver diseases. J Hepatol. 2016;**64**:1358–1364. DOI: 10.1016/j.jhep.2016.01.014.

[133] Diouf I, Magliano DJ, Carrington MJ, Stewart S, Shaw JE. Prevalence, incidence, risk factors and treatment of atrial fibrillation in Australia: the Australian diabetes, obesity and lifestyle (AusDiab) longitudinal, population cohort study. Int J Cardiol. 2016;**205**:127–132. DOI: 10.1016/j.ijcard.2015.12.013.

[134] Fukushima A, Lopaschuk GD. Cardiac fatty acid oxidation in heart failure associated with obesity and diabetes. Biochim Biophys Acta. 2016;**1861**:1525–1534. DOI: 10.1016/j.bbalip.2016.03.020.

[135] Merger SR, Kerner W, Stadler M, Zeyfang A, Jehle P, Müller-Korbsch M, Holl RW, For the DPV Initiative, and the German BMBF Competence Network Diabetes mellitus. Prevalence and comorbidities of double diabetes. Diabetes Res Clin Pract. 2016;**119**:48–56. DOI: 10.1016/j.diabres.2016.06.003.

[136] Orio F, Muscogiuri G, Nese C, Palomba S, Savastano S, Tafuri D, Colarieti G, La Sala G, Colao A, Yildiz BO. Obesity, type 2 diabetes mellitus and cardiovascular disease risk: an up to date in the management of polycystic ovary syndrome. Eur J Obstet Gynecol Reprod Biol. DOI: 10.1016/j.ejogrb.2016.08.026.

[137] Zlobine I, Gopal K, Ussher JR. Lipotoxicity in obesity and diabetes related cardiac dysfunction. Biochim Biophys Acta. 2016;**1861**:1555–1568. DOI: 10.1016/j.bbalip.2016.02.011.

[138] Huntley AL. The health benefits of berry flavonoids for menopausal women: cardiovascular disease, cancer and cognition. Maturitas. 2009;**63**:297–301. DOI: 10.1016/j.maturitas.2009.05.005.

[139] Linderborg KM, Järvinen R, Lehtonen HM, Viitanen M, Kallio HPT. The fiber and/or polyphenols present in lingonberries null the glycemic effect of the sugars present in the berries when consumed together with added glucose in healthy human volunteers. Nutr Res. 2012;**32**:471–478. DOI: 10.1016/j.nutres.2012.06.004.

[140] Mursu J, Virtanen JK, Tuomainen TP, Nurmi T, Voutilainen S. Intake of fruit, berries, and vegetables and risk of type 2 diabetes in Finnish men: the Kuopio Ischaemic heart disease risk factor study. Am J Clin Nutr. 2013. DOI: 10.3945/ajcn.113.069641.

[141] Castro Acosta ML, Smith L, Miller RJ, McCarthy DI, Farrimond JA, Hall WL. Drinks containing anthocyanin-rich blackcurrant extract decrease postprandial blood glucose, insulin and incretin concentrations. J Nutr Biochem. 2016. DOI: 10.1016/j.jnutbio.2016.09.002.

[142] Tsuda T. Recent progress in anti-obesity and anti-diabetes effect of berries. Antioxidants. 2016. DOI:10.3390/antiox5020013.

[143] Seymour EM, Tanone II, Urcuyo-Llanes DE, Lewis SK, Kirakosyan A, Kondoleon MG, Kaufman PB, Bolling SF. Blueberry intake alters skeletal muscle and adipose tissue peroxisome proliferator-activated receptor activity and reduces insulin resistance in obese rats. J Med Food. 2011;**14**:1511–1518. DOI: 10.1089/jmf.2010.0292.

[144] Schreckinger M, Lila M, Yousef G, Mejia E. Inhibition of glucosidase and amylase by *Vaccinium floribundum* and *Aristotelia chilensis* proanthocyanidins. In: Tunick and Gonzalez de Mejia; Hispanic Foods: Chemistry and Bioactive Compounds ACS Symposium Series; American Chemical Society. Washington, DC. 2012. 71–82 p. DOI: 10.1021/bk-2012-1109.ch006.

[145] Hidalgo J, Flores C, Hidalgo MA, Perez M, Yañez A, Quiñones L, Caceres DD, Burgos RA. Delphinol standardized maqui berry extract reduces postprandial blood glucose increase in individuals with impaired glucose regulation by novel mechanism of sodium glucose cotransporter inhibition. Panminerva Med. 2014;**56**:1–7.

[146] Crozier A, Jaganath IB, Clifford MN. Dietary phenolics: chemistry, bioavailability and effects on health. Nat Prod Rep. 2009;**26**:1001–1043. DOI: 10.1039/b802662a.

[147] Moreno PR, Purushothaman KR, Sirol M, Levy P, Fuster V. Neovascularization in human atherosclerosis. Circulation. 2006;**113**:2245–2252. DOI: 10.1161/CIRCULATIONAHA.105.578955.

[148] Lavie CJ, De Schutter A, Parto P, Jahangir E, Kokkinos P, Ortegac FB, Arena B, Milani RV. Obesity and prevalence of cardiovascular diseases and prognosis—the obesity paradox updated. Prog Cardiovasc Dis. 2016;**58**:537–547. DOI: 10.1016/j.pcad.2016.01.008.

[149] Cuende JI, Pérez de Diego IJ, Godoy D. Enfermedades cardiovasculares y enfermedades inflamatorias sistémicas. Clin Investig Arterioscler. 2016;**28**:94–101. DOI: 10.1016/j.arteri.2015.07.001.

[150] Kurata M, Nose M, Shimazu Y, Aoba T, Kohada Y, Yorioka S, Suehiro S, Fukuoka E, Matsumoto S, Watanabe H, Kumon Y, Okura T, Higaki J, Masumoto J. Microvasculature of carotid atheromatous plaques: hemorrhagic plaques have dense microvessels with fenestrations to the arterial lumen. J Stroke Cerebrovasc Dis. 2014;**23**:1440–1446. DOI: 10.1016/j.jstrokecerebrovasdis.2013.12.003.

[151] Ross R. The pathogenesis of atherosclerosisean update. N Engl J Med. 1986;**314**:488–500. DOI: 10.1056/NEJM198602203140806.

[152] Barger AC, Beeuwkes R 3rd, Lainey LL, Silverman KJ. Hypothesis: *vasa vasorum* and neovascularization of human coronary arteries. A possible role in the pathophysiology of atherosclerosis. N Engl J Med. 1984;**310**:175–177. DOI: 10.1056/NEJM198401193100307.

[153] Kordalewska M, Markuszewski MJ. Metabolomics in cardiovascular diseases. J Pharm Biomed Anal. 2015;**113**:121–136. DOI: 10.1016/j.jpba.2015.04.021.

[154] World Health Organization [Internet]. 2012. World Health Organization. Global Atlas on Cardiovascular Disease Prevention and Control. (2012) Available from: http://www.who.int/cardiovascular_diseases/en/ [Accessed: 2012-02-16].

[155] Erlund I, Koli R, Alfthan G, Marniemi J, Puukka P, Mustonen P, Mattila P, Jula A. Favorable effects of berry consumption on platelet function, blood pressure, and HDL cholesterol. Am J Clin Nutr. 2008;**87**:323–331.

[156] Basu A, Rhone M, Lyons TL. Berries: emerging impact on cardiovascular health. Nutr Rev. 2010;**68**:168–177. DOI:10.1111/j.1753-4887.2010.00273.x.

[157] Oudot C, Gomes A, Wang Z, Mateo P, Fischmeister R, Vieira H, Santos C, Brenner C. Beneficial effects of berries intake on survival and cardio-renal changes induced by high salt diet in Dahl/SS rat. Topic 27 – Hypertension, remodeling, arterial stiffness – B. Arch Cardiovasc Dis Suppl. 2015;**7**:185–186.

[158] Huang H, Chen G, Liao D, Zhu Y, Xue X. Effects of berries consumption on cardiovascular risk factors: a meta-analysis with trial sequential analysis of randomized controlled trials. Sci Rep. DOI: 10.1038/srep23625.

[159] Zhu Y, Miao Y, Meng Z, Zhong Y. Effects of vaccinium berries on serum lipids: a meta-analysis of randomized controlled trials. Evid Based Complement Altern Med. 2015. DOI: 10.1155/2015/790329.

[160] Rodriguez-Mateos A, Heiss C, Borges G, Crozier A. Berry (poly)phenols and cardiovascular health. J Agric Food Chem. 2014;**62**:3842–3851. DOI: 10.1021/jf403757g.

[161] Maxwell S, Lip G. Reperfusion injury: a review of the pathophysiology, clinical manifestations and therapeutic options. Int J Cardiol. 1997;**58**:95–117. DOI: 10.1016/S0167-5273(96)02854-9.

[162] Dhalla NS, Elmoselhi AB, Hata T, Makino N. Status of myocardial antioxidants in ischemia–reperfusion injury. Cardiovasc Res. 2000;**47**:446–456. DOI: 10.1016/S0008-6363(00)00078-X.

[163] Miranda-Rottman S, Aspillaga A, Pérez D, Vásquez L, Martinez A, Leighton F. Juice and phenolic fractions of the berry *Aristotelia chilensis* inhibit LDL oxidation *in vitro* and protect human endothelial cells against oxidative stress. J Agric Food Chem. 2002;**50**:7542–7547. DOI: 10.1021/jf025797n.

[164] Falkenberg SS, Tarnow I, Guzman A, Mølgaard P, Simonsen HT. Mapuche herbal medicine inhibits blood platelet aggregation. Evid Based Complement Altern Med. 2012;**2012**:1–9. DOI: 10.1155/2012/647620.

[165] Jofré I, Pezoa C, Cuevas M, Scheuermann E, Freires IA, Rosalen PL, de Alencar SM, Romero F. Antioxidant and vasodilator activity of *Ugni molinae* Turcz. (Murtilla) and its modulatory mechanism in hypotensive response. Oxid Med Cell Longev. 2016. DOI: 10.1155/2016/6513416.

5

Chilean Endemic/Native Plant Resources as Functional and Superfoods

Patricia Velásquez and Gloria Montenegro

Abstract

The current consumer demand for foods or food supplements with "super properties" is being covered by previously under-exploited ethnic products. The endemic flora from multiple continents serve as source of plant foods such as cereals or tropical fruits. Chile, one of the top five plant biodiversity hotspots on the planet, is a promising source of functional foods with little scientific and commercial research. The aim of this chapter is to summarize the findings related to the antioxidant and antibacterial potential of native/endemic plants and plant-derived compounds from Chile. Resources of these compounds may be found in honey, bee pollen, and berry-like fruits. These products, unknown to many parts outside the country, not only have the advantage of their functional properties but also possess denomination of origin, which gives added value and allows them to be used as food additives such as natural colorants, antioxidants, antibacterials, and antifungals. In the coming years, many of these products will be more commercially known and many of these plant species will be selected and improved, as have happened with products such as tofu or blueberries.

Keywords: biodiversity, endemism, berries, honey, bee pollen, antioxidant, antibacterial

1. Introduction

The latest food consumer trends point beyond fulfilling the function of providing nutrients to the body. It is intended that foods provide compounds capable of reducing the likelihood of developing diseases, improving or complementing the functions of the body, and even increasing life expectancy. A search for new food sources of these "healthy compounds" is underway to meet the needs of today's consumers. New analytical methods are being used with known foods to demonstrate properties they always had, but properties that had not been properly tested because of technological limitations; new foods or derived food compounds are also being found.

The diversity of food we know is derived from the biodiversity of plant and animal species we know. However, there are still many plant species which have not been explored or whose potential is just beginning to come to light. Many of these plants have been used by aboriginal groups around the world since ancient times. These species, which only grow in specific geographic locations (endemism), are rarely objects of scientific study or for industrial or commercial scaling.

Chile is one of the top five hotspots of plant biodiversity on the planet; here it is possible to find new food and food-derived resources of interesting compounds in the poorly explored flora. In addition, the biodiversity of plant species found in Chile have a high degree of endemism, indicating that they do not grow elsewhere. Leaves, stems, roots, or fruits can be sources of antioxidants and/or antibacterial compounds. Among the plant species with potential are the non-fruiting tree specimens such as quillay and ulmo; within fruit tree species, we may find maqui, murta, calafate, and others that are less known. All these products have high contents of polyphenolics, which have high antioxidant and antibacterial properties.

Polyphenolics are secondary metabolites from plants that have been associated with several healthy benefits such as the prevention of cancer, cardiovascular, inflammatory, and neurodegenerative diseases [1–3]; they are also associated with bioactive properties such as antioxidant and antibacterial properties [4–7].

Each phenolic/flavonoid compound has different antioxidant/antibacterial potency depending on its action mechanism. Phenolic compounds alter the permeability of bacterial cell membranes, which may result in the uncoupling of oxidative phosphorylation, the inhibition of active transport, and the loss of pool metabolites due to cytoplasmic membrane damage [8, 9]. Other authors explain the antibacterial activity of phenolics by the presence of more number of hydroxyl groups that may form hydrogen bonds with enzymes, altering their metabolism and also the lipid solubility and the degree of steric hindrance [10, 11]. In the case of flavonoids, antibacterial activity has been associated with its capacity to form complex bonds with proteins through non-specific forces such as hydrogen bonding and hydrophobic effects, as well as by covalent bond formation. Thus, it may inactivate microbial adhesins, enzymes, and cell envelope transport proteins. Lipophilic flavonoids may also disrupt microbial membranes [12, 13].

2. Effect of endemic/native Chilean plants on the functional activity of honeybee products

Honey has been recognized for many centuries as a healthy food, because of its positive effects such as healing [14], anti-inflammatory [15], antibacterial [16–20], and antioxidant [20–24] properties; and prebiotic capacity [24–28]. Meanwhile, the pollen has also been recognized by health claims. Scientific studies have been shown that bee pollen acts as an anti-anemic, tonic and restorative, hormonal and intestinal regulator, vasoprotector, hepatoprotective and detoxifying agent, and antioxidant and antibacterial [29, 30]. All these properties vary with the botanical and geographical origin (**Table 1**).

Honey	Phenolic compound	References
Heather	Benzoic acid, phenyl acetic acid	[31, 32]
Heather	Mandelic acid, B-phenyllactic acid	[32]
Honeydew	Protocatechuic acid	[32]
Rape	Hydrocinnamic acid	[32]
Buckwheet	4-hydroxybenzoic acid	[32]
Honeydew	Protocatechuic acid	[33]
Chestnut	Ferulic acid, p-coumaric acid	[33]
Chestnut	4-hydroxibenzoic acid, 4-hydroxyphenyllactic acid, phenylacetic acid	[34]
Heather	B-phenyllactic acid, benzoic acid, phenyl acetic acid	[34]
Sunflower	p-coumaric acid, phenyllactic acid, caffeic acid	[34]
Lime	3-hydroxybenzoic acid	[34]
Lavander	Caffeic acid, gallic acid	[34]
Strawberry	Homogentisic acid	[35]
Heather	Ellagic acid,abscisic acid	[36, 37]
Eucaliptus	Abscisic acid, ellagic acid	[38]
Citrus	Hesperetin	[39]
Rosemary	Kaempferol	[40]
Sunflower	Quercentin	[36, 37]
Eucaliptus	Myricetin, tricetin, luteolin, quercentin	[38, 41]
Manuka	Methylglyoxal	[42]
Heather	p-hydroxybenzoic, vanillic, chlorogenic, caffeic, syringic. p-coumaric, ferulic, m-coumaric, o-coumaric, ellagic, cinnamic acids	[43]
Lavander	Gallic, vanillic, chlorogenic, p-coumaric, ferulic, m-coumaric, cinnamic acids	[43]
Black locust	p-hydroxybenzoic, vanillic, p-coumaric, ferulic, trans-cinnamic acids. Vanillin, pinobanksin, apigenin, kaempherol, pinocembrin, crysina, acacetin	[44]

Honey	Phenolic compound	References
Acacia	Abscisic acid, p-hydroxybenzoic, vanillic, p-coumaric, Ferulic, trans-cinnamic acids. Vanillin, pinobanksin, apigenin, kaempherol, pinocembrin, crysina, acacetin	[44]
Rosemary	Pinobanksin, quercetin, luteolin, 8-methoxykaempferol, kaempferol, apigenin, isohamnetin. quercetin 3,3′-dimethyl ether, pinocembrin, quercetin 7,3′-dimethyl ether, quercetin 3,7-dimethylether, chrysin, galangin, tectochrysin	[40]
Eucalyptus	Quercentin, luteolin, myricetin	[45]
Lotus	Quercentin, luteolin, myricetin	[45]
Buckwheat	3-hydroxybenzoic acid, chlorogenic acid, 4-hydroxybenzoic acid, vanillic acid, caffeic acid, syringic acid, ferrulic acid, p-coumaric acid, rosmarinic acid, ellagic acid, myricetin, quercetin, kaempferol, chrysin, galangin	[45]
Sage	Myricetin, quercetin, luteolin, kaempferol, apigenin, isorhamnetin, chrysin, galangin, abscisic acid, caffeic acid, p-coumaric acid	[46]
Robinia	Myricetin, quercetin, luteolin, kaempferol, apigenin, chrysin, galangin	[47]
Eucalyptus	Myricetin, tricetin, quercertin, luteolin, quercertin-3-methyl ether, kaempferol, pinobanksin, chrysin, pinocembrin	[41]
Quillay	Chlorogenic, caffeic, coumaric, syringic, p-coumaric, vanillic and salicylic acids. Naringenin, quercetin, kaempferol	[48]
Ulmo	p-coumaric, ferulic, chlorogenic, caffeic, sinapic, syringic and salicylic acid Kaempferol luteolin	[49]

Table 1. Different polyphenolic compounds found in honeys with several botanical origins.

2.1. Honeys

Chilean honey has shown biological activity against bacteria and fungi. *Pseudomonas aeruginosa, Escherichia coli, Staphylococcus aureus, Streptococcus pneumoniae,* and *Vibrio cholerae* have been inhibited by hydroalcoholic extracts derived from honey [49, 50]. Meanwhile, *Candida albicans* has also shown sensitivity to Chilean honey. Chilean honey even has higher antimicrobial activity than Manuka honey, which has a standard antioxidant and antimicrobial activity potential [51]. The antimicrobial activity of honey is probably the result of the total number of active compounds and not the presence of any one of them (i.e., phenolics and flavonoids). This activity may be the result of synergism between flavonoids and phenolic compounds or between phenolic compounds and terpenes. Some phenolic compounds and flavonoids are present only in certain unifloral honeys. These results have allowed for the identification and certification of these honeys. References [48, 52, 53] identified chlorogenic, caffeic, coumaric, syringic, p-coumaric, vanillic and salicylic acids, naringenin, quercetin and kaempferol in the unifloral honey of Quillay (*Quillaja saponaria*). In the same report, [52] found p-coumaric, ferulic, and salicylic acids in the endemic unifloral honey of Ulmo (*Eucryphia cordifolia*). Pinobanksin and kaempferol are typically identified in Chilean honeys.

Other more recent Chilean honeys currently being studied are Avellano honey (*Gevuina avellana* Molina), Tiaca honey (*Caldcluvia paniculata* (Cav.) D. Don), and Corontillo honey (*Escallonia pulverulenta*), which have shown antibacterial and antioxidant properties [50].

2.2. Bee pollen

Bee pollen provides important ingredients to the human diet, such as carbohydrates, protein, fat, and other components in lesser amount such as minerals. Carbohydrates are mainly polysaccharides such as starch and sugars and represent between 13 and 55 g per 100 g of sample. With regard to protein content, bee pollen provides all essential amino acids to the human diet and their percentages vary between 10 and 40% of the test sample [55–63]. Referring to fats, a study reveals that 3% of the total lipids are free fatty acids and about half of them are omega-3 unsaturated oleic, linoleic (omega-6), and linolenic acids (omega-3) [55]. With reference to the mineral content, bee pollen contains potassium, phosphorus, calcium, magnesium, iron, copper, zinc, and selenium in amounts that satisfy the daily recommended intake per person [64].

Bee pollen classification	Total phenolics	FRAP	DPPH	β-Carotene	Licopene	Total flavonoids	References
Pollen multiflora	–	–	–	17.05–489.20 μg/g	–	–	[63]
Pollen multiflora	–	–	–	1–20 mg/100 g	–	1293–8243 mg/100 g	[23]
Pollen multiflora	–	–	–	–	–	530–3258 mg/100 g	[71]
Pollen multiflora	–	0.25–5.35 mM Fe^{2+}/g	0.27–2.8 mmol Trolox/g	–	–	–	[55]
Pollen multiflora	4.4–16.4 mg GAE/g	0.255–5.355 mM Fe^{2+}/g	0.274–2.814 mmol Trolox/g	–	–	2.8–13.6 mg eq quercetin/g	[74]
Pollen multiflora	817,33-138367 mg tannin/kg	-	47,97-86,25 % of inhibition	-	-	-	[75]
Quillaja saponaria	18.15 mg GAE/g	34.48 mM Fe^{2+}/g	2.97 mg ascorbic acid/g	0 μg/g	0 μg/g	–	[54, 76]
Azara petiolaris	16.43 mg GAE/g	14.58 mM Fe^{2+}/g	2.86 mg ascorbic acid/g	13.60 μg/g	60.40 μg/g	–	
Puya chilensis	11.83 mg GAE/g	28.24 mM Fe^{2+}/g	2.87 mg ascorbic acid/g	4.60 μg/g	14.70 μg/g	–	
Cryptocarya alba	11.74 mg GAE/g	28.32 mM Fe^{2+}/g	3.06 mg ascorbic acid/g	0 μg/g	0 μg/g	–	
Colliguaja odorifera	11.50 mg GAE/g	46.39 mM Fe^{2+}/g	2.87 mg ascorbic acid/g	7.10 μg/g	7.30 μg/g	–	
Schinus polygamus	7.12 mg GAE/g	24.85 mM Fe^{2+}/g	2.93 mg ascorbic acid/g	25.40 μg/g	28.80 μg/g	–	

Table 2. Main antioxidant parameters and pigments presented in bee pollen from different resources.

Several reports demonstrate the health benefits of bee pollen. Scientific studies have shown that bee pollen acts as an anti-anemic, tonic and restorative, hormone regulator, intestinal regulator, vasoprotector, and hepatoprotective, detoxifying, and antioxidant agent [28, 29, 65]. However, very few studies have identified the phenolic compounds of Chilean bee pollen. The information on bee pollen production for food applications and some reports concerning their antimicrobial and antioxidant activity [54, 66, 67].

Phenolic acids, flavonoids, and pigments such as β-carotene are mainly responsible for the healthy properties such as antioxidant and antibacterial properties exhibited by bee pollen [68–70]. The phenolic acids and flavonoid glycosides are present in the nectar of flowers visited by bees, which are hydrolyzed and transferred to bee pollen. The number and variety of phenolic acids and flavonoids are highly variable, since beekeepers mix bee pollen with different botanical origins from different plant species [22, 71]. A major flavonoid found in bee pollen is rutin [72]. The main group of pigments that compose bee pollen are carotenoids, especially β-carotene, whose concentration also depends on the botanical origin of the sample [63]. The β-carotene content is about 17% of total carotenoids. In some cases, it may contain 20 times less carotenoids that some foods [73]. In Chilean bee pollen, the carotenoid content varies with the botanical origin (**Table 2**).

The type and concentration of the polyphenolic compound influence the antibacterial and antioxidant activity exhibited by bee pollen. The most important polyphenolic compounds related to these activities are vanillic acid, protocatechuic acid, gallic acid, p-coumaric acid, hesperidin, rutin, kaempferol, apigenin, luteolin, quercetin, and isorhamnetin [70]. Bee pollen rich in these compounds has shown activity against specific pathogens such as *S. aureus*, which causes skin infections; *E. coli*, which causes diarrhea [67, 77], *Streptococcus pyogenes*, which causes acute bacterial pharyngitis [78], *P. aeruginosa*, which produces tissue damage and affects the immune system [79] and *S. pyogenes*, which causes skin wounds [16]. Another important study demonstrated the inhibition activity against *Salmonella* spp., as shown **Figure 1** [66].

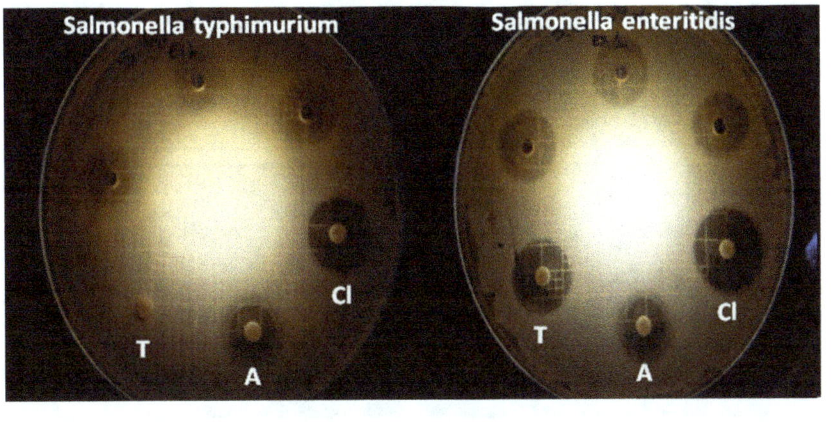

Figure 1. Antibacterial activity of Chilean multiflora bee pollen hydrophilic extracts evaluated by inhibition zone diameter against *Salmonella typhimurium* and *Salmonella enteriditis*. Tetracycline (T), ampicillin (A) and chloramphenicol (Cl) were used as controls.

3. Endemic/native berries

Chile is the main exporter of berries in the Southern Hemisphere and the fifth berry exporter worldwide because of its comparative advantages: geographic isolation of the country (desert in the north, the Pacific Ocean, the Andes mountains, and the Patagonian ice), which makes it an island from the health point of view, decreasing the incidence of pests and diseases; the Mediterranean climate is beneficial to obtain optimal raw material and production and in a counter-season and phased production [80, 81]. Maqui, murta, and others recently explored are included in the list of actual and future production (**Figures 2** and **3**).

Figure 2. *Luma apiculata* or "arrayán" fruits. These berry-like fruits have higher antioxidant activity than blueberries. Many unknown Chilean endemic/native fruits are potential functional foods.

Figure 3. *Myrceugenia obtusa* or "Rarán" fruits. These berrys have antioxidant and antibacterial activities (Orellana et al., 2017).

3.1. "Maqui" (*Aristotelia chilensis*)

Maqui is a berry with antioxidant and antihemolytic properties [82, 83], and it limits adipogenesis and inflammatory pathways in vitro [84, 85], protects against oxidative stress by reducing lipid peroxidation [86], inhibits LDL oxidation in vitro and protects human endothelial cells against oxidative stress [87] and has cardioprotective [88] and gastroprotective properties [89]. These healthy effects are produced by anthocyanins and many other bioactive compounds such as flavonoids, coumarins, phenolic acid (i.e. gallic, gentisic, sinapic, hydroxybenzoic, vanillic acids, makonine, 8-oxo-9 dehydrohobartine and 8-oxo-9 dehydromakomakine [90–93] present in the fruits. Recently, Maqui has been used to design new functional foods such as drinks and cakes with antioxidant properties for in vivo and clinical trials [94–96].

3.2. "Murta" or "murtilla" (*Ugni molinae*)

Murta fruits are berries which have a rich chemical composition of bioactive compounds associated with health properties [97]. They have shown analgesic *in vitro* activity [98], protective capacity against oxidative damage of human erythrocytes [99], antimicrobial activity [100], antioxidant activity [101, 102], and α-glucosidase/α-amylase inhibition [102] as the main beneficial effects.

3.3. Other berries and berry-like fruits

"Calafate" (taxonomically described as *Berberis buxifolia* and also *Berberis microphylla*) fruits are berries that are scarcely studied. However, the available information is very interesting and indicates its potential as an antioxidant, which may be related to its high anthocyanin and hydroxycinnamic acid levels [103, 104]. Most recently, exploratory studies have revealed new native/endemic berry-like fruits such as *Luma apiculata, Ribes punctatum, Ribes magellanicum, Ribes cucullatum* and *Ribes tribolum* [105, 106]. *Ribes* spp., *Rubus* spp., *Gaultheria* spp., and *Berberis* spp., among others, as promising crops of functional foods or food additives/supplements such as natural colorants (**Table 3**). Some other non-scientific studies have been related with functional properties of several non-fruiting plants with anticoagulant, antithrombin, and analgesic properties and related health effects [107].

Common name	Scientific name	Attributed properties	Description	References
Chaura	*Gaultheria pumila*	Antioxidant (antocyanin content)	The fruit is a berry, white or pink. ovoid shaped, 6 mm to 12 mm in diameter	[108]
Chaura	*Gaultheria mucronata*	Antioxidant	The fruit is a berry, between 6 and 9 mm in diameter, plum-shaped, passing from white to pink and finally to dark purple when ripe	[104, 109]
Chaura	*Gaultheria antarctica*	Antioxidant	The fruit is a berry, white or pink, ovoid shaped, 6 mm to 10 mm in diameter	[110]

Common name	Scientific name	Attributed properties	Description	References
Uva de cordillera, calafatillo	*Berberis empetrifolia*	Antioxidant	The fruit is a globose, blue-black,about 7 mm in diameter	[109, 110]
Calafate, chelia	*Berberis ilicifolia*	Antioxidant	Fruits are blue-black berries about 1 cm long, with four to six seeds, 5–6 mm in diameter	[110]
Calafate	*Berberis microphylla*	Antioxidant, antibacterial	The fruit is a spherical blue-black berry, about 1 cm. in diameter, and contains six angular seeds	[110–113]
Calafate	*Berberis buxifolia*	Antioxidant (anthocyanin content)	The fruit is a globose, blue-black, about 7–10 mm in diameter	[109, 114, 115]
Michay, mechay	*Berberis darwinii*	In vitro evidence for Alzheimer's disease therapy	The fruit is a globose, blue-black, about 7–10 mm in diameter	[116]
Copihue, Chilean bell national flower	*Lapageria rosae*	Antioxidant	The fruits are red berries, ovoid, between 3 and 6 cm long, with a thick skin containing numerous seeds	[117, 118]
Chilco, Chilca, Palo blanco	*Fuchsia magellanica*	Hypotensive and diuretic effect, antioxidant activity, significant inhibitory activity against B-glucuronidase enzyme	Fruit is a black berry, about 8–10 mm diameter	[104, 119–122]
Peumo	*Cryptocarya alba*	Significant inhibitory activity against B-glucuronidase enzyme, free radical scavenging activity, antibacterial activity	Red fruit with one large seed	[119, 123–125]
Daudapo, Huarapo, Zarapito, Té de la turba, naurapo, mirteola	*Myrteola nummularia*	Antioxidant (higher antioxidant content than blueberries), it may reduce colon cancer risk, source of natural colorant as anthocyanin	The fruit is up to 1 cm in diameter, it has a soft juicy flesh and a delicious slightly aromatic flavor	[104, 109, 126–128]
Copihuelo, Copihue chilote, Copihuelo, Coicopiu,	*Philesia buxifolia, Philesia magellanica*	Antioxidant	The fruit is a yellowish green ovoid berry, size up to 13 mm long	[121, 129]
Queule, keule,	*Gomortega nítida, Gomortega keule*	Antioxidant	The fruit is a drupe, yellow, about 34–45 mm (1.3–1.8 in) in diameter, usually with 1–2 seeds	[130]
Cauchao (from Luma or red	*Amomyrtus luma*	Antioxidant, inhibit	The fruit is a black to purplish-black berry when ripe, with about 1–1.5 cm in	[121, 131, 132]

Common name	Scientific name	Attributed properties	Description	References
luma tree)		platelet aggregation (anticoagulant effect), antibacterial	diameter, generally with 3 seeds, about 3–4.5 mm	
Cauchao (from Meli or White Luma tree)	*Amomyrtus meli*	Antioxidant, antibacterial	The fruit is a black or purplish black Berry, 5–8 mm in diameter, generally with 3 seeds, about 3–4.5 mm	[112, 113, 132]
Chequén, Arrayán blanco,	*Luma chequen*	Antioxidant	The berry-like fruit (drupe) is a dark purple, about 1 cm in diameter	[112, 113, 133]
Arrayán	*Luma apiculata*	Antioxidant, antibacterial, inhibit platelet aggregation (anticoagulant effect)	Berry rounded black fruit, about 1.3–1.5 cm. diameter, containing three seeds	[105, 112, 125, 131, 133]
Chilean strawberry, wild strawberry	*Fragaria chiloensis*	Antioxidant, free radical scavenging activity, anticancer cell Proliferation properties (human lung epithelial cancer cells	The fruit is whitish or pale pink	[134–139]
Chañar, chañal	*Geoffroea decorticans*	Antioxidant, antinoceptive, anti-inflammatory activities; antitussive and expectorant significant effect, antibacterial	The berry-like fruit is a drupe, ovoid, red-brown when ripe, about 1.7–3.5 cm to 1.5 cm. The pulp is white-yellowish and has 1 or 2 seeds	[140–142]
Maqui	*Aristotelia chilensis*	Inhibidor de la enzima xantina oxidasa (sintomatología de la gota); antimicrobial activity (wound treatment); in vitro and in vivo antidiabetic effects, antibacterial activity, cardioprotective effects, antioxidant	The fruit is a small fleshy edible berry (green when unripe and purple black when ripe), about 5 mm, with 2–4 seeds	[82, 83, 86, 88, 90, 103, 143–146]
Murtilla de Magallanes, brecillo, uvilla	*Empetrum rubrum*	Antioxidant	Globose and fleshy fruit, about –8 mm in diameter, dark red	[109]
Murta, murtilla, Murta blanca, Tautau	*Ugni molinae*	Antioxidant, vasodilator activity, antibacterial	The fruit is a bright red berry, around 5–15 mm in diameter	[83, 99, 103, 112, 113, 147–149]
Zarzaparrilla, parrilla, uvilla, mulul, milul, Chilean currant	*Ribes punctatum, Ribes cucullatum, Ribes magellanicum,*	Antioxidant, cytoprotective effect in human gastric cells	Fruits are red, black, or green	[104, 106]
Zarzaparrilla, parrilla, Chilean currant	*Ribes trilobum*	Antioxidant, cytoprotective effect in human gastric cells	The fruit is initially green and becomes glossy black when ripe	[106]

Common name	Scientific name	Attributed properties	Description	References
Zarzaparrilla, parrilla, Chilean currant	*Ribes valdivianum*	Antioxidant	Purple-black berry-like fruit	[150]
Zarzaparrilla, Miñe-miñe, strawberry of Magallanes, wild raspberry	*Rubus geoides*	Antioxidant, cytoprotective effect in human gastric cells	Berry-like fruit	[111, 151]

Table 3. Main functional properties of native/endemic berries and berry-like fruits.

4. Conclusions and future trends

In spite of the endemism, there are promising bee hive–derived products obtained from Chilean plants, as well as Chilean plant products in general. We are convinced that the main exponents of functional foods and super foods are in nature, which is where we have to explore to find them. However, they should be used and exploited in a sustainable way.

Acknowledgements

The authors would like to thank the Regional Innovation Fund for Competitiveness, ID 30126395-0 Región del Libertador Bernardo O'Higgins; UC-VRI interdisciplinary project N°13, 2014; to CONICYT Fellowship National Doctoral - Operating expenses N° 21110822; to Program Attraction and Integration of Advanced Human Capital (PAI-CONICYT) Doctoral Thesis in Business N° 781412002.

Author details

Patricia Velásquez[1] and Gloria Montenegro[2*]

*Address all correspondence to: gmonten@uc.cl

1 Department of Chemistry and Bioprocess, Faculty of Engineering, Pontifical Catholic University of Chile, Santiago, Chile

2 Department of Plant Science, Faculty of Agricultural and Forestry Sciences, Pontifical Catholic University of Chile, Santiago, Chile

References

[1] Huang, W.Y., Cai, Y.Z., Zhang, Y. Natural phenolic compounds from medicinal herbs and dietary plants: potential use for cancer prevention. Nutrition and Cancer. 2009;62(1):1–20.

[2] Aboul-Enein, Y., Berczynski, P., Kruk, I. Phenolic compounds: the role of redox regulation in neurodegenerative. Mini Reviews in Medicinal Chemistry. 2013;13(3):385–398.

[3] Rodríguez-Morató, J., Xicota, L., Fitó, M., Farré, M., Dierssen, M., De La Torre, R. Potential role of olive oil phenolic compounds in the prevention of neurodegenerative diseases. Molecules. 2015;20(3):4655–4680.

[4] Heleno, S.A., Martins, A., Queiroz, M.J.R., Ferreira, I.C. Bioactivity of phenolic acids: metabolites versus parent compounds: a review. Food Chemistry. 2015;173(1):501–513.

[5] Xu, Y., Burton, S., Kim, C., Sismour, E. Phenolic compounds, antioxidant, and antibacterial properties of pomace extracts from four Virginia-grown grape. Food Science and Nutrition. 2016;4(1):125–133. DOI: http://doi.org/10.1002/fsn3.264

[6] Heim, K.E., Tagliaferro, A.R., Bobilya, D.J. Flavonoid antioxidants: chemistry, metabolism and structure-activity relationships. The Journal of Nutritional Biochemistry. 2002;13(10):572–584.

[7] Pietta P.G. Flavonoids as antioxidants. Journal of Natural Products. 2000;63(7):1035–1042.

[8] Shan, B., Cai, Y.Z., Brooks, J.D., Corke, H. Antibacterial properties and major bioactive components of cinnamon stick (*Cinnamomum burmannii*): activity against food-borne pathogenic bacteria. Journal of Agricultural and Food Chemistry. 2007;55(14): 5484–5490.

[9] Maillard, J.Y. Bacterial target sites for biocide action. Journal of applied microbiology. 2002;92(S1):16s–27s.

[10] Salawu, S.O., Ogundare, A.O., Ola-Salawu, B.B., Akindahunsi, A.A. Antimicrobial activities of phenolic containing extracts of some tropical vegetables. African Journal of Pharmacy and Pharmacology. 2011;5(4):486–492.

[11] Ceylan, E., Fung, D.Y. Antimicrobial activity of spices. Journal of Rapid Methods and Automation in Microbiology. 2004;12(1):1–55.

[12] Kumar, S., Pandey, A.K. Chemistry and biological activities of flavonoids: an overview. The Scientific World Journal. 2013;2013(1):Article ID 162750. DOI: doi: 10.1155/2013/162750

[13] Cushnie, T.P.T., Hamilton, V.E.S., Chapman, D.G., Taylor, P.W., Lamb, A.J. Aggregation of *Staphylococcus aureus* following treatment with the antibacterial flavonol galangin. Journal of Applied Microbiology. 2007;103(5):1562–1567.

[14] Molan, P.C. The evidence supporting the use of honey as a wound. The International Journal of Lower Extremity Wounds. 2006;5(1):40–54.

[15] Taormina P.J., Niemira B.A., Beuchat L.R. Inhibitory activity of honey against foodborne pathogens as influenced by the presence of hydrogen peroxide and level of antioxidant power. International Journal of Food Microbiology. 2001;69(1):217–225.

[16] Basualdo C., Sgroy V., Finola M.S., Marioli J.M. Comparison of the antibacterial activity of honey from different. Veterinary Microbiology. 2007;124(1):375–381.

[17] Weston R.J. The contribution of catalase and other natural products to the antibacterial activity of honey: a review. Food Chemistry. 2000;71(1):235–239.

[18] Basim E., Basim H., Ozcan M. Antibacterial activities of Turkish pollen and propolis extracts against plant bacterial pathogens. Journal of Food Engineering. 2006;77(1): 992–996.

[19] Carpes, S.T., Begnini, R., de Alencar, S.M., Masson, M.L. Study of preparations of bee pollen extracts, antioxidant and antibacterial activity. Ciência e Agrotecnologia. 2007;31(6):1818–1825.

[20] Rauha, J.P., Remes, S., Heinonen, M., Hopia, A., Kähkönen, M., Kujala, T., Vuorela P. Antimicrobial effects of Finnish plant extracts containing flavonoids and other phenolic compounds. International Journal of Food Microbiology. 2000;56(1):3–12.

[21] Frankel, S., Robinson, G.E., Berenbaum, M.R. Antioxidant capacity and correlated characteristics of 14 unifloral honeys. Journal of Apicultural Research. 1998;37(1):27–31.

[22] Campos, M.G., Webby, R.F., Markham, K.R., Mitchell, K.A., Da Cunha, A.P. Age-induced diminution of free radical scavenging capacity in bee pollens and the contribution of constituent flavonoids. Journal of Agricultural and Food Chemistry. 2003;51(1):742–745.

[23] LeBlanc B.W., Davis O.K., Boue S., DeLucca A., Deeby T. Antioxidant activity of Sonoran Desert bee pollen. Food Chemistry. 2009;115(1):1299–1305.

[24] Ajibola, A., Chamunorwa, J.P., Erlwanger, K.H. Nutraceutical values of natural honey and its contribution to human. Nutrition and Metabolism. 2012;9(61):1.

[25] Aachary A.A., Prapulla S.G. Xylooligosaccharides (XOS) as an emerging prebiotic: microbial synthesis, utilization. Comprehensive Reviews in Food Science and Food Safety. 2011;10(1):6–16.

[26] Shin, H.S., Ustunol, Z. Carbohydrate composition of honey from different floral sources and their influence on growth of selected intestinal bacteria: an in vitro comparison . Food Research International. 2005;38(6):721–728.

[27] Riazi, A., Ziar, H. Effect of honey and starter culture on growth, acidification, sensory properties and bifidobacteria cell counts in fermented skimmed milk. African Journal of Microbiology Research. 2012;6(3):486–498.

[28] Yıldız, O., Can, Z., Saral, Ö. Yuluğ, E., Öztürk, F., Aliyazıcıoğlu, R., Kolaylı, S. Hepatoprotective potential of chestnut bee pollen on carbon tetrachloride-induced hepatic damages in rats. Evidence-based Complementary and Alternative Medicine. 2013;2013(1):Article ID 461478. DOI: 10.1155/2013/461478

[29] Negri, G., Teixeira, E.W., Florêncio Alves, M.L., Moreti, A.C., Otsuk, I.P., Borguini, R.G., Salatino, A. Hydroxycinnamic acid amide derivatives, phenolic compounds and antioxidant activities of extracts of pollen samples from southeast Brazil. Journal of Agricultural and Food Chemistry. 2011;59(10):5516–5522.

[30] Steeg, E., Montag A. Quantitative Bestimmung aromatischer carbon sauren in Honig fur Lebensmitteluntersuchung und-forschung. Zeitschrift. 1988;187(1):115–120.

[31] Andrade, P., Ferreres, F., Gil, M.I., Tomás-Barberán, F.A. Determination of phenolic compounds in honeys with different floral origin by capillary zone electrophoresis. Food Chemistry. 1997;60(1):79–84.

[32] Joerg, E., Sontag, G. Multichannel coulometric detection coupled with liquid chromatography for determination of phenolic esters in honey. Journal of Chromatography A. 1993;635(1):137–142.

[33] Dimitrova, B., Gevrenova, R., Anklam, E. Analysis of phenolic acids in honeys of different floral origin by solid-pase extraction and high-performance liquid chromatography. Phytochemical Analysis. 2007;18(1):24–32.

[34] Soler C.,Gil M.I.,García-Viguera C.,Tomás-Barberán F.A. Flavonoid patterns of French honeys with different floral origin. Apidologie. 1995;26(1):53–60.

[35] Cabras, P., Angioni, A., Tuberoso, C., Floris, I., Reniero, F., Guillou, C., Ghelli, S. Homogentisic acid: a phenolic acid as a marker of strawberry-tree (Arbutus unedo) honey. Journal of Agricultural and Food Chemistry. 1999;47(10):4064–4067.

[36] Ferreres F., Andrade P., Gil M.I., Tomas-Barberan F.A. Floral nectar phenolics as biochemical markers for the botanical origin of heather honey. Z Lebensm Unters Forsch. 1996;200(1):40–44.

[37] Yao, L., Datta, N., Tomás-Barberán, F., Ferreres, F., Martos, I., Singanusong, R. Flavonoids, phenolic acids and abscisic acid in Australian and New Zealand *Leptospermum* honeys. Food Chemistry. 2003;81(1):159–168.

[38] Ferreres F., García-Viguera C., Tomás-Lorente F., Tomás-Barberán F.A. Hesperetin: A marker of the floral origin of citrus honey. Journal of the Science of Food and Agriculture. 1993;61(1):121–123.

[39] Gil M.I., Ferreres F., Ortiz A., Subra E., Tomás-Barberán F.A. Plant phenolic metabolites and floral origin of rosemary honey. Journal of the Science of Food and Agriculture. 1995;43(1):2833–2838.

[40] Martos I., Ferreres F., Tomás-Barberán F.A. Identification of flavonoid markers for the botanical origin of eucalyptus honey. Journal of the Science of Food and Agriculture. 2000;48(1):1498–1502.

[41] Andrade, P., Ferreres, F., Amaral, M.T. Analysis of honey phenolic acids by HPLC, its application to honey botanical characterization. Journal of Liquid Chromatography and Related Technologies. 1997;20(14):2281–2288.

[42] Adams, C.J., Manley-Harris, M., Molan, P.C. The origin of methylglyoxal in New Zealand manuka (*Leptospermum scoparium*) honey. Carbohydrate Research. 2009;344(8): 1050–1053.

[43] Pyrzynska, K., Biesaga, M. Analysis of phenolic acids and flavonoids in honey. Trends in Analytical Chemistry. 2009;28(7):893–902.

[44] Iurlina, M.O., Saiz, A.I., Fritz, R., Manrique, G.D. Major flavonoids of Argentinean honeys. Optimisation of the extraction method and analysis of their content in relationship to the geographical source of honeys. Food Chemistry. 2009;115(3):1141–1149.

[45] Jasicka-Misiak, I., Poliwoda, A., Dereń, M., Kafarski, P. Phenolic compounds and abscisic acid as potential markers for the floral origin of two Polish unifloral honeys. Food Chemistry. 2012;131(4):1149–1156.

[46] Kenjerić, D., Mandić, M.L., Primorac, L., Čačić, F. Flavonoid pattern of sage (*Salvia officinalis* L.) unifloral honey. Food Chemistry. 2008;110(1):187–192.

[47] Kenjerić, D., Mandić, M.L., Primorac, L., Bubalo, D., Perl, A. Flavonoid profile of Robinia honeys produced in Croatia. Food Chemistry. 2007;102(3):683–690.

[48] Montenegro G., Gómez M., Casaubon G., Belancic A., Mujica A.M., Peña R.C. Analysis of volatile compounds in three unifloral native Chilean honeys. Phyton. 2009;78(1):61–65.

[49] Velásquez P., Retamal M., Giordano A., Valenzuela L, Montenegro G. Chemical composition and in vitro antibacterial activity of Chilean bee-pollen and honey blend-extracts. ICEF12 - 12th International Congress on Engineering and Food. 14-18 June 2015, Québec, Canada.

[50] Montenegro G., Santander F., Jara C., Nuñez G., Fredes C. Antioxidant and microbial activity of unifloral honeys of plants native to Chile. Boletín Latinoamericano y del Caribe de Plantas Medicinales y Aromáticas. 2013;12(1):257–268.

[51] Sherlock O., Dolan A., Athman R., Power A., Gethin G., Cowman S., Humphreys H. Comparison of the antimicrobial activity of Ulmo honey from Chile and Manuka honey against methicillin-resistant *Staphylococcus aureus, Escherichia coli* and *Pseudomonas aeruginosa*. BMC Complementary and Alternative Medicine. 2010;10(1):1–47.

[52] Muñoz O., Copaja S., Speisky H., Peña R.C., Montenegro G. Content of flavonoids and phenolic compounds in Chilean honeys. Química Nova 2007;30(1):848–851.

[53] Montenegro, G., Salas, F., Pena, R.C., Pizarro, R. Antibacterial and antifungic activity of the unifloral honeys of *Quillaja saponaria*, an endemic Chilean species. Phyton. 2009;78(1):141–146.

[54] Montenegro, G., Pizarro, R., Mejías, E., Rodríguez, S. Biological evaluation of bee pollen from native Chilean plants. Phyton. 2013;82(1):7–14.

[55] Bogdanov S. Pollen: Nutrition, Functional Properties, Health: A Review [Internet]. 2011. Available from: http://www.beehexagon.net/ [Accessed: October, 2015]

[56] Modro, A.F., Silva I.C., Luz C.F., Message D. Analysis of pollen load based on color, physicochemical composition and botanical source. Anais da Academia Brasileira de Ciências. 2009;81(2):281–285.

[57] Sommerville D.C. Nutritional value of bee collected pollens, Report DAN 134A. Rural Industries Research and Development Corporation (RIRDC). NSW Agriculture. Publication No. 01/047; 2001.

[58] Andrada, A.C., Tellería, M.C. Pollen collected by honey bees (*Apis mellifera* L.) from south of Caldén district (Argentina): floral origin and protein content. Grana. 2005;44(2):115–122.

[59] De Arruda, V.A., Santos Pereira, A.A., Estevinho, L.M., De Almeida-Muradian, L.B. Presence and stability of B complex vitamins in bee pollen using different storage conditions. Food Chemistry and Toxicology. 2013;51(1):143–148.

[60] De Arruda, V.A.S., Pereira, A.A.S., De Freitas, A.S., Barth, O.M., De Almeida-Muradian, L.B. Dried bee pollen: B complex vitamins, physicochemical and botanical composition. Journal of Food Composition and Analysis. 2013;29(2):100–105.

[61] Carpes, S.T., Mourao, G.B., Masson, M.L. Chemical composition and free radical scavenging activity of *Apis mellifera* bee pollen from Southern Brazil. Brazilian Journal of Food Technology. 2009;12(3):220–229.

[62] Human, H., Nicolson, S.W. Nutritional content of fresh, bee-collected and stored pollen of *Aloe greatheadii* var. *davyana* (*Asphodelaceae*). Phytochemistry. 2006;67(14):1486–1492.

[63] Almeida-Muradian, L.B., Pamplona, L.C., Coimbra, S., Barth, O.M. Chemical composition and botanical evaluation. Journal of Food Composition and Analysis. 2005;18(1): 105–111.

[64] Stanciu, O.G., Mărghitaş, L.A., Dezmirean, D., Campos, M.G. A comparison between the mineral content of flower and honeybee collected pollen of selected plant origin (*Helianthus annuus* L. and *Salix* sp.). Romanian Biotechnological Letters. 2011;16(4): 6291–6296.

[65] Cheng, N., Wang, Y., Gao, H., Yuan, J., Feng, F., Cao, W., Zheng, J. Protective effect of extract of Crataegus pinnatifida pollen on DNA damage response to oxidative stress. Food and Chemical Toxicology. 2013;59(1):709–714.

[66] Velásquez P., Rodríguez K., Retamal M., Giordano A., Valenzuela L, Montenegro G. Antimicrobial Activity of chilean bee-pollen extracts. XI Congreso Latinoamericano de Botánica. 19-24 Octubre 2014. Salvador de Bahía, Brasil.

[67] Cabrera C., Montenegro G. Pathogen control using a natural Chilean bee pollen extract of known botanical origin. Ciencia e Investigacion Agraria 2013;40(1):223–230.

[68] Meda A., Lamien C.E., Romito M., Millogo J., Nacoulma O.G. Determination of the total phenolic, flavonoid and proline contents in Burkina Fasan honey, as well as their radical scavenging activity. Food Chemistry. 2005;91(1):571–577.

[69] Aloisi, P.V., Ruppel, S. Bioactive and nutrition of bee pollen in the province of Chubut, Argentina. Revista de Investigaciones Agropecuarias. 2014;40(3):296–302.

[70] Aličić, D., Šubarić, D., Jašić, M., Pašalić, H., Ačkar, Đ. Antioxidant properties of pollen. Hrana u Zdravlju i Bolesti. 2014;3(1):6–12.

[71] Leja, M., Mareczek, A., Wyzgolik, G., Klepacz-Baniak, J., Czekonska, K. Antioxidative properties of bee pollen in selected plant species. Food Chemistry. 2007;100(1):237–240.

[72] Serra Bonvehí, J., Soliva Torrentó, M., Centelles Lorente, E. Evaluation of polyphenolic and flavonoid compounds in honeybee-collected pollen produced in Spain. Journal of Agricultural and Food Chemistry. 2001;49(4):1848–1853.

[73] Du Sert, P.P. Les pollens apicoles. Phytothérapie. 2009;7(2):75–82

[74] Marghitas L.A., Stanciu O.G., Dezmirean D.S., Bobis O., Popescu O., Bogdanov S., Campos M.G. *In vitro* antioxidant capacity of honeybee-collected pollen of selected floral origin harvested from Romania. Food Chemistry. 2009;115(1):878–883.

[75] Fatrcová-Šramková K., Nôžková J., Kačániová M., Máriássyová M., Rovná K., Stričík M. Antioxidant and antimicrobial properties of monofloral bee pollen. Journal of Environmental Science and Health, Part B: Pesticides, FoodContaminants, and Agricultural Wastes. 2013;48(2):133–138.

[76] Montenegro, G. Bee pollen Chilean: Differentiation and uses according to their properties and floral origin. 1st ed. Chile: Pontificia Universidad Católica de Chile; 2012.

[77] Libonatti, C., Soledad, V., Marina, B. Antibacterial activity of honey: a review of honey around the world. Journal of Microbiology and Antimicrobials. 2014;6(3):51–56.

[78] Esposito, S., Blasi, F., Bosis, S., Droghetti, R., Faelli, N., Lastrico, A., Principi, N. Aetiology of acute pharyngitis: the role of atypical bacteria. Journal of Medical Microbiology. 2004;53(7):645–651.

[79] Guo, S.A., DiPietro, L.A. Factors affecting wound healing. Journal of Dental Research. 2010;89(3):219–229.

[80] ASOEX (Asociación de Exportadores de Frutas de Chile, A.G.). Mercado de berries [Internet]. 2006 [Updated: 2016]. Available from: http://www.asoex.cl/ [Accessed: September, 2016]

[81] ProChile. Foods from Chile, Source of Life [Internet]. 2013. Available from: http://www.prochile.gob.cl/int/united-states/wpcontent/blogs.dir/21/files_mf/952FoodsfromChile.pdf [Accessed: August, 2016]

[82] Fredes, C., Montenegro, G., Zoffoli, J.P., Gómez, M., Robert, P. Polyphenol content and antioxidant activity of maqui (*Aristotelia chilensis* Molina Stuntz) during fruit development and maturation in Central Chile. Chilean Journal of Agricultural Research. 2012;72(4):582–589.

[83] Rubilar, M., Jara, C., Poo, Y., Acevedo, F., Gutierrez, C., Sineiro, J., Shene, C. Extracts of maqui (*Aristotelia chilensis*) and murta (*Ugni molinae* Turcz.): sources of antioxidant compounds and -glucosidase/-amylase inhibitors. Journal of Agricultural and Food Chemistry. 2011;59(5):1630–1637.

[84] Reyes-Farias, M., Vasquez, K., Ovalle-Marin, A., Fuentes, F., Parra, C., Quitral, V., Garcia-Diaz, D.F. Chilean native fruit extracts inhibit inflammation linked to the pathogenic interaction between adipocytes and macrophages. Journal of Medicinal Foods. 2015;18(5):601–608.

[85] Schreckinger, M.E., Wang, J., Yousef, G., Lila, M.A., Gonzalez de Mejia, E. Antioxidant capacity and *in vitro* inhibition of adipogenesis and inflammation by phenolic extracts of *Vaccinium floribundum* and *Aristotelia chilensis*. Journal of Agricultural and Food Chemistry. 2010;58(16):8966–8976.

[86] Céspedes, C.L., Valdez-Morales, M., Avila, J.G., El-Hafidi, M., Alarcón, J., Paredes-López, O. Phytochemical profile and the antioxidant activity of Chilean wild blackberry fruits, *Aristotelia chilensis* (Mol) Stuntz (*Elaeocarpaceae*). Food Chemistry. 2010;119(3):886–895.

[87] Miranda-Rottmann, S., Aspillaga, A.A., Pérez, D.D., Vasquez, L., Martinez, A.L., Leighton, F. Juice and phenolic fractions of the berry *Aristotelia chilensis* inhibit LDL oxidation *in vitro* and protect human endothelial cells against oxidative stress. Journal of Agricultural and Food Chemistry. 2002;50(26):7542–7547.

[88] Céspedes, C.L., El-Hafidi, M., Pavon, N., Alarcon, J. Antioxidant and cardioprotective activities of phenolic extracts from fruits of Chilean blackberry *Aristotelia chilensis* (*Elaeocarpaceae*), Maqui, Food Chemistry. 2008;107(2):820–829.

[89] Céspedes, C.L. Anti-inflammatory, antioedema and gastroprotective activities of *Aristotelia chilensis* extracts. Boletín Latinoamericano y del Caribe de Plantas Medicinales y Aromáticas. 2011;9(6):432–439.

[90] Fredes, C., Yousef, G.G., Robert, P., Grace, M.H., Lila, M.A., Gómez, M., Montenegro, G. Anthocyanin profiling of wild maqui berries (*Aristotelia chilensis* [Mol.] Stuntz) from different geographical regions in Chile. Journal of the Science of Food and Agriculture. 2014;94(13):2639–2648.

[91] Romanucci, V., D'Alonzo, D., Guaragna, A., Di Marino, C., Davinelli, S., Scapagnini, G., Zarrelli, A. Bioactive compounds of *Aristotelia chilensis* Stuntz and their pharmacological effects. Current Pharmaceutical Biotechnology. 2016;17(6):513–523.

[92] Muñoz, O., Christen, P., Cretton, S., Backhouse, N., Torres, V. Chemical study and anti-inflammatory, analgesic and antioxidant activities of the leaves of *Aristotelia chilensis* (Mol.) Stuntz, *Elaeocarpaceae*. Journal of Pharmacy and Pharmacology. 2011;63(1):849–859.

[93] Suwalsky, M., Vargas, P., Avello, M., Villena, F., Sotomayor C.P. Human erythrocytes are affected in vitro by flavonoids of *Aristotelia chilensis* (Maqui) leaves. International Journal of Pharmaceutics. 2008;363(1):85–90.

[94] Wu, B.H., Wang, W.C., Kuo, H.Y. Effect of multi-berries drink on endogenous antioxidant activity in subjects who are regular smokers or drinkers. Journal of Food and Nutrition Research. 2016;4(5):289–295.

[95] Lee, H.J. Antioxidant activity and properties characteristics of pound cakes prepare using freeze dried maquiberry (*Aristotelia chilensis* [Mol.]) powder. The Korean Journal of Food and Nutrition. 2014;27(6):1067–1077.

[96] Gironés-Vilaplana, A., Valentão, P., Moreno, D.A., Ferreres, F., García-Viguera, C., Andrade, P.B. New beverages of lemon juice enriched with the exotic berries maqui, açaı, and blackthorn: bioactive components and *in vitro* biological properties. Journal of Agricultural and Food Chemistry. 2012;60(26):6571–6580

[97] Schreckinger, M.E., Lotton, J., Lila, M.A., de Mejia, E.G. Berries from South America: a comprehensive review on chemistry, health potential, and commercialization. Journal of Medicinal Food. 2010;13(2):233–246.

[98] Delporte, C., Backhouse, N., Inostroza, V., Aguirre, M.C., Peredo, N., Silva, X., Miranda, H.F. Analgesic activity of *Ugni molinae* (murtilla) in mice models of acute pain. Journal of Ethnopharmacology. 2007;112(1):162–165.

[99] Suwalsky, M., Orellana, P., Avello, M., Villena, F. Protective effect of *Ugni molinae* Turcz against oxidative damage of human erythrocytes. Food and Chemical Toxicology. 2007;45(1):130–135.

[100] Shene, C., Reyes, A.K., Villarroel, M., Sineiro, J., Pinelo, M., Rubilar, M. Plant location and extraction procedure strongly alter the antimicrobial activity of murta extracts. European Food Research and Technology. 2009;228(3):467–475.

[101] Peña-Cerda, M., Arancibia-Radich, J., Valenzuela-Bustamante, P., Pérez-Arancibia, R., Barriga, A., Seguel, I., Delporte, C. Phenolic composition and antioxidant capacity of *Ugni molinae* Turcz leaves of different genotypes. Food Chemistry. 2017;215(1):219–227.

[102] Rubilar, M., Pinelo, M., Ihl, M., Scheuermann, E., Sineiro, J., Nuñez, M.J. Murta leaves (*Ugni molinae* Turcz) as a source of antioxidant polyphenols. Journal of Agricultural and Food Chemistry. 2006;54(1):59–64.

[103] Ruiz, A., Hermosín-Gutiérrez, I., Mardones, C., Vergara, C., Herlitz, E., Vega, M., von Baer, D. Polyphenols and antioxidant activity of calafate (*Berberis microphylla*) fruits and other native berries from Southern Chile. Journal of Agricultural and Food Chemistry. 2010;58(10):6081–6089.

[104] Ruiz, A., Mardones, C., Vergara, C., Hermosín-Gutiérrez, I., von Baer, D., Hinrichsen, P., Dominguez, E. Analysis of hydroxycinnamic acids derivatives in calafate (*Berberis microphylla* G. Forst) berries by liquid chromatography with photodiode array and mass spectrometry detection. Journal of Chromatography A. 2013;1281(1):38–45.

[105] Fuentes, L., Valdenegro, M., Gómez, M.G., Ayala-Raso, A., Quiroga, E., Martínez, J.P., Figueroa, C.R. Characterization of fruit development and potential health benefits of arrayan (*Luma apiculata*), a native berry of South America. Food Chemistry. 2016;196(1):1239–1247.

[106] Jiménez-Aspee, F., Theoduloz, C., Vieira, M.N., Rodríguez-Werner, M.A., Schmalfuss, E., Winterhalter, P., SchmedaHirschmann, G. Phenolics from the Patagonian currants *Ribes* spp.: isolation, characterization and cytoprotective effect in human AGS cells. Journal of Functional Foods. 2016;26(1):11–26.

[107] Fredes, C., Montenegro, G. Chilean Plants as a Source of Polyphenols. In: Céspedes C.L., Sampietro D.A., Seigler D.S., Mahendra Rai, editors. Natural Antioxidants and Biocides from Wild Medicinal Plants. CABI (CAB International); 2013. p. 116–136

[108] Villagra, E., Campos-Hernandez, C., Cáceres, P., Cabrera, G., Bernardo, Y., Arencibia, A.,. Morphometric and phytochemical characterization of chaura fruits (*Gaultheria pumila*): a native Chilean berry with commercial potential. Biological Research. 2014;47(1):1–8.

[109] INIA. Tierra adentro, N5, agosto-septiembre [Internet]. 2011. Available from: http://www2.inia.cl/medios/tierraadentro/TierraAdentro95.pdf [Accessed: September, 2016]

[110] Ruiz, A., Zapata, M., Sabando, C., Bustamante, L., von Baer, D., Vergara, C., Mardones, C. Flavonols, alkaloids, and antioxidant capacity of edible wild *Berberis* species from patagonia. Journal of Agricultural and Food Chemistry. 2014;62(51):12407–12417.

[111] Manosalva, L., Mutis, A., Urzúa, A., Fajardo, V., Quiroz, A. Antibacterial activity of alkaloid fractions from Berberis microphylla G. Forst and study of synergism with ampicillin and cephalothin. Molecules. 2016;21(1): 76–85. DOI: 10.3390/molecules6

[112] Ramirez, J.E., Zambrano, R., Sepúlveda, B., Kennelly, E.J., Simirgiotis, M.J. Anthocyanins and antioxidant capacities of six Chilean berries by HPLC–HR-ESI-ToF-MS. Food Chemistry. 2015;176(1):106–114.

[113] Brito, A., Areche, C., Sepúlveda, B., Kennelly, E.J., Simirgiotis, M.J. Anthocyanin characterization, total phenolic quantification and antioxidant features of some Chilean edible berry extracts. Molecules. 2014;19(8):10936–10955.

[114] Albrecht, C., Pellarin, G., Rojas, M.J., Albesa, I., Eraso, A.J. Beneficial effect of *Berberis buxifolia* Lam, *Zizyphus mistol griseb* and *Prosopis alba* extracts on oxidative stress induced by chloramphenicol. Medicina. 2010;70(1):65–70.

[115] INIA. Domesticacion del calafate para fines agroindustriales. Ministerio de Agricultura. Gobierno de Chile. [Internet]. 2001. Available from: http://repositoriodigital.corfo.cl/bitstream/handle/11373/4289/777.076_IF.pdf?sequence=3 [Accessed: September, 2016]

[116] Habtemariam, S. The therapeutic potential of *Berberis darwinii* stem-bark: quantification of berberine and *in vitro* evidence for Alzheimer's disease therapy. Natural Product Communications. 2011;6(8):1089–1090.

[117] Vergara, C., Von Baer, D., Hermosín, I., Ruiz, A., Hitschfeld, M.A., Castillo, N., Mardones, C. Anthocyanins that confer characteristic color to red copihue flowers (*Lapageria rosea*). Journal of the Chilean Chemical Society. 2009;54(2):194–197.

[118] Pino, J.A., Abril, D., Contreras, D., Lamí-Izquierdo, L., Lorenzo-Izquierdo, M. Volatiles from *Lapageria rosea* Ruiz et Pav. red flower. Journal of Essential Oil Research. 2013;25(3):224–226.

[119] Schmeda-Hirschmann, G., Loyola, J.L., Razmilic, I., Reyes, S., Rodríguez, J., Pacheco, P., Teoduloz, C. La Farmacopea Mapuche, una fuente de productos biológicamente activos. Revista Universum de la Universidad de Talca. Chile. 1993;8(1):153–179.

[120] Rodriguez, J., Pacheco, P., Razmilic, I., Loyola, J.I., Schmeda-Hirschmann, G., Theoduloz, C. Hypotensive and diuretic effect of *Equisetum bogotense* and *Fuchsia magellanica* and micropropagation of *E. bogotense*. Phytotherapy Research. 1994;8(3):157–160.

[121] Strzałka, K., Świeżewska, E., Suwalsky, M. Tocochromanols, plastoquinone and polyprenols in selected plant species from Chilean Patagonia. Acta Biologica Cracoviensia. Series Botanica. 2009;51(1):39–44.

[122] Pérez-Cruz, F., Cortés, C., Atala, E., Bohle, P., Valenzuela, F., Olea-Azar, C., Speisky, H., Aspée, A., Liss, E., López-Alarcón, C., Bridi, R. Use of pyrogallol red and pyranine as

probes to evaluate antioxidant capacities towards hypochlorite. Molecules. 2013;12(8): 1638–1652.

[123] Schmeda-Hirschmann, G., Razmilic, I., Gutierrez, M.I., Loyola, J.I. Proximate composition and biological activity of food plants gathered by Chilean Amerindians. Economic Botany. 1999;53(2):177–187.

[124] Simirgiotis M.J. Antioxidant capacity and HPLC-DAD-MS profiling of Chilean Peumo (*Cryptocarya alba*) fruits and comparison with German Peumo (*Crataegus monogyna*) from Southern Chile. Molecules. 2013;18(1):2061–2080.

[125] Franco, W., Fuentes L., Valdenegro M., Figueroa C. Antimicrobial and Antioxidant Capacity of Peumo (*Cryptocarya alba*) and Arrayan (*Luma apiculata*) Leaves and Fruits [Internet]. 2013. Available from: https://iafp.confex.com/iafp/2013/webprogram/Paper3946.html [Accessed: September, 2016]

[126] Arancibia-Avila, P., Toledo, F., Werner, E., Suhaj, M., Leontowicz, H., Leontowicz, M., Martinez-Ayala A.L., Paweł P., Gorinstein, S. Partial characterization of a new kind of Chilean Murtilla-like berries. Food Research International. 2011;44(7):2054–2062.

[127] Flis, S., Jastrzebski, Z., Namiesnik, J., Arancibia-Avila, P., Toledo, F., Leontowicz, H., Gorinstein, S. Evaluation of inhibition of cancer cell proliferation in vitro with different berries and correlation with their antioxidant levels by advanced analytical methods. Journal of Pharmaceutical and Biomedical Analysis. 2012;62(1):68–78.

[128] Gorinstein, S., Arancibia-Avila, P., Toledo, F., Namiesnik, J., Leontowicz, H., Leontowicz, M. Kyung-Sik Ham, Seong-Gook Kang, Kann Vearasilp, Suhaj, M. Application of analytical methods for the determination of bioactive compounds in some berries. Food Analytical Methods. 2013;6(2):432–444.

[129] Suwalsky, M., Avello, M., Obreque, J., Villena, F., Szymanska, R., Stojakowska, A., Strzalka, K. Protective effect of *Philesia Magellanica* (coicopihue) from Chilean Patagonia against oxidative damage. Journal of the Chilean Chemical Society. 2015;60(2): 2935–2939.

[130] Simirgiotis, M.J., Ramirez, J.E., Hirschmann, G.S., Kennelly, E.J. Bioactive coumarins and HPLC-PDA ESI-ToF-MS metabolic profiling of edible queule fruits (*Gomortega keule*), an endangered endemic Chilean species. Food Research International. 2013;54(1):532–543.

[131] Falkenberg, S.S., Tarnow, I., Guzman, A., Mølgaard, P., Simonsen, H.T. Mapuche herbal medicine inhibits blood platelet aggregation. Evidence-Based Complementary and Alternative Medicine. 2012;2012(1):Article ID 647620. DOI: 10.1155/2012/647620

[132] Cerón, M.A.O. Acción antimicrobiana de extractos crudos de especies de plantas nativas sobre *Escherichia coli* y *Salmonella* spp. [thesis]. Chile. 2012.

[133] Simirgiotis, M.J., Bórquez, J., Schmeda-Hirschmann, G. Antioxidant capacity, polyphenolic content and tandem HPLC–DAD–ESI/MS profiling of phenolic compounds from

the South American berries *Luma apiculata* and *L. chequen*. Food Chemistry. 2013;139(1): 289–299.

[134] Cheel, J., Theoduloz, C., Rodríguez, J., Saud, G., Caligari, P.D., Schmeda-Hirschmann, G. E-cinnamic acid derivatives and phenolics from Chilean strawberry fruits, *Fragaria chiloensis* ssp. *chiloensis*. Journal of Agricultural and Food Chemistry. 2005;53(22):8512–8518.

[135] Cheel, J., Theoduloz, C., Rodríguez, J.A., Caligari, P.D., Schmeda-Hirschmann, G. Free radical scavenging activity and phenolic content in achenes and thalamus from *Fragaria chiloensis* ssp. *chiloensis*, F. vesca and F. x *ananassa* cv. Chandler. Food Chemistry. 2007;102(1):36–44.

[136] Wang, S.Y., Lewers, K.S., Bowman, L., Ding, M. Antioxidant activities and anticancer cell proliferation properties of wild strawberries. Journal of the American Society for Horticultural Science. 2007;132(5):647–658.

[137] Simirgiotis, M.J., Theoduloz, C., Caligari, P.D., Schmeda-Hirschmann, G. Comparison of phenolic composition and antioxidant properties of two native Chilean and one domestic strawberry genotypes. Food Chemistry. 2009;113(2):377–385.

[138] Simirgiotis, M.J., Schmeda-Hirschmann, G. Determination of phenolic composition and antioxidant activity in fruits, rhizomes and leaves of the white strawberry (*Fragaria chiloensis* spp. *chiloensis* form chiloensis) using HPLC-DAD–ESI-MS and free radical quenching techniques. Journal of Food Composition and Analysis. 2010;23(6):545–553.

[139] Schmeda-Hirschmann, G., Simirgiotis, M., Cheel, J. Chemistry of the Chilean Strawberry (*Fragaria chiloensis* spp. *chiloensis*). Gene, Genome and Genomics. 2011;5(1):85–90.

[140] Salvat, A., Antonacci, L., Fortunato, R.H., Suárez, E.Y., Godoy, H.M. Antimicrobial activity in methanolic extracts of several plant species from northern Argentina. Phytomedicine. 2004;11(2):230–234.

[141] Costamagna, M.S., Zampini, I.C., Alberto, M.R., Cuello, S., Torres, S., Pérez, J., Quispe J., Schmeda-Hirschmann G., Isla, M.I. Polyphenols rich fraction from *Geoffroea decorticans* fruits flour affects key enzymes involved in metabolic syndrome, oxidative stress and inflammatory process. Food Chemistry. 2016;190(1):392–402.

[142] Reynoso M.A., Sánchez Riera A., Vera N.R. Nutraceutical properties and safety evaluation of fruits and arrope of *Geoffroea decorticans* (Chañar). Journal of Nutrition and Food Sciences. 2016;6(2):485. DOI: 10.4172/2155-9600.1000485

[143] Mølgaard, P., Holler, J.G., Asar, B., Liberna, I., Rosenbæk, L.B., Jebjerg, C.P., Simonsen, H.T. Antimicrobial evaluation of Huilliche plant medicine used to treat wounds. Journal of Ethnopharmacology. 2011;138(1):219–227.

[144] Rojo, Leonel E., Ribnicky, David, Logendra, Sithes, Poulev, Alex, Rojas-Silva, Patricio, Kuhn, Peter, Dorn, Ruth, Grace, Mary H., Ann Lila, Mary, Raskin, Ilya. In vitro and in

vivo antidiabetic effects of anthocyanins from Maqui Berry (*Aristotelia chilensis*). Food Chemistry. 2012;131(1):387–396.

[145] Alfaro, S., Mutis, A., Palma, R., Quiroz, A., Seguel, I., Scheuermann, E. Influence of genotype and harvest year on polyphenol content and antioxidant activity in murtilla (*Ugni molinae* Turcz) fruit. Journal of Soil Science and Plant Nutrition. 2013;13(1):67–78.

[146] Genskowsky, E., Puente, L.A., Pérez-Álvarez, J.A., Fernández-López, J., Muñoz, L.A., Viuda-Martos, M. Determination of polyphenolic profile, antioxidant activity and antibacterial properties of maqui [*Aristotelia chilensis* (Molina) Stuntz] a Chilean blackberry. Journal of the Science of Food and Agriculture. 2016;96(12):4235–4242.

[147] Augusto, Thalita Riquelme, Salinas, Erick Sigisfredo Scheuermann, Alencar, Severino Matias, D'arce, Marisa Aparecida Bismara Regitano, Camargo, Adriano Costa de, Vieira, Thais Maria Ferreira de Souza. Phenolic compounds and antioxidant activity of hydroalcoholic extracts of wild and cultivated murtilla (*Ugni molinae* Turcz.). Food Science and Technology (Campinas). 2014;34(4):667–679.

[148] Junqueira-Gonçalves, M.P., Yáñez, L., Morales, C., Navarro, M., A Contreras, R., Zúñiga, G.E. Isolation and characterization of phenolic compounds and anthocyanins from Murta (*Ugni molinae* Turcz.) fruits. Assessment of antioxidant and antibacterial activity. Molecules. 2015;20(4):5698–5713.

[149] Jofré I., Pezoa C., Cuevas M., Scheuermann E., Almeida Freires I., Rosalen P.L., Matias de Alencar, Romero F. Antioxidant and vasodilator activity of *Ugni molinae* Turcz. (Murtilla) and its modulatory mechanism in hypotensive response. Oxidative Medicine and Cellular Longevity. 2016;2016(1):Article ID 6513416. DOI: 10.1155/2016/6513416

[150] Moyer, R.A., Hummer, K.E., Finn, C.E., Frei, B., Wrolstad, R.E. Anthocyanins, phenolics, and antioxidant capacity in diverse small fruits: *Vaccinium, Rubus*, and *Ribes*. Journal of Agricultural and Food Chemistry. 2002;50(3):519–525.

[151] Jiménez-Aspee, F., Theoduloz, C., Ávila, F., Thomas-Valdés, S., Mardones, C., von Baer, D., Schmeda-Hirschmann, G. The Chilean wild raspberry (*Rubus geoides* Sm.) increases intracellular GSH content and protects against H_2O_2 and methylglyoxal-induced damage in AGS cells. Food Chemistry. 2016;194(1):908–919.

6

Açaí (*Euterpe oleracea*) and Bacaba (*Oenocarpus bacaba*) as Functional Food

Wanessa Almeida da Costa,
Mozaniel Santana de Oliveira,
Marcilene Paiva da Silva, Vânia Maria Borges Cunha,
Rafael Henrique Holanda Pinto,
Fernanda Wariss Figueiredo Bezerra and
Raul Nunes de Carvalho Junior

Abstract

This chapter reviews two oleaginous fruits that are widely consumed by people in the Amazon region: Bacaba (*Oenocarpus bacaba*) and Açaí (*Euterpe oleracea*). Besides their food and the folk medicinal uses, studies suggest that substances present in both berries have antioxidative effects, antimicrobial, and therapeutic properties such as hypocholesterolemic and neuroprotection effects. These therapeutic effects are related to phenolic compounds, anthocyanins, and fatty acids, which can prevent serious problems such as coronary heart disease, hypertension, and depression. The use of supercritical fluid technology is described as a technique to obtain the best extracts of bacaba and açaí, as well as their valuable constituents. Indubitably, this technology is a great tool for human health and all with the advantage of presenting nontoxic solvents such as carbon dioxide or water. Açaí and bacaba fruits represent not only food but also a source of compounds that can work in both prevention and treatment of diseases.

Keywords: Amazon, açaí, bacaba, bioactive compounds, antioxidants, functional food

1. Introduction

The Brazilian Amazon represents one of the richest biomes found in the world. It presents many sources of plants, including native ones not yet explored, but that have potential for use. The economic importance that the aromatic plants represent to the Amazon region is associated with the application of their vegetable oils and aromas in technological and industrial processes. Because of this, there is a greater investment in such plants extraction sector, causing an expansion of the domestic and international markets.

Because of this biodiversity, there is a wide variety of oleaginous species, as is the case of andiroba (*Carapa guianensis*), tucumã (*Astrocaryum vulgare*), buriti (*Mauritia flexuosa*), palm (*Elaeis guineensis, Jacq*), açaí (*Euterpe oleracea*), and bacaba (*Oenocarpus bacaba*). These species experimentally have a high yield in vegetable oils, with the potential for production of biologically active natural products, the so-called bioactive compounds, which have a high value added. Among these, the fat-soluble vitamins carotenoids (provitamin A), tocopherols (provitamin E and antioxidant), dyes, and flavonoids (anthocyanins, which are dyes with antioxidant effects) can be highlighted.

The characteristics of the Amazon region are conducive to the proliferation of palm trees, among which there are the oleaginous ones that are commercially cultivated with already fully established management technology, as is the case of açaí and bacaba, which can be considered new "superfruits." The consumption of these fruits pulps has been increasing, mainly due to the benefits that are being showed by scientific papers. Açaí, for example, has a high economic potential, mainly due to its use in the preparation of açaí beverages, which are exported all over the world as an energetic drink [1].

Besides the folk use as a drink, studies suggest that substances present in both berries have therapeutic properties such as hypocholesterolemic and neuroprotection effects. These therapeutic effects are related to fatty acids, which can prevent serious problems such as coronary heart disease, hypertension, and depression [2, 3]. The presence of phenolic compounds in their composition also gives them properties such as antimicrobial and antioxidant effects [4, 5].

Another group of compounds with significant presence in açaí and bacaba is anthocyanins. Anthocyanins are plant-derived compounds belonging to the flavonoids subgroup of phenolic compounds. Besides antioxidative properties, anthocyanins are the focus of studies for application on humans against diseases such as cancer and Alzheimer's [6–8].

Among the various methods of obtaining natural extracts, the process of supercritical fluid extraction has become appropriate and of great interest to the food industry, pharmaceutical, and cosmetic technology. It provides the obtainment of products free of residual solvents and with superior quality, while preserving the organoleptic properties of the material. The most used solvent in the supercritical technology is carbon dioxide (CO_2), which is inert, nontoxic, has a high solubility, and allows performing low-temperature processes, which are perfect for the extraction of thermosensible compounds, as is the case of, for example, anthocyanins.

2. Açaí and bacaba as functional food

The food industry has high expectations in food products that meet the consumers' demand for a healthy lifestyle. In this context, functional food plays a specific role, which is not only to satisfy hunger but also to provide humans the necessary nutrients. It also prevents nutrition-related diseases and increases their physical and mental well-being [9].

In Brazil, there are two kinds of functional food: açaí and bacaba (see **Figure 1**), which are oleaginous fruits, present black-violet color, and are from typical palm trees in the Amazon region. They belong to the *Arecaceae* family and when processed with water, form an emulsion. Both are commercially exploited for the production of foods and beverages. The juices of bacaba and açaí are considered tasty and much appreciated by the Amazonian population. In the period between harvests of açaí, from December to April, bacaba has the highest sales potential, in a relay system [10, 11].

The functional quality of bacaba oil was analyzed by Pinto [12] through the determination of atherogenicity index (AI) and thrombogenicity index (TI) proposed by Ulbricht and Southgate [13] and hypocholesterolemic/hypercholesterolemic ratio (h/H) suggested by Santos-Silva et al. [14]. The results of AI, TI, and h/H were satisfactory. Although the values of AI and TI were low, h/H was high in levels that show bacaba oil could be regarded as cardioprotective, suggesting the direct consumption of it in the form of table oil, similar to olive oil, or in encapsulated form as a phytopharmaco. In the same study, bacaba oil was used for coating iron oxide for the synthesis of Fe_3O_4 magnetic nanoparticles (MNP). The results showed that the oil well replaced the oleic acid, with the formation of MNP with morphological and desirable

(a) (b)

Figure 1. Bacaba (**a**) and açaí (**b**) berries.

magnetic characteristics. MNP have therapeutic features, being used as drug carriers in the treatment of cancer by magnetic induction, reducing collateral effects to patients.

Açaí, being a source of fibers and rich in antioxidants, has considerable potential for nutritional applications and in the health field as a functional food or dietary supplement [15]. The work conducted by Barbosa et al. [2] evaluated the effect of a diet with daily consumption of açaí pulp in the prevention of oxidative damage by measuring the activity of antioxidant enzymes and the use of protein biomarkers in healthy women. The results showed that the açaí intake increased the activity of catalase, an intracellular enzyme which is also known as hydroperoxidase, able to decompose the hydrogen peroxide (H_2O_2), which is associated with various pathologies connected to oxidative stress; the results also showed an increase in total antioxidant capacity and a reduction in the production of reactive oxygen species. These studies reveal the antioxidant effect of açaí, increasing the understanding of its beneficial health properties.

The antioxidants found in açaí and bacaba are necessary to prevent the formation and oppose the actions of reactive oxygen species (ROS) and reactive nitrogen species (RNS), which are continuously formed in the human body. Mechanisms of free radicals such as these are related to various human diseases, including cancer, atherosclerosis, malaria, rheumatoid arthritis, and neurodegenerative diseases. Many components of the diet such as carotenoids and plant pigments are suggested as important antioxidants; however, the interest in phenolic compounds of plants, particularly flavonoids, is also increasing. Thus, diets based on functional foods rich in antioxidants are important for the maintenance of human health [16–19].

3. Chemical composition of açaí and bacaba

The nutritional properties of Amazonian palm trees are related to the composition of fatty acids and phytochemical compounds, the so-called bioactive compounds. Açaí and bacaba are some of the species of fruits that have become quite attractive, not only for lipid content they present, but also for their composition of bioactive compounds.

The fatty acids present in fruit species such as these are considered one of the most important constituents in living organisms due to their structural role in cell membranes and as metabolic energy sources [20]. Those considered essential to life are known as essential unsaturated fatty acids and must be supplied by food. The main representatives are omega-9 (ω-9), omega-6 (ω-6), and omega-3 (ω-3). Of these groups, the α-linolenic acid (n-3), the linoleic and arachidonic acids (n-6), and the oleic acid (n-9) can be highlighted [21]. The vegetable oils, such as bacaba and açaí, are good sources of these components and fat-soluble vitamins such as vitamins A, D, E, and K [22].

According to Martin et al. [23], the availability of ω-3 and ω-6 fatty acids in the human species depends on the food supply, and moreover, it is important to know what are the sources capable of supplying these needs. **Table 1** shows some sources of monounsaturated and polyunsaturated fatty acids of fruits that come from palm trees and are considered as dietary sources of fatty acids.

Batista et al. [8] obtained the fatty acids profile of lyophilized açaí pulp extracts obtained by extraction with supercritical CO_2, as shown in **Table 2**.

Fruits that come from palm trees	Part of the fruit analyzed	(C12:0) lauric (%)	(C14:0) myristic (%)	(C16:0) palmitic (%)	(C18:1) oleic (%)	(C18:2) linoleic (%)	(C18:3) linolenic (%)
Babaçu (*Orbignya phalerata* Martius)[1]	Kernel	44.0	17.0	8.0	14.0	2.0	–
Buriti (*Mauritia flexuosa L.f.*)[2]	Mesocarp	–	–	18.0	73.5	2.7	2.1
Dendê (palm) (*Elaeis olifera*)[3]	Endocarp	47.9	16.1	8.4	16.2	2.7	Traces
Pupunha (*Bactris gasipaes*)[4]	Mesocarp	–	–	35.20	51.7	4.9	1.2
Tucumã (*Astrocaryum vulgare*)[5]	Epicarp + mesocarp	–	0.10	24.6	65.1	2.6	0.2
Bacaba (*Oenocarpus bacaba*)[6]	Mesocarp	0.18	0.59	32.27	40.82	9.78	1.93
Bacaba (*Oenocarpus bacaba*)[7]	Mesocarp	–	–	30.6	47.3	20.6	–
Patauá (*Jessenia bataua*)[5]	Mesocarp	–	0.10	13.3	76.7	3.9	0.1
Açaí (*Euterpe oleracea*)[3]	Mesocarp	–	–	25.9	54.9	11.5	1.1

Sources: [1]Lima et al. [24], [2]Tavares et al. [25], [3]Rogez. [26], [4]Yuyama et al. [27], [5]Rodrigues et al. [28], [6]Montúfar et al. [29], [7]Santos et al. [30].

Table 1. Content of the main fatty acids present in palm tree fruits consumed in the human diet.

Foods rich in fatty acids, such as bacaba and açaí, can play an important role in human food base, because the linolenic, linoleic, and oleic acids present in these raw materials are considered functional and exhibit inflammation-reducing and immunity-increasing properties in the human body, as demonstrated by Wallace et al. [31], Schwab and Serhan [32], Serhan et al. [33], and Calder [34].

In addition to fatty acids, various bioactive compounds can be found in these fruits. Yamaguchi et al. [1] report that about 90 substances have been found in açaí, of which approximately 31% consist of flavonoids, followed by 23% of phenolic compounds, 11% of lignoids, and 9% of anthocyanins. These are compounds that are correlated with high biological activity.

Content of fatty acids in % g/100mg									
Fatty	**50°C**	**50°C**	**50°C**	**60°C**	**60°C**	**60°C**	**70°C**	**70°C**	**70°C**
Acid	**150 bar**	**220 bar**	**350 bar**	**190 bar**	**270 bar**	**420 bar**	**220 bar**	**320 bar**	**490 bar**
C8:0	0.69	1.26	0.83	0.77	1.58	0.40	0.33	2.27	0.02
C10:0		0.03	0.02	0.02	0.04	0.03	–	–	–
C12:0	0.07	0.17	0.17	0.13	0.19	0.25	0.07	0.33	0.14
C13:0	–	–	–	–	–	–	0.02	0.21	–
C14:0	0.13	0.24	0.16	0.19	0.21	0.30	0.13	0.42	0.18
C15:0	–	–	–	–	–	–	–	–	–
C16:0	28.15	30.91	23.47	26.29	29.20	28.58	25.41	90.86	27.81
C16:1	4.95	0.03	5.49	6.14	7.08	6.83	4.16	0.08	5.81
C17:0	–	0.04	0.14	0.03	–	–	0.05	0.19	0.03
C18:0	1.05	1.25	1.02	0.80	1.14	1.16	1.43	5.35	1.33
C18:1	64.86	65.81	52.73	50.78	60.42	62.41	55.71	0.23	64.65
C18:2	–	–	15.54	14.80	–	–	12.59	–	–
C18:3	–	–	–	–	–	–	–	–	–
C20:0	0.08	–	–	–	0.10	–	–	–	–
C22:0	–	0.22	0.38	–	–	–	0.04	–	–
SFA	30.18	34.15	26.22	28.25	32.48	30.74	27.53	99.67	29.53
MUFA	69.81	65.84	58.23	56.93	67.51	69.25	59.87	0.31	70.46
PUFA	–	–	15.54	14.80	–		12.59	–	–
S/U	0.43	0.52	0.35	0.39	0.48	0.44	0.38	321.52	0.42

C8:0 (caprylic acid); C10:0 (capric acid); C12:0 (lauric acid); C13:0 (tridecanoic acid); C14:0 (myristic acid); C15:0 (pentadecanoic acid); C16:0 (palmitic acid); C16:1 (palmitoleic acid); C17:0 (margaric acid); C18:0 (stearic acid); C18:1 (oleic acid); C18:2 (linoleic acid); C18:3 (linolenic acid); C20:0 (arachidic acid); C22:0 (behenic acid); SFA (saturated fatty acids); MUFA (monounsaturated fatty acids); PUFA (polyunsaturated fatty acids).

Table 2. Content of fatty acids in açaí pulp extracts obtained by extraction with supercritical CO_2.

Of these components, anthocyanins have received great attention due to their potential benefits in preventing chronic diseases, including cancer and Alzheimer [8]. They are glycosides of anthocyanins and have, at their core, the flavylium cation. They belong to the group of flavonoids and subgroup of phenolic compounds. These compounds are responsible for defining the color of a variety of vegetables, including purple color in açaí [1]. They are hydrophilic, stable at acid pH, sensitive to light exposure, elevated temperatures, and presence of O_2.

To obtain bioactive substances such as anthocyanins, different extraction techniques have been developed with the aim of reducing the extraction time and the solvent consumption, increasing the extraction yield and improving the quality of the extracts. Among these techniques are included: ultrasound assisted extraction, microwave assisted extraction, supercritical fluid extraction, and accelerated solvent extraction [35].

The choice of a method for extracting anthocyanins depends largely on the purpose of extraction and the nature of the constituent molecules of these compounds [36]. Therefore, as these pigments are very soluble in water, they are easily extracted by polar solvents. Their extraction typically involves the use of aqueous acidified solutions of ethanol, methanol, or acetone [37]. However, these solvents have also used limitations such as lower extraction efficiency compared to other solvents, as well as a lower durability of their extracts [38, 39].

References	Application	Anthocyanins quantification
Finco et al. [40]	Characterization and analysis of total phenolic compounds and total flavonoids of bacaba extract (*Oenocarpus bacaba* Mart.) by HPLC-DAD-MS	The total content of monomeric anthocyanin was evaluated by a differential pH method described by Sellappan et al. [41]. The anthocyanin cyanidin-3-glucoside was used as pattern
Gouvêa et al. [42]	Isolation of anthocyanins patterns (cyanidin-3-O-glucoside and cyanidin-3-O-rutinoside) of lyophilized açaí (*Euterpe oleracea* Mart.) by HPLC	The isolation of anthocyanins was carried out by HPLC. The anthocyanin identification in the lyophilized açaí was done by mass spectrometry. They used the anthocyanins patterns: cyanidin-3-*O*-glucoside and cyanidin-3-*O*-rutinoside
Santos et al. [43]	This study evaluated the encapsulation of anthocyanin extract obtained from jabuticaba (*Myrciaria cauliflora*) using supercritical CO_2 as solvent and ethanol as co-solvent	In the extraction of jabuticaba anthocyanin, supercritical CO_2 was used together with the co-solvent ethanol in certain conditions of pressure, temperature, and flow ratio
Paes et al. [44]	Extraction of anthocyanins and phenolic compounds of blueberry (*Vaccinium myrtillus* L.) using supercritical CO_2 and water and ethanol as co-solvents	HPLC and mass spectrometry. Pelargonidin was used as pattern for the identification of anthocyanins
Neves et al. [45]	The objective of this study was to follow the physicochemical and functional alterations of açaí and bacaba pulps processed by hand	For the determination of total anthocyanins, the method of Francis [46] was used
Novello et al. [47]	This study aimed to evaluate the influence of organic solvents on the extraction of anthocyanins from açaí. The anthocyanins, the fatty acids profile, and the antioxidant activity of the extract were analyzed by HPLC	The anthocyanins were determined by spectrophotometry using differential pH method described by Giusti and Wrolstad [48]. The identification and quantification of anthocyanins of lyophilized açaí extract were performed by HPLC-DAD. The identified anthocyanins were cyanidin-3-*O*-glucoside and cyanidin-3-*O*-rutinoside
Batista et al. [8]	This study determined the phenolic compounds and anthocyanins of lyophilized açaí pulp after extraction with supercritical CO_2	The anthocyanins were determined by UV-visible spectrophotometry using the Folin-Ciocalteu reagent, according to the method described by Singleton and Rossi [49]

Table 3. Overview of anthocyanin extraction applications.

The main anthocyanins found in açaí are cyanidin-3-*O*-glucoside and cyanidin-3-*O*-rutinoside. In bacaba, it is cyanidin-3-glucoside. This information is presented in **Table 3**, as well as an overview of some anthocyanin extraction applications of açaí, bacaba, and other raw materials. Their chemical structures are presented in **Figure 2**.

(a) Cyanidin-3-O-rutinoside (b) Cyanidin-3-glucoside

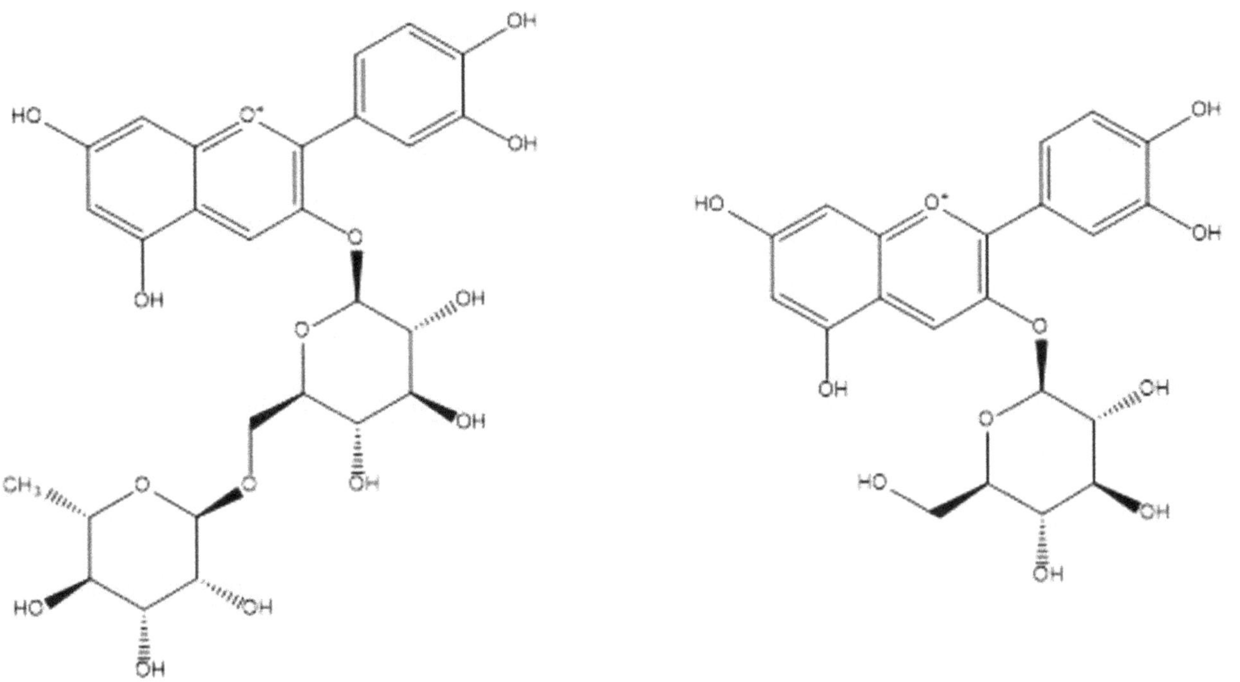

Figure 2. Chemical structures of the main anthocyanins found in açaí and bacaba (**a**): 2-(3,4-dihydroxyphenyl)-5,7-dihydroxy-3-chromeniumyl6-*O*-(6-deoxy-α-L-mannopyranosyl)-β-D-glucopyranoside; (**b**): 2-(3,4-dihydroxyphenyl)-5,7-dihydroxy-3-chromeniumyl β-D-glucopyranoside [nomenclatures according to IUPAC].

In addition to anthocyanins, other bioactive compounds have been identified in açaí and bacaba. Pacheco-Palencia et al. [50] analyzed two species of açaí and identified several flavones, including homoorientin, orientin, deoxyhexose taxifolin, and isovitexin; flavanol derivatives, including (+)—catechin, (−)—epicatechin, procyanidin dimers and trimers, and phenolic acids such as protocatechuic, p-hydroxybenzoic, vanillic, syringic, and ferulic. Phenolic compounds are also reported to be potentially protective against cardiovascular disease and cancer [51]. Also, large amounts of phenolic compounds such as phenolic acids, flavanols, and flavonols can be found, which act as cofactors to improve the biological action of anthocyanins [52].

Santos et al. [53] evaluated the content of bioactive compounds and total antioxidant capacity of native fruits of the Amazon palm trees, including the species *O. bacaba*. Their results showed a high content of total polyphenols, presence of carotenoids, higher levels of anthocyanins, and antioxidant capacity in the bacaba extracts. In the study of Finco et al. [40], the phenolic classes: C-glycoside, flavonoid, C-hexoside, C-glycosylflavone, isorhamnetin hexoside, quercetin hexoside, quercetin diglycoside, quercetin glycoside, and isorhamnetin glycoside, were identified.

4. Methods for obtaining vegetable oils

The economic importance that aromatic plants have in the Amazon region is associated with the application of their vegetable oils and use of their aromas in technological and industrial processes. Because of this, there is a greater investment in such plants extraction sector, causing an expansion of the domestic and international markets.

The soil and climate of the Amazon region are conducive to the proliferation of palm trees, among which there are the oleaginous ones cultivated with commercial purpose. This is the case of açaí and bacaba, whose extraction already constitutes a significant economic activity in the state of Pará-Brazil. There are other native palm trees in the region that provide oleaginous fruits rich in provitamins A and E, yet poorly explored, such as pupunha (*Guilielmaspeciosa*) and tucumã (*Astrocaryumvulgare*). These and other vegetable raw materials present in their composition have a high content of lipids, with significant potential for extraction.

Extraction is a unit operation widely used in the food industry and can be used for the production of coffee, sugar, caffeine extraction, vegetable oils, flavorings, and essential oils [54]. Obtaining these extracts may be accomplished by different methods such as mechanical pressing extraction, solvent extraction, supercritical fluids extraction, or others, depending on their content [55–57].

4.1. Mechanical pressing extraction

The extraction by mechanical pressing is one of the oldest methods of obtaining oil and fats from seed and fruits. For this kind of extraction, the packaged material enters through a feed shaft in the press. The press consists of a basket formed of spaced rectangular steel bars, through blades, whose thickness varies depending on the raw material. In the center of the basket, there is a screw that rotates and moves the material forward, compressing it at the same time. The pressure is regulated via an outlet cone [58, 59].

Souza et al. [60] and Pighinelli et al. [61] report that although the mechanical pressing extraction is less efficient than other methods, it is a more workable system on a small scale, for not being dependent on facilities and safety that are characteristics of the solvent processing, besides being fast, easy to handle and presents low cost of installation and maintenance.

One of the disadvantages of the mechanical pressing method is its low oil yield recovery: even in the most efficient presses, there is still a range of 3–5% of remaining oil in the cake. This residual oil present in the cake can be recovered by a two-step process: pre-extraction (with the screw-press) and solvent extraction, thus maximizing efficiency. Furthermore, the solvent extraction is recommended only in raw materials with <25% of fat content [62–69].

4.2. Solvent extraction

This type of extraction occurs by partitioning a solute between two immiscible or partially miscible phases. The mass transfer occurs from the solutes in the food matrix to the solvent. First, the solute is dissolved in the solvent, then the penetration of the particle solution in the food surface occurs, and finally the solution is dispersed in the solvent. According to Ghosh [64], solvent extraction can be classified into four types depending on the phase of the matrix: (i) solid-liquid extraction; (ii) liquid-liquid extraction; (iii) vapor extraction; and (iv) supercritical fluids extraction.

The solvent choice is of fundamental importance in the aspects that aim at efficiency, economy, and preservation of the physicochemical and nutritional characteristics of oils. In conventional extraction, some solvents used for obtaining oils from plants are hexane, n-hexane, pentane, ethanol, and petroleum ether [59, 63, 65–70].

In the solvent extraction, there can be a reduction in the product quality because of the several steps necessary to recover the solvent, elevated temperature, long periods of thermal exposure, high oxygen concentration, and extraction of other compounds considered undesirable [63, 71].

4.3. Supercritical fluids extraction

The supercritical fluids (SCFs) extraction is a unit operation by contact that is based on the balance and on the physicochemical properties of the SCFs, being dependent on operating conditions such as temperature, pressure, solvent flow, the material morphology, prior treatment of the porous solid matrix, and the physical properties of the packed bed, such as porosity, distribution and particle size, initial content of solute in the solid matrix, and the fixed bed height [72].

The SCFs present intermediate characteristics between liquids and gases. The diffusion coefficient (DC) of SCFs is high and close to the gases DC, thus increasing the diffusivity when they are in the liquid state, providing a rapid and efficient mass transfer. The density of SCFs is greater than that of a gas, having a higher solvating power due to the high compressibility. Furthermore, they exhibit low viscosity and the absence of surface tension, which promote greater penetration into the solid matrix [73, 74].

Carbon dioxide (CO_2) is widely used as SCF due to having low critical temperature and pressure (73.74 bar and 304.12 K, respectively), besides being: nontoxic, nonflammable, odorless, and easily separated from the extract. Due to its low critical temperature, it is possible to use it to extract reactive and thermosensitive compounds. CO_2 is suitable for extracting apolar compounds, but when polar organic solvents such as ethyl acetate, ethanol, or methanol are added, the polarity is modified, being possible to extract other compounds. These aggregate solvents are called co-solvents [75].

Batista et al. [8] obtained açaí extracts fractions with supercritical CO_2 and analyzed the allelopathic effects of these extracts on two species of invasive plants: *Mimosa pudica* and *Senna obtusifolia*. They observed that depending on the operating conditions of temperature and pressure used, the pattern of phytotoxic responses can change: in some cases, the effect may be stimulatory to seed development. Studies on allelopathy have direct influence on human health, because the use of chemicals such as pesticides, which can cause diseases such as cancer, can be avoided [76–78]. However, other studies must be conducted to isolate the specific metabolites for each role assigned to the açaí.

Pinto [12] also obtained bacaba extracts fractions with supercritical CO_2 at different conditions of temperature and pressure. In his work, bacaba is mentioned as a rich source of natural antioxidants and dyes. However, there is a need for further studies to elucidate bacaba's behavior in different processes.

4.4. Other extraction methods

The methods of soxhlet, hydrodistillation, solid-liquid, and ultrasound-assisted extraction do not present a performance as good as the one presented by the extraction with supercritical fluids: it has a high selectivity, low or no organic solvent consumption, operates at temperature close to room, no request for subsequent purification steps, and reduces post-processing costs as there is no longer need to eliminate solvent extracts [75, 79, 80].

5. Anthocyanins extraction by SFE

Anthocyanins are the most abundant flavonoid constituents of fruits and vegetables. Their use into food and/or medical fields has proven to be a technological challenge since these compounds have low stability and are susceptible to degradation through factors such as the presence of light, pH, temperatures usually higher than 60–80°C, the presence of sulfite, ascorbic acid, enzymes (such as glycosidases and phenolases), among other factors [43, 81, 82].

In the literature, the recovery of phytochemicals from solid wastes has been reported using conventional and alternative technologies. According to Paes et al. [44], conventional methods are Soxhlet extraction, maceration extraction, extraction by infusion and vapor distillation. Alternative techniques such as supercritical fluid extraction (SFE) and pressurized liquid extraction (PLE) eventually assisted with ultrasound are also reported.

Supercritical fluids processes have proved to be an excellent alternative to extract natural pigments due to the use of environmentally friendly solvents, such as carbon dioxide. According to Vatai et al. [83], extractions with supercritical CO_2 result in non-deteriorated reactions, due to low process temperatures. The CO_2 is readily available, relatively cheap, and accepted as a solvent in the food industry. SFE with CO_2 is an excellent isolation method for natural materials and gives an alternative to replace the nonpolar organic solvents, which can be a source of food contamination.

Supercritical fluid extraction (SFE) using carbon dioxide (CO_2) has been applied for the pretreatment of natural materials, as shown in the works of Paula et al. [84], Ghafoor et al. [85], and Floris et al. [86]. Operating conditions (temperatures varying from 40 to 50°C and pressures above 200 bar) and the use of co-solvents such as ethanol and water were used in their studies as modifiers to obtain the maximum extract yield. According to Seabra et al. [87], even though the choice of the appropriate polar solvent is a key factor for the success of the anthocyanin extraction procedure, its influence on the extract's characteristics is not always clear, due to the diverse structure and composition of plant materials and also the relation material-solvent.

6. Conclusion

Açaí (*Euterpe oleracea*) and bacaba (*Oenocarpus bacaba*) are highly consumed fruits in Amazon that come from common palm trees and have remarkable properties. There are many benefits that help increasing their role in the growing market for nutraceuticals. Their extracts have a range of bioactive and polyphenolic components with antioxidant properties that make them new "superfruits"; however, further studies still need to be conducted in order to elucidate all the roles that these fruits can play. Açaí and bacaba represent not only food, but also a real source of health for humans.

Author details

Wanessa Almeida da Costa, Mozaniel Santana de Oliveira, Marcilene Paiva da Silva, Vânia Maria Borges Cunha, Rafael Henrique Holanda Pinto, Fernanda Wariss Figueiredo Bezerra and Raul Nunes de Carvalho Junior*

*Address all correspondence to: raulncj@ufpa.br

1 LABEX/FEA, Faculty of Food Engineering, Federal University of Pará, Belém, Brazil

References

[1] Yamaguchi KKL, Pereira LFR, Lamarão CV, Lima ES, Veiga-Junior VF. Amazon acai: chemistry and biological activities: a review. *Food Chemistry*. 2015; 179: 137–151.

[2] Barbosa PO, Pala D, Silva CT, de Souza MO, do Amaral JF, Vieira RAL, et al. Acai (*Euterpe oleracea* Mart.) pulp dietary intake improves cellular antioxidant enzymes and biomarkers of serum in healthy women. *Nutrition*. 2016; 32(6): 674–680.

[3] Kris-Etherton PM, Harris WS, Appel LJ, Nutrition Committee. Fish consumption, fish oil, omega-3 fatty acids, and cardiovascular disease. *Circulation*. 2012; 106(21): 2747–2757.

[4] Xie C, Kang J, Burris R, Ferguson ME, Schauss AG, Nagarajan S, et al. Acaí juice attenuates atherosclerosis in ApoE deficient mice through antioxidant and anti-inflammatory activities. *Atherosclerosis*. 2011; 216(2): 327–333.

[5] Santos VS, Bisen-Hersh E, Yu Y, Cabral ISR, Nardini V, Culbreth M, et al. Anthocyanin-rich Açaí (*Euterpe oleracea* Mart.) Extract attenuates manganese-induced oxidative stress in rat primary astrocyte cultures. *Journal of Toxicology and Environmental Health, Part A*. 2014; 77(7): 390–404.

[6] Kang J, LI Z, Wu T, Jensen GS, Schauss AG, Wu X. Anti-oxidant capacities of flavonoid compounds isolated from acai pulp (*Euterpe oleracea* Mart.). *Food Chemistry*. 2010; 122(3): 610–617.

[7] Hogan S, Chung H, Zhang L, Li J, Lee Y, Dai Y, et al. Antiproliferative and antioxidant properties of anthocyanin-rich extract from açai. *Food Chemistry*. 2010; 118(2): 208–214.

[8] Batista CCR, Oliveira MS, Araújo ME, Rodrigues AMC, Botelho JSR, Filho APSS, et al. Supercritical CO_2 extraction of açaí (*Euterpe oleracea*) berry oil: global yield, fatty acids, allelopathic activities, and determination of phenolic and anthocyanins total compounds in the residual pulp. *The Journal of Supercritical Fluids*. 2016; 107: 364–369.

[9] Menrad K. Market and marketing of functional food in Europe. *Journal of Food Engineering*. 2013; 56(2): 181–188.

[10] Mendonça MS, De Araújo MGP. A semente de bacaba (*Oenocarpus bacaba* Mart. Arecacea): Aspectos morfológicos. Revista Brasileira de Sementes. 1999; 21(1): 122–124.

[11] Mertens-Talcott SU, Rios J, Jilma-Stohlawetz P, Pacheco-Palencia LA, Meibohm B, Talcot ST, Derendorf H. Pharmacokinetics of anthocyanins and antioxidant effects after the consumption of anthocyanin-rich acai juice and pulp (Euterpe oleracea Mart.) in human healthy volunteers. *Journal of agricultural and food chemistry*. 2008; 56(17), 7796–7802.

[12] Pinto RHH. *Extração do óleo de bacaba (Oenocarpus bacaba) com CO_2 supercrítico: Parâmetros de processo, perfil de ácidos graxos e aplicação na síntese de nanopartículas de Fe_3O_4*. Dissertação de Mestrado, Instituto de Tecnologia, Programa de Pós-Graduação em Ciência e Tecnologia de Alimentos, Universidade Federal do Pará. 2016: 16–54.

[13] Ulbricht TLV, Southgate DAT. Coronary heart disease: seven dietary factors. *The Lancet*. 1991; 338(8773): 985–992.

[14] Santos-Silva J, Bessa RJB, Santos-Silva F. Effect of genotype, feeding system and slaughter weight on the quality of light lambs. II. Fatty acid composition of meat. *Livestock Production Science*. 2002; 77(2): 187–194.

[15] Pérez-Jiménez J, Arranz S, AlvesRE, de Brito ES, Oliveira MS, Saura-Calixto F. Açaí (*Euterpe oleraceae*) 'BRS Pará': a tropical fruit source of antioxidant dietary fiber and high antioxidant capacity oil. *Food Research International*. 2011; 44(7): 2100–2106.

[16] Halliwell B. Antioxidants in human health and disease. *Annual Review of Nutrition*. 1996; 16(1): 33–50.

[17] Aruoma OI. Free radicals, oxidative stress, and antioxidants in human health and disease. *Journal of the American Oil Chemists' Society*. 1998; 75(2): 199–212.

[18] Devasagayam TPA, Tilak JC, Boloor KK, Sane KS, Ghaskadbi SS, Lele RD. Free radicals and antioxidants in human health: current status and future prospects. *Japi*. 2004; 52(794804): 4.

[19] Zafra-Stone S, Yasmin T, Bagchi M, Chatterjee A, Vinson JA, Bagchi D. Berry anthocyanins as novel antioxidants in human health and disease prevention. *Molecular Nutrition & Food Research*. 2007; 51(6): 675–683.

[20] Simopoulos, AP. The importance of the ratio of omega-6/omega-3 essential fatty acids. *Biomedicine & Pharmacotherapy*. 2002; 56(8): 365–379.

[21] Mataix J. *Lipiodos alimentarios*. In: Mataix J, Gil A (eds). Libro blanco de los Omega-3. Madrid: Instituto Omega-3. 2002: 14–32.

[22] Calder PC. Mechanisms of action of (n-3) fatty acids. *The Journal of Nutrition*. 2012; 142(3): 592S–599S.

[23] Martin CA, Almeida VV, Ruiz MR, Visentainer JEL, Matshushita M, Souza NE, et al. Ácidos graxos poliinsaturados ômega-3 e ômega-6: importância e ocorrência em alimentos. *Revista de Nutrição*. 2006; 19(6): 761–770.

[24] Lima JRO, Silva RB, Silva CM. Bidiesel de babaçu (*Orgignya* sp.) obtido por via etanólica. *Química Nova*. 2007; 30(3): 600–603.

[25] Tavares M, Aued-Pimentel S, Lamardo LCA, Campos NC, Jorge LIF, et al. Composição química e estudo anatômico dos frutos de buriti do Município de Buritizal, Estado de São Paulo. *Revista Instituto Adolfo Lutz*. 2003; 3(62): 227–232.

[26] Rogez H. *Açaí: Preparo, Composição e Melhoramento da Conservação*. EDUFPA. 2000: 313.

[27] Yuyama LKO, Aguiar JPL, Yuyama K, Clement CR, MacedoSHM, Fávaro DIT., et al. Chemical composition of the fruit mesocarp of three peach palm (*Bactrisgasipaes*) populations grown in Central Amazonia, Brazil. *International Journal of Food Sciences and Nutrition*. 2003; 54(1): 49–56.

[28] Rodrigues AMC, Darnet S, Silva LHM. Fatty acid profiles and tocopherol contents of buriti (*Mauritia flexuosa*), patawa (*Oenocarpus bataua*), tucuma (*Astrocaryum vulgare*), mari (*Poraqueiba paraensis*) and inaja (*Maximiliana maripa*) fruits. *Journal of the Brazilian Chemical Society*. 2010; 21(10): 2000–2004.

[29] Montúfar R, Laffargue A, Pintaud J-C, Hamon S, Avallone S, Dussert S. *Oenocarpusbataua* Mart. (Arecaceae): rediscovering a source of high oleic vegetable oil from Amazonia. *Journal of the American Oil Chemists' Society*. 2010; 87(2): 167–172.

[30] Santos MFG, Alves RE, Ruíz-Méndez MV. Minor components in oils obtained from amazonian palm fruits. *Grasas y Aceites*. 2013; 64(5): 531–536.

[31] Wallace FA, Miles EA, Calder PC. Comparison of the effects of linseed oil and different doses of fish oil on mononuclear cell function in healthy human subjects. *British Journal of Nutrition*. 2003; 89(5): 679–689.

[32] Schwab JM, Serhan CN. Lipoxins and new lipid mediators in the resolution of inflammation. *Current Opinion in Pharmacology*. 2006; 6(4): 414–420.

[33] Serhan CN, Hong S, Gronert K, Colgan SP, Devchand PR, Mirick G, et al. Resolvins: a family of bioactive products of omega-3 fatty acid transformation circuits initiated by aspirin treatment that counter proinflammatory signals. *The Journal of Experimental Medicine*. 2002; 196(8): 1025–1037.

[34] Calder PC. Omega-3 fatty acids and inflammatory processes. *Nutrients*. 2010; 2(3): 355–374.

[35] Wang L, Weller CL. Recent advances in extraction of nutraceuticals from plants. *Trends in Food Science & Technology*. 2006; 17(6): 300–312.

[36] Cipriano, PA. *Antocianinas de açaí (Euterpe oleracea Mart.) e casca de jabuticaba (Myrciaria jaboticaba) na formação de bebidas isotônicas*. Dissertação (Mestrado)—Universidade Federal de Viçosa, Viçosa, MG, 2011.

[37] Castaneda-Ovando A, de Lourdes Pacheco-Hernández M, Páez-Hernández ME, Rodríguez JÁ, Galán-Vidal C. A. Chemical studies of anthocyanins: a review. *Food Chemistry*. 2009; 113(4): 859–871.

[38] Nicoué EE, Savard S, BelkacemiK. Anthocyanins in wild blueberries of Quebec: extraction and identification. *Journal of Agricultural and Food Chemistry*. 2007; 55(14): 5626–5635.

[39] Terci DBL. Aplicações analíticas e didáticas de antocianinas extraídas de frutas. Tese (Doutorado em Química Analítica)—Instituto de Química da UNICAMP, Universidade Estadual de Campinas, Campinas, SP, 2004: 231.

[40] Finco FDBA, Kammerer DR, Carle R, TsengWH. Antioxidant activity and characterization of phenolic compounds from bacaba (*Oenocarpus bacaba* Mart.) fruit by HPLC-DAD-MSn. *Journal of Agricultural and Food Chemistry*. 2010; 60(31): 7665–7673.

[41] Sellappan S, Akoh CC, Krewer G. Phenolic compounds and antioxidant capacity of Georgia-grown blueberries and blackberries. *Journal of Agricultural and Food Chemistry*. 2002; 50(8): 2432–2438.

[42] Gouvêa ACMS, AraujoMCP, Schulz DF, Pacheco S, Godoy RLO, CabralLMC. Anthocyanins standards (cyanidin-3-*O*-glucoside and cyaniding-3-*O*-rutinoside) isolation from freeze-dried açaí (*Euterpe oleracea* Mart.) by HPLC. *Ciência e Tecnologia de Alimentos*. 2012; 32(1): 43–46.

[43] Santos DT, Albarelli JQ, Beppu MM, Meireles MAA. Stabilization of anthocyanin extract from jabuticaba skins by encapsulation using supercritical CO_2 as solvent. *Food Research International*. 2013; 50(2): 617–624.

[44] Paes J, Dotta R, Barbero GF, Martínez J. Extraction of phenolic compounds and anthocyanins from blueberry (*Vaccinium myrtillus* L.) residues using supercritical CO_2 and pressurized liquids. *The Journal of Supercritical Fluids*. 2014; 95: 8–16.

[45] Neves LTBC, Campos DCDS, Mendes JKS, Urnhani CO, Araújo KG. Quality of fruits manually processed of açaí (*Euterpe oleracea* Mart.) and bacaba (*Oenocarpus bacaba* Mart.). *Revista Brasileira de Fruticultura*. 2015; *37*(3): 729–738.

[46] Francis FJ. Analysis of anthocyanins. In Pericles Markakis. (ed.). *Anthocyanins as food colors*. New York: Academic Press, 1982. p. 181–207.

[47] Novello AA, Conceição LL, Dias MMS, Cardoso LM, Castro CA,Ricci-Silva ME, et al. Chemical characterization, antioxidant and antiatherogenic activity of anthocyanin-rich extract from *Euterpe edulis* Mart. in mice. *Journal of Food and Nutrition Research*. 2015; 54(2): 101–112.

[48] Giusti MM, Wrolstad RE. Anthocyanins. *Characterization and Measurement with UV-Visible Spectroscopy*. In: Wrolstad, R. E. (Ed.). Current Protocols in Food Analytical Chemistry Unit. F1.2.1-13. New York: John Wiley & Sons, 2001.

[49] Singleton VL, Rossi, JA. Colorimetry of total phenolics with phosphomolybdic–phosphotungstic acid reagents. *American Journal of Enology and Viticulture*. 1965; 16(3): 144–158.

[50] Pacheco-Palencia LA, Duncan CE, Talcott ST. Phytochemical composition and thermal stability of two commercial açai species, *Euterpe oleracea* and *Euterpe precatoria*. *Food Chemistry*. 2009; 115(4): 1199–1205.

[51] Schreckinger ME, Lotton J, Lila MA, de Mejia EG. Berries from South America: a comprehensive review on chemistry, health potential, and commercialization. *Journal of Medicinal Food*. 2010; 13(2): 233–246.

[52] Portinho JA, Zimmermann LM, Bruck MR. Efeitos Benéficos do Açaí. *International Journal of Nutrology*. 2012; 5(1): 15–20.

[53] Santos MFG, Mamede RVS, Rufino MSM, Brito ES, Alves RE. Amazonian native palm fruits as sources of antioxidant bioactive compounds. *Antioxidants*. 2015; 4(3): 591–602.

[54] Fellows PJ. *Tecnologia do processamento de alimentos: princípios e prática*. 2. ed. Porto Alegre: Artmed, 2006. 602 p.

[55] Canuto GAB, Xavier AAO, Neves LC, Benassi MDT. Caracterização físico-química de polpas de frutos da Amazônia e sua correlação com a atividade anti-radical livre. *Revista Brasileira de Fruticultura*. 2010; 32(4): 1196–1205.

[56] Muniz MAP, dos Santos MNF, da Costa CEF, Morais L, Lamarão MLN, Ribeiro-Costa RM, et al. Physicochemical characterization, fatty acid composition, and thermal analysis of *Bertholletia excelsa* HBK oil. *Pharmacognosy Magazine*. 2015; 11(41): 147–151.

[57] Silva LB, Queiroz MB, Fadini AL, da Fonseca RC, Germer SP, Efraim P. Chewy candy as a model system to study the influence of polyols and fruit pulp (açai) on texture and sensorial properties. *LWT-Food Science and Technology*. 2016; 65: 268–274.

[58] Moretto E, Fett R. *Tecnologia de óleos e gorduras vegetais na indústria de alimentos*. São Paulo: Varela, 1998.

[59] Ramalho HF, Suarez PA. A química dos óleos e gorduras e seus processos de extração e refino. *Revista Virtual Química*. 2013; 5: 2–15.

[60] Souza ADV, Fávaro SP, Ítavo LCV, Roscoe R. Caracterização química de sementes e tortas de pinhão manso, nabo forrageiro e crambe. *Pesquisa Agropecuária Brasileira*. 2009; 44(10): 1328–1335.

[61] Pighinelli AL, Park KJ, Rauen AM, Oliveira RAD. Otimização da prensagem de grãos de girassol e sua caracterização. *Revista Brasileira de Engenharia Agrícola e Ambiental*. 2009; 13(1): 63–67.

[62] Berk Z, Marcondes Borge J, Pena S, Alvarez Arancedo M, Spagnolo R, Wilmart A. Technology of production of edible flours and protein products from soybeans. Technologie de production de farines alimentaires et de produits protéiques issus du soja (No. FAO ASB-97) FAO, Roma (Italia). 1992. Available from: http://www.fao.org/docrep/t0532E/t0532e00.htm [Accessed 13th August 2016].

[63] Pradhan RC, Meda V, Rout PK, Naik S, Dalai AK. Supercritical CO_2 extraction of fatty oil from flax seed and comparison with screw press expression and solvent extraction processes. *Journal of Food Engineering*. 2010; 98(4): 393–397.

[64] Ghosh R. *Principles of bioseparations engineering*. World Scientific, McMaster University, Canada. 2006: 282.

[65] Amarn iF, Kadi H. Kinetics study of microwave-assisted solvent extraction of oil from olive cake using hexane: comparison with the conventional extraction. *Innovative Food Science & Emerging Technologies*. 2010; 11(2): 322–327.

[66] Danlami JM, Arsad A, Zaini MAA. Characterization and process optimization of castor oil (*Ricinuscommunis* L.) extracted by the soxhlet method using polar and non-polar solvents. *Journal of the Taiwan Institute of Chemical Engineers*. 2015; 47: 99–104.

[67] Pessoa AS, Podestá R, Block JM, Franceschi E, Dariva C, Lanza M. Extraction of pequi (*Caryocarcoriaceum*) pulpoil using subcritical propane: determination of process yield and fatty acid profile. *The Journal of Supercritical Fluids*. 2015; 101: 95–103.

[68] Kartika IA, Evon P, Cerny M, Suparno O, Hermawan D, Ariono D, et al. Simultaneous solvent extraction and transesterification of jatropha oil for biodiesel production, and potential application of the obtained cakes for binder less particle board. *Fuel*. 2016; 181: 870–877.

[69] Nascimento ADP, Soares LAL, Stragevitch L, Danielski, L. Extraction of *Acrocomia intumescens* drude oil with supercritical carbon dioxide: process modeling and comparison with organic solvent extractions. *The Journal of Supercritical Fluids*. 2016; 111: 1–7.

[70] Toda TA, Sawada MM, Rodrigues, CE. Kinetics of soybean oil extraction using ethanol as solvent: experimental data and modeling. *Food and Bioproducts Processing*. 2016; 98: 1–10.

[71] Ribeiro MC, Vilas Boas EVB, Riul TR, Pantoja L, Marinho HA, Santos AS. Influence of the extraction method and storage time on the physicochemical properties and carotenoid levels of pequi (*Caryocar brasiliense* Camb.) oil. *Ciência e Tecnologia de Alimentos*. 2012; 32(2): 386–392.

[72] Brunner G. Gas extraction: an introduction to fundamentals of supercritical fluids and the application to separation process. New York: Springer. 1994: 386.

[73] Carrilho E, Tavares MCH, Lanças FM. Fluidos supercríticos em química analítica. I. cromatografia com fluido supercrítico: conceitos termodinâmicos. *Química Nova*. 2001; 24(4): 509–515.

[74] Castro MDL, Jurado-López A, Luque-Garcia JL. Drug extraction. In: York P, Kompella UB, Shekunov BY (eds). *Supercritical fluid technology for drug product development*. New York: Marcel Dekker. 2004: 498–531.

[75] Maul AA, Wasicky R, Bacchi EM. Extração por fluido supercrítico. *Revista Brasileira de Farmacognosia*. 1996; 5(2): 185–200.

[76] Dich J, Zahm SH, Hanberg A, Adami HO. Pesticides and cancer. *Cancer Causes & Control*. 1997; 8(3): 420–443.

[77] Zahm SH, Ward MH, Blair A. Pesticides and cancer. *Occupational Medicine*. 1997; 12(2): 269.

[78] Sufen G, Xiaojun T. Pesticides and cancer. *World Sci-tech R & D*. 2005; 2: 006.

[79] Piantino CR, Aquino FWB, Follegatti-Romero LA, Cabral FA. Supercritical CO_2 extraction of phenolic compounds from *Baccharis dracunculifolia*. *The Journal of Supercritical Fluids*. 2008; 47(2): 209–214.

[80] Herrero M, Mendiola JA, Cifuentes A, Ibáñez E. Supercritical fluid extraction: recent advances and applications. *Journal of Chromatography A*. 2010; 1217(16): 2495–2511.

[81] Patras A, Brunton NP, O'Donnell C, Tiwari BK. Effect of thermal processing on anthocyanin stability in foods; mechanisms and kinetics of degradation. *Trends in Food Science & Technology*. 2010; 21(1): 3–11.

[82] de Rosso VV, Mercadante AZ. The high ascorbic acid content is the main cause of the low stability of anthocyanin extracts from acerola. *Food Chemistry*. 2007; 103(3): 935–943.

[83] Vatai T, Škerget M, Knez Ž. Extraction of phenolic compounds from elder berry and different grape marc varieties using organic solvents and/or supercritical carbon dioxide. *Journal of Food Engineering*. 2009; 90(2): 246–254.

[84] Paula JT, Paviani LC, Foglio MA, Sousa IM, Duarte GH, Jorge MP, et al. Extraction of anthocyanins and luteolin from *Arrabidaea chica* by sequential extraction in fixed bed using supercritical CO_2, ethanol and water as solvents. *The Journal of Supercritical Fluids*. 2014; 86: 100–107.

[85] Ghafoor K, Park J, Choi Y H. Optimization of supercritical fluid extraction of bioactive compounds from grape (*Vitis labrusca* B.) peel by using response surface methodology. *Innovative Food Science & Emerging Technologies*. 2010; 11(3): 485–490.

[86] Floris T, Filippino G, Scrugli S, Pinna MB, Argiolas F, Argiolas A, et al. Antioxidant compounds recovery from grape residues by a supercritical antisolvent assisted process. *The Journal of Supercritical Fluids*. 2010; 54(2): 165–170.

[87] Seabra IJ, Braga MEM, Batista MT, Sousa HC. Effect of solvent (CO_2/ethanol/H_2O) on the fractionated enhanced solvent extraction of anthocyanins from elderberry pomace. *The Journal of Supercritical Fluids*. 2010; 54(2): 145–152.

7

Phytochemicals in Fruits and Vegetables

Ayse Tulin Oz and Ebru Kafkas

Abstract

Fruits and vegetables are the most important sources of phytochemicals. Phytochemicals use for both human diets and natural antimicrobial agents in food preservation. Their benefits for health are mainly due to high antioxidant activity. Antimicrobials of plant origin are known as secondary metabolites that could play a role not only individually or jointly against food-borne pathogens but also contribute to food flavor. Phytochemicals have a strong effect on control and prevention of natural spoilage processes and growth of microorganisms, including pathogens causing food safety issues. Microorganisms are always associated with harvested plants and slaughtered animals, the basic unprocessed materials of the food industry. Since foods consumed by humans undergo several processing treatments, it is important to understand the effect of such treatments on the phytochemical composition of foods.

Keywords: phytochemicals, food preservatives, food spoilage, food phenolics

1. Introduction

Fruits and vegetables are consumed as fresh or processed and known to be among the most important sources of phytochemicals for the human diet. About 200,000 phytochemicals are known so far and 20,000 of them have been identified as originating from fruits, vegetables and grains [1]. Phytochemicals has many health effects as antioxidants against many diseases or antibacterial, antifungal, antiviral, cholesterol-lowering, antithrombotic, or anti-inflammatory effects [2]. Phytochemicals are used for various purposes such as pharmaceuticals, agrochemicals, flavors, fragrances, coloring agents, biopesticides and food additives [1]. Their chemical structures composed of phytochemicals such as phenolics, alkaloids, saponins and terpenoids [1]. These compounds are known as secondary metabolites having various identifiable structures, although a benzene ring with one or more hydroxyl groups is a common feature. They are commonly classified as flavonoids (anthocyanins, flavan-3-ols, flavonols, proanthocyanidins or flavones, non-hydrolyzable tannins, isoflavones and flavanones) and non-flavonoids (hydroxycinnamic, hydroxybenzoic acid, hydrolyzable tannins, benzoic acids and stilbenes) [3]. Sugars, acids and polysaccharides are an important source of phytochemicals, secondary metabolites of plants also known as their antioxidant activity and other properties [4]. Lately, there are many investigations on plant "antimicrobial," "antiviral," or "antibacterial" effects [1]. In addition, phytochemicals are some of the most important natural preservation structures to reduce and inhibit pathogenic microorganism growth and preserve the overall quality of food products [5]. These antimicrobials can protect food products, extending the shelf life naturally [5]. Chilling, fermentation, freezing, acidification, nutrient restriction, water activity reduction, synthetic antimicrobials and pasteurization have been used in food preservation technology and phytochemicals such as flavonoids, polyphenols, anthocyanins and carotenoids are also used to preserve and control microbial spoilage in foods traditionally [6]. In general, food antimicrobials can be classified as natural and synthetic substances depending on their origin. Synthetic antimicrobials are found in fruits naturally such as benzoic acid in cranberries, tartaric acid in grapes, sorbic acid in rowanberries, malic acid in apples and citric acid in lemons [6]. Secondary metabolites are in close contact through sophisticated communication involving metabolic attacks by plants on their pathogens [7]. Fruits and vegetables have phenolics which are biologically active compounds. Fruits and vegetables have a special phytochemical group which protect plants from their environment stress such as pollution, pathogens, or various abiotic stresses [2]. Even if secondary metabolites having different structures, they can have similar functions. First, plant-defensive metabolites include phytoalexins biosynthesized to respond to biotic and abiotic stresses with the effect of both protecting the plant and controlling the pathogen growth [7]. Secondly, most of these metabolites are responsible for the organoleptic and qualitative properties of foods originating from such plants. For example, anthocyanins constitute a pigment group responsible for the color of a great variety of fruits, flowers and leaves [8] and flavan-3-ols are polyphenols involved in the bitterness and astringency of tea, grapes and wine [9, 10]. Thirdly, these compounds are unique sources of industrial material in the form of food additives, pharmaceuticals and flavors [11]. Finally, they are considered to be beneficial for health, mainly due to their antioxidant activity. Many studies have suggested that a high intake of polyphenol-rich foods may have cardiovascular benefits and provides some level of cancer chemopreventive activities and beneficial effects against other less prevalent but devastating illnesses, such as urinary bladder dysfunctions and Alzheimer's disease [12]. Furthermore, food scientists and nutrition specialists suggest that phytochemicals offer many health benefits when consumed as part of the usual human diet [13].

2. Commonly used methods of treating plant foods

Many fresh fruits especially small berries and vegetables are highly perishable after harvest. During the harvest, bruising can reduce shelf life, influencing both color and texture of fresh products. The freshness of fruits and vegetables can be maintained in storage through reduction of temperature and/or oxygen levels, increase in carbon dioxide levels, use of modified atmosphere packaging or edible coatings, or treatment with gamma irradiation or high pressure. These can also be combined with treatments of 1-MCP, ozone and ultraviolet (UV) irradiation to further prevent losses. One of the most basic treatments used to lengthen the shelf life of fresh commodities during storage is to store in a low temperature and high relative humidity conditions. It has been known and used to extend the shelf life of fruits and vegetables since antiquity [14]. Moreover, exposure to low temperature during storage optimizes produce appearance and has the additional benefit of protecting nonappearance quality attributes, such as texture, nutrition, aroma and flavor [14]. There are many chemical and natural preservative treatments used to reduce postharvest losses and extend the shelf life of fresh commodities. Using plant extracts with known antimicrobial properties can be of great importance in food preservation. There are some chemical substances in plants that produce a definite action on the microbiological, chemical and sensory quality of foods and these phytochemicals have been grouped in several categories including polyphenols, flavonoids, tannins, alkaloids, terpenoids, isothiocyanates, lectins, polypeptides, or their oxygen-substituted derivatives [6]. On the other hand, alternative sources of natural products, such as plant extracts, either as pure compounds or as standardized extracts, provide unlimited opportunities for control of microbial growth owing to their chemical diversity. The use of natural antimicrobials as phytochemicals is organic acids, essential oils, or plant extracts and could be a good alternative to ensure food safety [6]. To inactivate or inhibit the growth of spoilage and pathogenic microorganisms during preservation of food, there are several processing techniques used including the use of chemical preservatives and synthetic antimicrobials [5]. However, these techniques have not been considered natural antimicrobial agents in food preservation. But, naturally derived compounds in plant extracts can be good control agents for pathogenic microorganisms. The use of synthetic chemicals is increasingly restricted in many countries. Nowadays, the recent trend has been for the use of natural preservatives due to the adverse health effects of synthetic ones. The alternative methods preserve foods and reduce pest and pathogen injury, with the use of resistant varieties or integrated cropping strategies in which plant secondary metabolites may improve crop protection [6]. The major goals of such natural antimicrobials are to protect the food from food poisoning and spoilage microorganisms that cause off-odors, off-flavors and discoloration quality losses [6]. Antimicrobials are called traditional when they have been used for many years and many countries approve them for inclusion in foods. Although many synthetic antimicrobials are found naturally (benzoic acid, sorbic acid, citric acid, malic acid, tartaric acid), the perception of natural has become important for many consumers. The safety and shelf life of food ingredients can also be improved by application of novel technologies to avoid or delay microbial growth like packaging in modified atmosphere, nonthermal treatments, activated films, irradiation, etc. [6]. The use of fruits and vegetables as a source of certain phytochemicals, such as ascorbic acid (AA), carotenoids, phenols and flavonoids, has not only health-promoting effects but also widely used to restrict oxidation-induced degenerative changes in cell physiology and aging [15] and is well known due to their significant impact on the food industry [16]. Both glucosinolates and leaf surface waxes are important phytochemicals that also play an important role in protecting plants from pest and pathogen injury [17]. These factors that positively affect plant protection also minimize crop damage by pests and pathogens. B-Carotene, lycopene, lutein and zeaxanthin are known to exhibit antioxidant activity. Increasing oxidative stress results in produce losing keeping quality, not only in terms of microbial contamination, excessive softening and browning but also in terms of significant depletion of phytochemicals, such as phenolics, flavonoids, ascorbic acid and carotenoids. The addition of AA minimizes oxidative deterioration in processed fruits and vegetables. Exogenous treatment of AA in minimally processed fruits and vegetables reduces or stops enzymatic browning and oxidation-susceptible degenerative changes such as the deterioration of carotenoids, phenolics and flavonoids [16].

Phenolics and their metabolites are common constituents of fruits and vegetables that play an important role as to provide astringency and aroma constituents [15]. Polyphenolic compounds are important as food preservatives that inactivate free radicals giving them an important role fighting against pathogenicity, infestation and photooxidation [1, 15]. In general, antibacterial activity of phenolic acids is stronger against Gram-positive bacteria than Gram-negative bacteria [1]. The main problems for such antimicrobials are food-poisoning microorganisms and spoilage microorganisms that are metabolic end products causing off-odors, off-flavors, texture problems and discoloration of food [1]. Phytochemicals, such as phenolic compounds, are of great importance as antithrombotic, anticarcinogenic and anti-inflammatory agents. However, due to the possible negative effects of synthetic antioxidants, food industries prefer natural ones and can be used as food additives or pharmaceutical supplements [13]. First of all, they protect plants from biotic and abiotic stress factors. Indeed, such phenolic compounds are only induced when stress factors are present and so-called phytoalexins are specifically involved in defense mechanisms and are synthesized after pathogen or predator attack or injury [18].

3. Phytochemical and secondary metabolites present in plant foods

Secondary metabolites present in plant foods, such as alkaloids, phenolic compounds (flavonoids, isoflavonoids and anthocyanins) and terpenoids, have gained importance because of their antioxidant, antiviral, antibacterial and anticancer effects [19]. These phytochemicals are mixtures of several components, including phytophenolics in herbs and spices, phenolics, flavonoids and acids in fruits and glucosinolates in cruciferous vegetables (mustard) [5]. As shown in **Table 1**, *Rubus* (cloudberry and raspberry) extracts have antimicrobial effects against food spoilage and poisoning bacteria [21]. In general, antioxidant compounds have important protection effects from fruit insects and microbial organisms during storage [23]. Secondary metabolites are very important in medicine and agricultural science due to the activity of chemotherapeutic compounds or pesticides. Phenolics and flavonoids provide very important defense mechanisms against postharvest diseases [23]. For example, walnut seed coats contain gallic acid which is a phenolic compound that prevents aflatoxin biosynthesis by *Aspergillus flavus* (**Table 1**) [23]. Therefore, polyphenol compounds have antiviral activities to some various types of viruses [23]. One of the other very important phytochemical groups are flavonoids which have anti-allergenic, antiviral, antifungal and anti-inflammatory activity. It is abundant in most of the fruits and vegetables such as apples, grapes, lemons, tomatoes, onions, lettuce, broccoli, etc. [23]. Flavonoids are also known as one of the largest groups of natural phenolic compounds in plants [27]. These natural compounds have important effects against a variety of microorganisms [27]. In addition, flavonoids either have protective effects from microbial attacks or respond as phytoalexins against them. Volatiles are phytochemicals that are either inhibitory or stimulatory to fungal growth. Acetaldehyde, a volatile compound that occurs in fruit during ripening, has a fungicidal effect on postharvest pathogens [23]. As shown in **Table 1**, "Isabella" (*Vitis labrusca* L.) grape variety volatiles have a strong effect on *B. cinerea* development [19].

Name plant	Effects	References
The cranberry juice	Inhibition of *E. coli*	Howell et al. [20]
Some berry extracts (*Rubus*)	Food spoilage and poisoning bacteria	Rauha et al. [21]
Grape seed or rosemary extracts	Food preservatives	Blasa et al. [22]
Riesling grape	Gram-positive food-borne pathogens	Tajkarimi and Ibrahim [5]
Flavonoids in plant tissue	Antimicrobial	Saxena et al. [15]
Flavonoids	Antiallergenic, antiviral, and antifungal	Ippolito and Nigro [23]
Bergamot fruit extract	*Saccharomyces cerevisiae*	Mandalari et al. [24]
Citrus species and grape	*Fusarium oxysporum*	Okwu et al. [25]
Glucosinolates and leaf surface	Pest and pathogen injury	Björkman et al. [17]
The "Isabella" (*Vitis labrusca* L.) grape	(*B. cinerea*)	Makkar et al. [19]
The walnut seed coats (gallic acid)	Aflatoxin	Ippolito and Nigro [23]
Pomegranate fruit	*P. italicum*, *R. stolonifer* and *B. cinerea*	Tehranifar et al. [26]

Table 1. The effects of fruit and vegetable extracts on food pathogens.

4. Phytochemicals as natural preservatives and antimicrobials

Natural preservatives derived from plant extracts such as phytochemicals and essential oils are used against fungal development in many fruits and vegetables after harvest [28]. The efficiency of an antimicrobial treatment depends on many factors, such as type, genus, species and strain of the main microorganism, in addition to environmental factors such as pH, water activity, temperature, atmospheric composition and an initial microbial load of the food materials [6]. Therefore, other important subject to know is type of the microorganism(s) owing to usually combinations of antimicrobials are more effective than adding just one. The natural antimicrobial preservative activity is not clear since there are many influencing factors, one of the most important being the interaction between phytochemicals and growth of microorganisms [6]. Processing of foods containing phytochemicals is expected to result in some changes in their phytochemical content. Phytochemicals present in many food stuffs are lost by heat processing such as sterilization, pasteurization and dehydration [6].

Many investigations have evaluated phytochemical effects on antifungal activity. The potential use of plant extracts as natural antimicrobial agents in food preservation forms the basis for many applications such as grape seed or rosemary extracts that have been used as food preservatives [22]. Researchers reported that grape extracts of Riesling *Vitis vinifera* L. grapes showed strong preservative effects against some of the Gram-positive food-borne pathogens [5]. The alkaloids, steroids, tannins, flavonoids, saponins and glycosides which were secondary metabolites showed various biological activities and act in plant defense mechanisms. Flavonoids usually occur as glycosides and aglycones in plant tissue which have significant antioxidant properties and antimicrobial and insect-repellent properties as well [15]. Flavonoids and their antimicrobial effect are useful as a food preservative to extend the shelf life and safety of foods. Flavonoids play important roles in biological activities, including antiallergenic, antiviral and antifungal effects [23]. It is also present in various common fruits and vegetables (apples, grapes, lemons, tomatoes, onions, lettuce and broccoli). The following flavonoids are antifungal agents in plants: isoflavonoids, flavans flavanones. However, the antifungal activity of flavonoid compounds plays an important role between plant-microorganism and host plant's defensive systems [8]. Saponin and flavonoids are found in fruits and vegetables and in general they form a soapy lather after extracted from parts of plants [5]. Mandalari et al. [24] reported that Bergamot fruit extract which is rich in flavonoid has an effective on the yeast *Saccharomyces cerevisiae* (**Table 1**).

Okwu et al. [25] also showed that the antifungal activity of both citrus species and grape has an important effect against *Fusarium oxysporum* (**Table 1**). Thiosulfinates come from hydrolysis products of garlic and onion. They have a strong potential of producing antimicrobial effects against pathogenic microorganisms [5]. Broccoli, Brussels' sprouts, cabbage mustard and horseradish have glucosinolates that also have a wide range of antibacterial effects. Moreover, olive leaves (*Olea europaea*) are rich in phenolic compounds, with demonstrated strong antimicrobial effects and can be potentially used in food processing [5]. Al-Zoreky [29] reported that phenolics and flavonoids present in pomegranate fruit peels demonstrated strong antimicrobial activity against some food pathogen microorganisms. In addition, *Psidium guajava* has phenolic, flavonoid, carotenoid, terpenoid and triterpenes that demonstrated strong antimicrobial activity [30]. Salas et al. [31] reported that flavonoids extracted from citrus species, not only naringin, hesperidin and neohesperidin but also enzymatically modified derivatives of these compounds, have strong antifungal activity [31]. The limonoid compounds have important antibacterial and antiviral activity as shown in **Table 1** [1]. Vikram et al. [32] reported that seeds of grapefruits have significant inhibitory effect on pathogenic *Escherichia coli* O157:H7. Black raspberry and Chardonnay seed extracts have also antibacterial activity to inhibiting growth of some food pathogen microorganisms [33]. Tehranifar et al. [26] reported that high percentage of phenolic content in the peel and seed of pomegranate fruit has high antifungal activity especially on postharvest fungi (*Penicillium italicum*, *Rhizopus stolonifer* and *Botrytis cinerea*). Another study showed that berry extracts exhibit selective inhibitory properties against intestinal bacteria [1]. Recently, antifungal activity has been found in all tissue types of strawberry fruit due to the phenolic compounds that inhibit the growth of fungi [23].

5. Other beneficial properties of plant phytochemicals

In the last decade, the results of many research have shown the positive effects of phytochemicals in human health. There is a strong correlation of antioxidant consumption with lower risk of many diseases such as cardiovascular cancer, diabetes and hypertension diseases as well as other medical conditions [34, 35]. Fruits and vegetables have phenolic compounds, pigments and natural antioxidants; these compounds protect many diseases like cancer and heart disease [36]. The importance of antioxidant effects on cardiovascular diseases and cancer is especially important [23] and these antioxidants can be found in various fruits, vegetables and herbs. Phenolics as flavonoids have an important effects such as antimicrobial, anti-inflammatory, antioxidant, antiviral, antiallergic, anticancer, antiulcer, antidiabetic, antiplasmodial, antihypertensive, anticonvulsant and all reducing risks for severe human diseases [27]. Antioxidants in fruits and vegetables have defensive effects and are three main groups: vitamins, phenolics and carotenoids [35]. Vitamin C (L-ascorbic acid, AA) and the oxidized form (dehydroascorbic acid, DHAA), carotenoids and phenolic compounds prevent cardiovascular disease, cancer and cataracts which are associated with the oxidative damage of lipids, DNA and proteins [4]. Moreover, some carotenoids also have antioxidant activity (AOA) and shown beneficial effects on the reduction of cardiovascular diseases [4]. The fruits and vegetables that have phytochemicals are also not only low in fat and saturated fat, cholesterol and calories but also are rich in potassium and sodium, fiber, folic acid and AA [34]. One of the most important flavonols is quercetin, which is higher in onion (red and yellow), broccoli, kale, French beans, apple, red grapes and cherries. Quercetin is anticarcinogenic and inhibits low-density lipoprotein (LDL) oxidation activities [34].

6. Conclusion

In a conclusion, potentially a great number of phytochemicals including some of the vitamins, flavonoids, terpenoids, carotenoids, phenolics, phytoestrogens, minerals and antioxidants in plant materials are used as alternative preservative agents for controlling postharvest physiological disorders or microbial pathogen injuries of both fresh fruit and vegetables in the food industry. Many publications have focused on the potential protective nature of these natural phytochemical compounds against fungal and bacterial attacks. Moreover, these natural compounds have become interesting candidates not only for plant protection but also human and animal health protection from fungal and bacterial diseases because of their lower toxicity or absence of toxicity.

Author details

Ayse Tulin Oz[1]* and Ebru Kafkas[2]

*Address all correspondence to: aysetulinoz@osmaniye.edu.tr

1 Food Engineering Department, Engineering Faculty, Osmaniye Korkut Ata University, Osmaniye, Turkey

2 Horticulture Department, Agriculture Faculty, Cukurova University, Adana, Turkey

References

[1] Patra AK. Dietary Phytochemicals and Microbes. ISBN 978-94-007-3925-3 ISBN 978-94-007-3926-0 (eBook). 2012; DOI 10.1007/978-94-007-3926-0 Springer Dordrecht Heidelberg New York London.

[2] Schreiner M, Huyskens-Keil S. Phytochemicals in Fruit and Vegetables. Crit. Rev. Plant Sci. 2006;25:267–278.

[3] Waterhouse AL. Wine Phenolics. Ann. N. Y. Acad. Sci.. 2002;957:21–36.

[4] Escobedo-Avellaneda Z, Janet G.-U., Aurora Valdez-Fragosoa J, Antonio T., Welti-Chanesa J. Phytochemicals and antioxidant activity of juice, flavedo, albedo and comminuted orange. J. Funct. Foods. 2014;6:470–481.

[5] Tajkarimi M, Ibrahim SA. Phytochemicals as Anti-microbial Food Preservatives. Chapter 2012;7:207–225. In: Patra AK. Dietary Phytochemicals and Microbes. ISBN 978-94-007-3925-3 ISBN 978-94-007-3926-0 (eBook). DOI 10.1007/978-94-007-3926-0 Springer Dordrecht Heidelberg New York London.

[6] Negi PS. Plant Extracts for the Control of Bacterial Growth: Efficacy, Stability and Safety Issues for Food Application. Int. J. Food Microbiol. 2012;156:7–17.

[7] Soledade M, Pedras C. Fungal Attack and Cruciferous Defenses: Tricking Plant Pathogens. Chapter. 2011;9:127 pp. in: Gang DR. The Biological Activity. of Phytochemicals. Vol. 41, DOI 10.1007/978-1-4419-7299-6. ISBN 978-1-4419-6961-3 e-ISBN 978-1-4419-7299-6. Springer New York Dordrecht Heidelberg London.

[8] Harborne JB, Williams CA. Advances in Flavonoids Research Since 1992. Phytochemistry. 2000;55:481–504.

[9] Noble AC. Bitterness in Wine. Physiol. Behav. 1994;56:1251–1255.

[10] Halsam E, Lilley TH. Natural Astringency in Foodstuffs—A Molecular Interpretation. Crit. Rev. Food Sci. Nutr. 1988;27:1–40.

[11] Zhao J, Davis LC, Verpoorte R. Elicitor Signal Transduction Leading to Production of Plant Secondary Metabolites. Biotechnol. Adv. 2005;23:283–333.

[12] De Pascual-Teresa S, Moreno DA, Garcia-Viguera C. Flavanols and Anthocyanins in Cardiovascular Health: A Review of Current Evidence. Int. J. Mol. Sci. 2010;11:1679–1703.

[13] Tlili N, Mejri H, Yahia Y, Saadaoui E, Rejeb S, Khaldi A, Nasri N. Phytochemicals and antioxidant activities of Rhus tripartitum (Ucria) fruits depending on locality and different stages of maturity.Food Chemistry, 2014;160:98–103.

[14] Paull RE. Effect of Temperature and Relative Humidity on Fresh Commodity Quality. Postharvest Biol. Technol. 1999;15:263–277.

[15] Saxena A, Bawa AS, Raju PS. Phytochemical Changes in Fresh-cut Jackfruit (*Artocarpus heterophyllus* L.) Bulbs During Modified Atmosphere Storage. Food Chem. 2009;115:1443–1449.

[16] Devi RR, Jayalekshmy A, Arumughan C. Antioxidant Efficacy of Phytochemical Extracts from Defatted Rice Bran in the Bulk Oil System. Food Chem. 2007;104:658–664.

[17] Björkman M, Klingen I, Birch ANE, Bones AM, Bruce TJA, Johansen TJ, Meadow R, Mølmann J, Seljåsen R, Smart LE, Stewart D. Phytochemicals of Brassicaceae in Plant Protection and Human Health—Influences of Climate, Environment and Agronomic Practice. Phytochemistry. 2011;72:538–556.

[18] Cantos E, Espín JC, Fernández MJ, Oliva J, Tomás-Barberán A. Postharvest UV-C-Irradiated Grapes as a Potential Source for Producing Stilbene-Enriched Red Wines. J. Agric. Food Chem. 2003;51:1208–1214.

[19] Makkar HPS, Siddhuraju P, Becker K. Plant Secondary Metabolites. Institute for Animal Production in the Tropics and Subtropics, University of Hohenheim, Stuttgart, Germany. Humana Press Inc., a division of Springer Science+Business Media. 2007. LLC 999 Riverview Drive, Suite 208 Totowa, New Jersey 07512.

[20] Howell AB, Vorsa N, Marderosian AD, Foo LY. Inhibition of the Adherence of P-fimbriated *Escherichia coli* to Uroepithelial Surfaces by Proanthocyanidin Extracts from Cranberries. New Engl. J. Med. 1998;339:1085–1086.

[21] Rauha JP, Remes S, Heinonen M, Hopia A, Kähkönen M, Kujala T, Pihlaja K, Vuorela H, Vuorela P. Antimicrobial Effects of Finnish Plant Extracts Containing Flavonoids and Other Phenolic Compounds. Int. J. Food Microbiol. 2000;56:3–12. doi: 10.1016/S0168-1605(00)00218-X.

[22] Blasa M, Gennari L, Angelino D, Ninfali, P. Fruit and vegetables antioxidants in health, U: Bioactive foods in promoting health (fruits and vegetables), Watson, R.R.,Preedy, V.R. (Ed.), Elsevier Inc., New York, pp. (2010); 37-58.

[23] Ippolito A, Nigro F. Natural Antimicrobials for Preserving Fresh Fruit and Vegetables. in: Wim, J. Improving the Safety of Fresh Fruit and Vegetables. 2005.

[24] Mandalari G, Bennett R, Bisignano G, Trombetta D, Saija A, Faulds C, Gasson M, Narbad A. Antimicrobial Activity of Flavonoids Extracted from Bergamot (Citrus bergamia Risso) Peel, a by Product of the Essential Oil Industry. J. Appl. Microbiol. 2007;103(6):2056–2064.

[25] Okwu DE, Awurum AN, Okoronkwo JJ. Phytochemical Composition and in vitro Antifungal Activity Screening of Extracts from Citrus Plants against Fusarium oxysporum of Okra Plant (*Hibiscus esculentus*). Afr. Crop Sci. Conf. Proc. 2007;8:1755–1758.

[26] Tehranifar A, Selahvarzi Y, Kharrazi M, Bakhsh VJ. High Potential of Agro-Industrial by Products of Pomegranate (*Punica granatum* L.) as the Powerful Antifungal and Antioxidant Substances. Ind. Crop Prod. 2011;34(3):1523–1527.

[27] De Conti Lourenço RMC, P. da Silva Melo, De Almeida ABA. Flavonoids as Antifungal Agents. Chapter 10. in: M. Razzaghi-Abyaneh and M Rai (eds.), Antifungal Metabolites from Plants. 2013; DOI: 10.1007/978-3-642-38076-1_10, Springer-Verlag Berlin Heidelberg.

[28] Sharma R. Ozone Decontamination of Fresh Fruit and Vegetables. in: Wim, J Improving the Safety of Fresh Fruit and Vegetables. 2005.

[29] Al-Zoreky NS. Antimicrobial Activity of Pomegranate (*Punica granatum* L.) Fruit Peels. Int. J. Food Microbiol.. 2009;134(3):244–248.

[30] Gutierrez RMP, Mitchell S, Solis RV. *Psidium guajava*: A Review of its Traditional Uses, Phytochemistry and Pharmacology. J. Ethnopharmacol. 2008;117(1):1–27. doi: 10.1016/j.jep. 2008.01.025.

[31] Salas MP, Celiz G, Geronazzo H, Daz M, Resnik SL. Antifungal Activity of Natural and Enzymatically-Modified Flavonoids Isolated from Citrus Species. Food Chem. 2011;124(4):1411–1415.

[32] Vikram A, Jesudhasan PR, Jayaprakasha GK, Pillai BS, Patil BS. Grapefruit Bioactive Limonoids Modulate *E. coli* O157:H7 TTSS and Biofilm. Int. J. Food Microbiol. 2010;140:109–116.

[33] Luther M, Parry J, Moore J, Meng JH, Zhang YF, Cheng ZH, Yu LL. Inhibitory Effect of Chardonnay and Black Raspberry Seed Extracts on Lipid Oxidation in Fish Oil and their Radical Scavenging and Antimicrobial Properties. Food Chem. 2007;104(3):1065–1073.

[34] Craig WJ. Phytochemicals: Guardians of our Health. J. Am. Diet. Assoc. 1997;97(10):199–204.

[35] Thaiponga K, Boonprakoba U, Crosbyb K, Cisneros-Zevallosc L, Byrnec DH. Comparison of ABTS, DPPH, FRAP and ORAC Assays for Estimating Antioxidant Activity from Guava Fruit Extracts. J. Food Compos. Anal. 2006;19:669–675.

[36] Rodríguez-Fragoso L, Osuna-Martínez U, Gonzaga-Morales AI, Reyes-Esparza J. Potential Antioxidant Benefits of Commonly Used Fruits and Vegetables around the World. in: Pereira, DAM. Medicinal Plants: Antioxidant Properties, Traditional Uses and Conservation Strategies. Nova Science Publishers, Inc., Hauppauge, NY, USA: 2013.

8

Foods with Functional Properties and their Potential uses in Human Health

Ricardo Tighe-Neira, Miren Alberdi,
Patricio Arce-Johnson, Jesús L. Romero-Romero,
Marjorie Reyes-Díaz and
Claudio Inostroza-Blancheteau

Abstract

Vegetables and fruits have been a part of human diet since ancient times; nevertheless, as countries develop, its population's feeding habits change and tend to have a diet poor in vegetables and fruits, with well-known consequences. Several food plant products with massive consumption and within the reach of the population are products such as artichoke, leek, hot chili pepper, coriander, kiwifruit, sweet orange, highbush blueberry, and maracuyá to name a few. They have many beneficial properties principally by its content of phytochemicals with high impact on human health, beyond nutritional support. The phytochemicals are bioactive compounds such as vitamins, carotenoids, phenolic acid, and flavonoids, which contribute to antioxidant capacity and as a whole prevent chronic nontransmissible diseases such as: diabetes, high blood pressure, high cholesterol in blood, cardiovascular risks, among others. This relationship between food plant for human consumption and its impacts on human health is discussed in this chapter, highlighting coriander and kiwifruit by its wide range of benefits.

Keywords: phytochemicals, antioxidant capacity, polyphenols, chronical diseases, healthy feeding habits, life style

1. Introduction

Foods are neither inherently good nor bad. Rather, good or bad eating habits, as well as other factors such as smoking and physical activity, all influence human health. If we desire a healthy lifestyle and wish to avoid chronic nontransmissible disorders such as diabetes, high levels of cholesterol, cardiovascular diseases, etc., foods, especially those that are functional, are only a part of the solution [1].

Despite a wealth of information, there is no universal definition about what constitutes a functional food. However, there is consensus concerning central concepts, which are associated with their benefit for human health beyond their traditional nutrients [2]. Along the same lines, the importance of phytochemicals as a class of biologically active metabolites in plants is accepted [3]. When discussing "potential use for human health" to refer to a particular plant, preliminary evidence on its outstanding phytochemical content must already exist, which means it can be used in the future as a source to investigate more profoundly its beneficial implications in human health.

Some processes such as cooking alter the content and composition of phytochemicals present in vegetables, reducing their concentrations by thermal degradation or augmenting their concentration with respect to the raw material. However, these effects are varying with the cooking method and type of phytochemical [4]. These, together with the growing consumption of fiber, are the principal reasons to recommend the regular intake of fresh vegetables [5].

Functional foods may be plant or animal products, that are fresh, semi-processed or processed, but in this chapter we will refer mainly to fresh plants and their properties beyond their nutritional characteristics. In addition, we will also discuss the existence of several common horticultural and fruit plants that are widely available and consumed by the human population, whose functional properties have yet to be systematized and categorized. Vegetables with high functional interest such as artichoke, leek, hot chili pepper, and coriander, as well as fruit plants such as kiwifruit, sweet orange, and highbush blueberry are considered [6–9]. Within the species discussed in this chapter, only artichokes must be consumed cooked, whilst the others may be eaten fresh.

We believe that the updated information about plants with characteristics as functional foods responds to a need of the population and scientists to learn more about healthy habits and how consumption of natural foods can improve their quality of life.

2. Horticultural species with functional properties and their potential use in human health

The properties and features of some horticultural species that have beneficial effects on human health are mentioned here. It is important to note that the artichoke, leek, hot chili pepper, and coriander are considered in this section, because these species are widely consumed by the human population.

2.1. Artichoke (*Cynara cardunculus* var. scolymus and *Cynara scolymus*)

These species belong to the *Asteraceae* family (**Figure 1**) are native to the Mediterranean region [10] and grow in many parts of the world [11]. *Cynara scolymus* is cultivated due to its large immature flower heads, which have special functional and nutraceutical characteristics, and a high antioxidant capacity, with more than 9000 μmol trolox equivalents (TE) 100 g^{-1} FW [12]. This antioxidant capacity is due to its high content of total polyphenols like caffeoylquinic acids and flavonoids [7]. These polyphenols are present in flower heads, with values ranging from 4.8 to 29.8 mg g^{-1} FW in different Italian varieties [7, 13, 14]. It is important to mention that these values are not only dependent on the genetic background, as the interaction genotype-environment also has an influence [7]. Therefore, antioxidant content may be affected by agricultural practices, because in different locations, the same variety may have different antioxidant capacities.

Figure 1. Immature flower head of artichoke.

Choleretic and hypocholesterolemic activities, due to the presence of chlorogenic acid, cynarin, and lutein, have been reported in clinical studies, which demonstrated the effect of leaf extracts on the inhibition of the biosynthesis of cholesterol in rat hepatocytes [15]. Furthermore, these extracts prevented necrosis in rat hepatocytes provoked by hydroperoxides indicated for treatment of dyspepsia, or dyskinesia of the bile ducts, as well as disorders in the assimilation of fats in humans [16–18]. The nutraceutical and therapeutic actions of several metabolites are summarized in **Table 1**. The positive effects of ingesting *C. cardunculus* flower heads have been widely demonstrated; nevertheless, artichoke leaves and external parts of bracts may be used in industrial processes to obtain functional metabolites for use in human health.

Species	Molecule	Part of plant	Specific function for human health	References
Cynara cardunculus L. var. scolymus L. / *Cynara scolymus*	Caffeoylquinic acids, flavonoids, and cynarin	Flower head and leaves	Antioxidant capacity	[7, 12–14]
	Chlorogenic acid and cynarin	Flower head and leaves	Choleretic, hepatoprotective	[168, 169, 170]
	Luteolin	Flower head and leaves	Hypocholesterolemic	[171]
	Inulin	Flower head, stem and leaves	Stimulates the intestinal flora, hypocholesterolemic	[170]
Allium ampeloprasum Var. porrum / *Allium porrum*	Polyphenols	White part and green leaves	Antioxidant capacity	[28]
	Ascorbate	White part and green leaves	Growth and tissue repair, antioxidant	[28, 30]
	Steroidal saponines	White part	Antiinflammatory, gastroprotective and antiulcerogenis	[31, 32]
	Methanolic extracts	White part and green leaves	Antimicrobial	[35]
	Steroidal saponines/ polyphenols	White part	Cytotoxic activity: a potential anticancer agent	[33, 35]
	Hydroalcoholic extracts	White part and green leaves	Hypolipidemic/ hypocholesterolemic	[172, 173]
Capsicum annuum L.	Polyphenols	Fruit	Antioxidant capacity	[37, 174]
	Dietary fiber	Fruit	Anticarcinogenic, prevents diabetes, hypocholesterolemic	[9, 175]
Coriandrum sativum L.	Essential oil, aqueous, methanolic and ethanolic extracts	Fruit and leaves	Prevention of several chronic degenerative disorders	[8, 44, 48]
	Essential oils	Fruit and leaves	Anticancer activity	[176, 177]
	Extracts	Seeds	Hypocholesterolemic	[51]
	Extracts	Seeds	Antidiabetic	[49]
	Diethyl ether and aqueous extracts	Fruit, seeds and leaves	Anxiolytic, sedatives, antidepressant	[51, 54, 56]
	Aqueous, ethanolic and chloroformic extracts	Aerial parts	Analgesic	[55]
	Ethanolic extract	Root	Antimicrobial	[48]
	Extracts	Aerial parts	Antioxidant activity	[47]
	Hydroalcoholic extract	Aerial parts	Anticonvulsant	[15]
	Ethanolic extract	Seeds	Cognitive effects (improves learning in the long term)	[53]

Table 1. Selected plant species and their compounds that are beneficial for human health.

Inositol and inulin are soluble carbohydrates, which are present in external bracts of artichokes. In the case of inositol (chiro-scyllo- and myo-inositol), values fluctuate from 6.7 to 9.3 mg g^{-1} DW while for inulin they fluctuate from 69.8 to 114.6 mg g^{-1} DW [16]. These values are higher than the ones reported by Hernández-Hernández et al. [17] in edible bracts of artichoke. Regarding the beneficial properties, inositols have been used in treatments against diabetes mellitus [18]. In this sense, Crawford et al. [19] reported that inositol prevents diabetes mellitus in pregnant women and concluded that myo-inositol shows promising results by preventing the onset of the disease. Furthermore, inulin has been associated with some beneficial functional properties, as it can be a good source of carbohydrates and fiber, associated with positive effects in the prevention of colon cancer [20, 21]. In addition, prebiotic properties and effects on the absorption of calcium have been reported [20, 21]. In this sense, research on mineral absorption of calcium and magnesium concluded that inulin can reduce risk for osteoporosis by increasing their absorption [22].

2.2. Leek (*Allium ampeloprasum* var. porrum or *Allium porrum*)

It has been reported that this species has a substantial nutraceutical and functional properties (see **Table 1**) and is cultivated in Asia, America, and Europe, especially in the Mediterranean region [23, 24]. *A. ampeloprasum*, belongs to subgenus *Allium* section *Allium* (**Figure 2**), is considered a "complex species" due to different ploidy levels and genome constitution, and diploid, tetraploid, and octoploid accessions have been described [25, 26]. In fact, Hirschegger et al. [27] suggested that molecular evidence could be used to consider this species as a tetraploid horticultural group together with other *Allium* genus species. This condition is particularly important in accounting for its nutritional, functional, and nutraceutical characteristics [28].

Figure 2. Green leaves and white part of leek.

The principal beneficial properties of the *Allium* genus are mainly due to the presence of many sulfur compounds containing bioactive constituents, including: dimethyl disulfide, methyl propenyl disulfide, propyl propenyl disulfide, dimethyl trisulfide, methyl propyl trisulfide, methyl propenyltrisulfide, S-methyl cysteine sulfoxide, S-propyl cysteine sulfoxide, S-propenyl cysteine sulfoxide, and N-(γ-glutamyl)-S-(E-1-propenyl) cysteine [6].

In the evaluation of 30 leek cultivars, the content of total phenols varied from 5 to 15 mg gallic acid equivalents (GAE) g^{-1} DW for whole plants [28]. Other studies reported values of 5.5–6.0 mg GAE g^{-1} DW in whole leeks [29]. These differences in the total phenolic content could be attributed to the genetic variability of this species and agricultural systems [28]. Moreover, in the same study, the oxygen radical absorbance capacity (ORAC) was evaluated, where the green leaves possessed 82–135 μmol TE g^{-1} DW, whereas the white part contained just 27–88 μmol TE g^{-1} DW. Additionally, Vandekinderen et al. [30] determined the total vitamin C content (ascorbic acid + dehydroascorbic acid) whose values reached 9.65 mg 100 g^{-1} FW in whole leeks. Bernaert et al. [28] reported 5.54 mg ascorbic acid (AA) g^{-1} DW in whole leeks and higher values in green leaves than in white parts (2.77–8.52 mg AA g^{-1} DW and 0.89–3.55 mg AA g^{-1} DW, respectively). Values of polyphenols, AA, and antioxidant activity may be influenced by the season of year, genetic characteristics, and biotic and abiotic factors during vegetative growth, as well as agricultural practices.

The antiinflammatory, gastroprotective, and cytotoxic activities of organosulfate compounds, saponins, particularly steroidal saponines, have been well documented in *A. ampeloprasum* var. porrum [31–33].

Another property that is exclusive to the *Allium* genus is their antimicrobial activity (see **Table 1**). This has been reported since ancient times, and leeks have been used to treat wounds and respiratory diseases, as well as acting as an antibacterial agent due to the presence of alliin-containing structures [34, 35]. Polyphenols of methanolic extracts of green leaves and white parts of *A. porrum* are potent against Gram-positive (*Staphylococcys aureus* ATCC 25923 and *Bacillus subtilis* ATCC 6633) and Gram-negative (*Klebsiella pneumoniae* ATCC 13883, *E. coli* ATCC 25922, *Proteus vulgaris* ATCC 13315 and *Proteus mirabilis* ATCC 14153) bacteria, as well as fungal species (*Candida albicans* ATCC 10231 and *Aspergillus niger* ATCC 16404) [28]. These authors affirm a negative effect of methanolic extracts of *A. porrumon* Hep2c, L2OB, and RD cell cultures. This cytotoxic activity could indicate a future use of these natural biological compounds in human health.

2.3. Hot chili pepper (*Capsicum annuum* L. var. longum)

The fruit of this species (**Figure 3**), immature or mature and leaves, contains at least two groups of bioactive compounds of significance for human health, polyphenols and carotenoids. The polyphenol content is variable but reaches over 20 mg GAE g^{-1} DW in mature and dried fruits [36], and 40 mg GAE g^{-1} DW in leaves [37]. The polyphenols of fruits have a total antioxidant capacity of 26.6–44.4 μmol TE g^{-1} DW, depending on the variety [36].

According to Serrano et al. [38], the small intestine has around 25% of bioavailability of total polyphenols.

Figure 3. Immature fruit of red hot chili pepper.

Regarding the total carotenoids present in different varieties of red hot chili peppers, Hervert-Hernández et al. [36] indicated values from 87.6 to 373.3 mg 100 g^{-1} DW. In addition, the same authors determined that the bioavailability of chili carotenoids in the small intestine ranges from 20 to 50% of the total content, depending on the variety [36]. In addition, *C. annuum* is also a good source of vitamin C (up to 26.5 mg g^{-1} DW) [39]. Minerals, mainly potassium and magnesium, as well as dietary fiber, reducing sugars (around 20% DW), and antimicrobial activity useful for functional food production are also characteristic of this species [9]. Regarding the antimicrobial activity, De et al. [40] identified three pathogens that are susceptible to ethyl alcohol extracts of chili: *Bacillus subtilis* ATCC6633 (minimum inhibitory concentration (MIC) 10–25 mg mL^{-1}), *Escherichia coli* ATCC10536 (MIC 25–50% mg mL^{-1}), and *Saccharomyces cerevisiae* ATCC 9763 (MIC 2–5% mg mL^{-1}). In addition, an antimicrobial activity in *Staphylococcus aureus* ATCC14154, *Escherichia coli* ATCC-1698, *Pseudomonas aeruginosa* ATCC-23993, *Candida albicans* ATCC-14053, and *Sarcina lutea* (Collection of Microbiology Laboratory of Chemical Engineering Department, Institut Teknologi Bandung) has also been reported using ethanolic extracts of chili [41]. These authors indicate that capsaicin may be one of the main responsibles of microorganism inhibition. Likewise, Huang et al. [42] affirm that developments of *Staphylococcus aureus, Saccharomyces cerevisiae,* and *Aspergillus niger* were negatively affected by capsaicin.

2.4. Coriander (*Coriandrum sativum* L.)

This species of the *Apiaceae* family (**Figure 4**) is usually cultivated throughout the year in diverse edafoclimatic areas [43]. It is native to Italy and is currently propagated in several Mediterranean regions of Europe, as well as in America and Asia [44]. It is consumed principally fresh, either alone or in salads. This species has multiple human health benefits and has considerable potential as a functional horticulture species (see **Table 1**). *C. sativum* contains essential oils in seeds and in the pericarp whose content and composition appear to be dependent on biological and geographical traits [35]. The oil content is approximately 1% of seed weight of which linalool is the major component (73%) [44–46]. In the stem and immature leaves, the most important compounds are essential oils, flavonoids (quercetin, kaempferol, and acacetin), phenolic acids (vanillic acid, p-coumaric acid, syringic acid, p-OH benzoic acid, cis-ferulic acid, and trans-ferulic acid), and polyphenols [8]. The principal phytochemicals are β-carotene (5.1 mg g^{-1} DW), AA (1.16 mg 100 g^{-1} FW), total phenolics (2.05 mg GAE 100 g^{-1} DW), and antioxidants (1.12 mg GAE 100 g^{-1} DW) [47].

Figure 4. Stem and immature leaves of coriander.

According to Kumar et al. [48], ethanolic extracts of fresh coriander roots contain alkaloids, flavonoids, terpenoids, sterols, carbohydrates, saponins, and phenolic compounds. This extract and its fractions possess significant antibiotic activity against *Staphylococcus aureus, Bacillus cereus, Salmonella typhi*, and *Klebsiella*.

The antihyperglycemic (antidiabetic) activity of coriander has been studied by several authors. Deepa and Anuradha [49] analyzed the effects of coriander seed extracts in rats, which showed decreases by 44% in blood glucose and by 58% in glycosylated hemoglobin levels with respect to untreated rats. At the same time, the insulin level in plasma increased to 40%. They also reported beneficial effects in kidney and pancreas. Moreover, *C. sativum* seeds have hypocholesterolemic properties in rats, increasing hydroxymethyl glutaryl CoA (HMG-CoA) reductase and plasma lecithin cholesterol acyl transferase activities [50, 51].

Additional properties such as analgesic, anticonvulsive, anxiolytic, sedative, antidepressant, and cognitive effects of coriander have been tested *in vivo* in mice. In many cases, its effect was comparable with equivalent standardized doses of the typical drugs used to treat these diseases, elegantly demonstrating the beneficial properties of coriander in human health [15, 52–56].

3. Fruit species with functional properties and their potential use in human health

Fruits, in addition to horticultural species, constitute a group of foods for humans with important functional characteristics. In this chapter, we consider kiwifruit, sweet orange, and highbush blueberry given their extensive geographical distribution, consumption, and richness in biocompounds with nutraceutical properties.

3.1. Kiwifruit (*Actinidia deliciosa* [A. Chev] C.F Liang et A.R Ferguson/*Actidinia chinensis* [Planch])

This species originated in Asia [56] and belongs to the *Actinidiaceae* family (**Figure 5**). The *Actinidia* genus includes 66 species, but only four are cultivated for fruit production. Of these, *A. deliciosa* and *A. chinensis* are the most accepted by consumers worldwide [58, 59]. Here, we will refer to both species indistinctly. Kiwifruits have multiple sensory, nutritional, and phytochemical properties and are rich in dietary fiber, acids, phenols, and vitamins [57, 60] (see **Table 2**). These contribute to antioxidant activity, which varies with the variety and the part of fruit consumed. For example, Soquetta et al. [57] found higher values of antioxidant activity measured by the ferric reducing ability of the plasma (FRAP) method as well as carotenoids, flavonoids, and vitamin C in flour of the Monty variety compared to the Bruno variety. Moreover, the same authors found the highest content of these compounds in flours from kiwifruit skin compared to flour from kiwi fruit bagasse, reporting values from 59 to 189 mg AA in 100 g of kiwifruit flour, almost double that found in oranges and strawberries [61], and 200–1200 mg GAE 100 g^{-1} for phenolic compounds. D'Evoli et al. [60] indicated that in the total fresh kiwifruit, the content of oxalic acid was 8 mg 100 g^{-1} FW, while citric and malic acid contents were 1.2 and 0.24 g 100 g^{-1} FW, respectively. Furthermore, the same authors indicated that kiwifruits contain 90 mg GAE 100 g^{-1} FW of the total polyphenols, 0.2 mg 100 g^{-1} FW of lutein, and 0.06 mg 100 g^{-1} FW of β-carotene, in addition to α-tocopherol, γ-tocopherol, and γ-tocotrienol which represent 0.9, 0.04, and 0.12 mg 100 g^{-1} FW, respectively. All these compounds give kiwifruit strong antioxidant properties that contribute to protect cells against oxidative damage [62]. Its antioxidant capacity (ORAC) varies from 0.06 to 1.4 μmol TE 100 g^{-1} FW of fruit, depending on the hydrophilic or lipophilic fraction used in the analysis [60]. Lee et al. [63], using the same method but expressed as vitamin C equivalents (VCE), reported values from 595.7 to 2662.7 VCE 100 g^{-1} FW. Clinical studies indicate that the uptake of vitamin C derived from kiwifruit reaches 40% in humans, similar to the rate of synthetic vitamin C uptake [64, 65]. This was tested by Vissers et al. [66] in a mouse model, where they found highly effective delivery to tissues. Together with vitamin C uptake, several other nutrients and beneficial phytochemicals are consumed, with synergistic effects, among them, of iron (Fe). A clinical study with women [67, 68] indicated that a breakfast fortified with Fe, when consumed with kiwifruit, can improve Fe content in women with low Fe stores; this may be related with the high values of AA, lutein, and zeaxanthin of kiwifruit.

Figure 5. Fruits of *Actinidia deliciosa* commonly name kiwifruit.

The dietary fiber content of kiwifruit is 2–3.39 g 100 g^{-1} FW [60, 69] of which 25–30% is found in the flour skin and fruit bagasse [57]. These values are higher than several of the widely consumed fruits such as orange, apple, banana, strawberries, and blueberries [61]. Thus, kiwifruits contain sufficient fiber thus improving digestive performance, ameliorating digestive transit, alleviating constipation, and irritable bowel syndrome [60, 70–72]. Studies performed in rat have reported that kiwifruits improve digestion of the principal proteins of beef muscle, soy protein, gelatin, and gluten [73]. This is related with its content of actinidin, a very active proteolytic enzyme, which acts in concert with the gastric and intestinal proteases, pepsin, and pancreatin, generating an increment in protein digestion in the gastric and intestinal tracts [74, 75].

Species	Molecule	Part of plant	Specific function for human health	References
Actinidia deliciosa (A. Chev) C.F Liang et A.R Ferguson/*actidinia chinensis* Planch.	Dietary fiber	Fruit and flour skin and bagasse of fruit	Alleviate constipation/ laxative	[57, 60, 71]
	Actinidin	Whole fruit	Increment on protein digestion	[73 – 75]
	Polyphenols, flavonoids Vitamin C, E, lutein, zeaxanthin and other phytochemicals	Fruit and flour skin and bagasse of fruit	Antioxidant activity/ reduces cellular damage from oxidative stress	[57, 62, 71, 86]
	Vitamin C	Fruit	Positive synergic effect with nutrients and phytochemicals	[61]
	Essential amino acids, linolenic acid, folic acid, vitamins A, B6, B12, C, and E, minerals like Zn, Cu, Fe, and Se.	Fruit	Strengthen the immune system	[82, 84]
	Phytosterols and ursolic acid/independent or synergic actions of polyphenols, vitamin C, and E	Fruit	Improve lipid profile in men/reduction of risk of cardiovascular diseases	[61, 76, 77, 79, 86]
	Unidentified	Fruit	Antihypertensive	[79, 80]
	Unidentified	Fruit	Antithrombotic	[78]
	Vitamin C and inositol and other unidentified compounds	Fruit	Antiinflammatory activities	[71, 178]
	Phytosterols and ursolic acid	Fruit	Inhibit carcinogenesis processes	[86]
	Fiber	Fruit	Antiirritable bowel syndrome	[72]
Citrus sinensis (L.) Osbeck	Unidentified	Juice and ethanolic peel extract	Antihyperglycemic activity	[89, 104, 105]
	Flavonoids and other polyphenols	Juice	Antihypercholesterolemic activity	[107, 110, 111]
	Flavonoids other than anthocyanins	Juice	Antithrombotic	[112]
	Flavonoids and other polyphenols	Juice	Cardioprotective	[108, 111]
	Flavones	Juice	Enhanced antioxidant defense system	[114]

Species	Molecule	Part of plant	Specific function for human health	References
	Unidentified	Juice	Antiinflammatory activity	[119, 120]
	Flavones, flavonoids and flavonols	Juice	Microbial activity	[101, 122]
Vaccinium corymbosum L.	Unidentified	Juice	Hypoglycemic activity	[135]
	Unidentified	Juice	Antiinflammatory	[140]
	Interaction phenolic compounds	Fruit and leaf aqueous extract	Antimicrobial	[138]
	Anthocyanins and other phytochemicals	Fruit	Modulation of vascular function	[178, 179]
	Anthocyanins and other polyphenols	Extract hydroalcoholic of fruit	Cytotoxic activity	[142]
	Antioxidant action	Fruit	Antiatherogenic effect/hypocholesterolemic	[144]

Table 2. Selected fruit species and their compounds that are beneficial for human health.

Kiwifruits possess hypocholesterolaemic activity in hypercholesterolemic men [61]. This property may be related with the expression of the *Taq1B* gene in response to the consumption of kiwifruits, which modulates the content of lipids in blood plasma and has been associated with a reduced risk of cardiovascular diseases [76, 77]. However, other researchers did not find the same effect on cholesterol levels, but concluded that the consumption of two or three fruits per day can reduce levels of blood triglycerides by 15%, compared with the control [78]. Similar clinical studies have demonstrated that kiwifruits (two or three per day) can reduce blood pressure in male smokers [79], possibly related with 11% reduction in angiotensin-converting enzyme (ACE) activity. This finding is considered relevant because it is very difficult modulate hypertension by diet [79, 80]. In addition, a clinical study revealed that daily consumption of kiwifruit produces a reduction of 15% in platelet aggregation, which can be understood as antithrombotic activity [79, 81]. Nonetheless, Brevik et al. [81] discussed this effect because it may be influenced for the rate of kiwifruit consumption.

Additionally, Hunter et al. [82] and Skinner [83] affirm that kiwifruits have an important function in the modulation of the immune system. In this context, Hunter et al. [62] indicate that kiwifruit contribute significantly to lessening upper respiratory tract infections, head congestion, and sore throats in older individuals. Even though, there is a large source of variation in immune function, the nutrient status of this fruit is crucial. The most important phytochemicals present in kiwifruit include essential amino acids, linolenic acid, folic acid, vitamins A, B6, B12, C, and E, and minerals such as zinc, copper, iron, and selenium [84]. Given the type and content of phytochemicals, beneficial immune effects are not unexpected [82], although the mechanisms and the specific molecules underlying these effects are unknown. Moreover, preliminary studies under *in vitro* and *ex vivo* conditions found a protective effect of kiwifruit over oxidative damage of DNA, which may be interpreted as inhibition of the carcinogenesis process [85, 86]. Subsequently, Collins et al. [87] determined that kiwifruit consumption could protect against oxidative DNA damage protection in humans and *ex vivo* by both increasing the antioxidant status in the plasma, and stimulating DNA repair.

3.2. Sweet orange (*Citrus sinensis* (L.) Osbeck.)

The sweet orange is one of most economically important fruits in worldwide [88, 89]. It is believed that the *Citrus* genus is native to Asia, specifically from Southern China (Yunnan), which may be the origin and point of distribution of several contemporaneous *Citrus* species [90]. *Citrus sinensis* (L.) Osbeck belongs to the *Rutaceae* family and is believed to be a backcross hybrid between pummelo and mandarin (**Figure 6**) [91, 92]. The sweet orange species have several cultivated varieties, some of which are mentioned by Grosso et al. [93], but in this chapter we will discuss this species without distinguishing between varieties.

Figure 6. Fruits of *Citrus sinensis* commonly name sweet orange.

The sweet orange harbors several interesting phytochemical compounds that play an important role in human health (see **Table 2**). These include vitamins and polyphenols such as hesperidin, gallic acid, sinapic acid, caffeic acid, p-hydroxybenzoic acid, vanillic acid, narirutin, naringin, p-cumaric, and ferulic acid [93–96]. Hesperidin is the major polyphenol of sweet oranges, accounting for over 77% of the flavonol content [98–100]. These compounds are present in the edible fruit, juice, and/or peel, and here we concentrate on the juice and the edible fresh fruit, due to their direct implications in human health. As a functional food, Letaief et al. [97] determined that the AA content in orange juice fluctuates from 551 to 614 mg L^{-1}, total phenolics range from 413 to 417 mg GAE L^{-1}, and flavonoids from 25 to 60 mg catechin equivalents (CE) g^{-1} DW. Roussos [94] measured total phenols (964–1215 mg TAE L^{-1}). The percentage of antioxidant activity of the juice, evaluated by the DPPH method, fluctuated from 36.4% to 56.6% [94, 97]. The antioxidant activity of sweet orange juice is dependent on the state of maturity of the fruit. Indeed, Adu et al. [101] noted higher levels of antioxidant activity in fruits of 3–6 months (over 80%) than in fruits of 10–12 months (around 70%). Fiber and amino acids are also important in juice. In this sense, Aschoff et al. [96] informed 1.4 g 100 g^{-1} of dietary fiber, and Roussos [94] mentioned that juice contains 18 amino acids (included the essentials amino acids), especially proline, arginine, asparagine, glycine, serine, and γ-aminobutyric acid. Other authors have determined some of these and other phytochemicals in homogenate orange segments without peel, where Aschoff et al. [96] reported 36.2 mg 100 g^{-1} FW of AA, 271.5 μg100 g^{-1} FW of carotenoids, and 13.6 g 100 g^{-1} FW of dietary fiber. Recently, Molan et al. [102] evaluated some compounds and properties of sweet orange seeds, such as total polyphenols (10.9–39.4 mg GAE g^{-1} DW) and antioxidant activity (around 50%) by the DPPH method.

It has been reported that to maintain sufficient antioxidant protection, an estimated average consumption of 60 and 75 mg d^{-1} of vitamin C is required for young women and men, respectively; however, it is suggested an increase of 35 mg d^{-1} for smokers [103]. This is important because orange consumption provides other phytochemicals with multiple benefits to human health. Several clinical studies confirm this assertion. For example, sweet orange juice also harbors antidiabetic activity, as determined in rats by metabolome analysis [104]. This agrees with research performed by Kumar and Bhaskar [105] in rats, using ethanolic orange peel extract, where blood glucose decreased around 60% with respect to the control after 3 weeks of treatment, similar to the drug, glibenclamide. Furthermore, Mallick and Khan [89] suggest a combination of juice of *C. sinensis* and *C. paradisi* in order to reduce the level of glucose and improve the insulin level in the plasma of diabetic rats.

Hypocholesterolemic activity was demonstrated in women with aerobic exercise and a consumption of 500 mL of sweet orange juice daily [106]. These authors found a 15% decrease of low-density lipoprotein (LDL-C) in serum and an 18% increase of high-density lipoprotein (HDL-L), whereas the ratio LDL/HDL-cholesterol decreased by 27%. Furthermore, they also noted an improved performance during physical activity, by a reduction of blood lactate. Moreover, a long-term study (twelve months) showed that consumption of orange juice (480 mL daily) triggered reductions of 11% in total cholesterol, 18% in LDL-cholesterol, 12% in apolipoprotein B, and 12% in the LDL/HDL ratio in comparison to nonconsumers [107]. In addition, an increase in antiatherogenic activity levels with the consumption of sweet orange juice was found [108–110]. Recently, it was informed that in rats, antihyperlipidemic activity is due to phytochemical compounds like flavonoids and other polyphenols with antioxidant capacity present in the juice of sweet oranges [111]. Therefore, the juice of sweet oranges may play an important cardioprotective role by preventing thrombosis [111]. In humans, orange juice intake also decreases procoagulant activity, possibly due to flavonoids, like anthocyanins, or other juice components [112].

Another feature of sweet orange juice that supports its cardioprotective role is its effect on diastolic blood pressure, which was significantly lower in men after the daily consumption of 500 mL orange juice for 4 weeks, and an enhancement of endothelium-dependent microvascular reactivity [113]. These authors also suggest that hesperidin could be related to the beneficial effect of orange juice in cardioprotection. Likewise, Rangel-Huerta et al. [114] related the reduction of blood pressure in obese adults with the consumption of at least 300 mg flavanones over 12 weeks. On the contrary, Schär et al. [115] found a relatively high flavanone and phenolic metabolite content in plasma, but no effects were observed on blood pressure and cardiovascular risk biomarkers. Additionally, Giordano et al. [116] reported that a daily intake of 1 L of orange juice for 4 weeks was not effective in reducing cellular markers associated with cardiovascular risks. Nevertheless, in general, more evidence of positive rather than neutral or negative effects on cardiovascular risk of sweet orange juice consumption exists. In fact, risk factors are mainly associated with metabolic syndromes such as cholesterol, blood pressure, and blood coagulation, and frequent intake of orange juice may be a useful delaying strategy [117].

The antiinflammatory activity of sweet orange juice has been reported by Mohanty et al. [118] where glucose induced an acute increase in ROS and inflammation, and orange juice intake prevented meal-induced oxidative and inflammatory stress [119]. Recent studies in rats revealed the positive effect of orange juice over histological and biochemical changes related with a progress in colonic oxidative status [120]. Besides, the antimicrobial activity of sweet orange juice has been reported by several authors. Recently, Adu et al. [101] indicated an inhibitory effect of orange juice from fruits at different stages of development against Gram-positive and Gram-negative bacteria and fungi, like *B. subtilis* NCTC 10073, *C. albicans* ATCC 10231, *E. coli* ATCC 25922, *P. vulgaris* NCTC 4175, *Pseudomonas aeruginosa* ATCC 27853, and *S. aureus* ATCC. Similar results on bacteria and fungi were found by Javed et al. [121] using essential oils of orange peel. This positive effect appears to be related with flavones, flavonoids, and flavonols [122].

3.3. Highbush blueberry (*Vaccinium corymbosum* L.)

The highbush blueberry is a species that belongs to the *Ericaceae* family (**Figure 7**) [123] exhibiting a high level of morphological diversity [124]. It is native to eastern United States and was domesticated during the twentieth century [125, 126]. Its distribution and consumption is extensive due to the human health benefits (antioxidant and mineral characteristics) of fruits and leaves [123] (see **Table 2**). This fruit has a wide range of phenolic compounds, especially flavonols, such as quercetin, as well as anthocyanins [127]. Some values of the main phytochemicals that contribute to antioxidant capacity are: phenolic compounds (261–585 mg g^{-1} FW), flavonoids (50 mg g^{-1} FW), and anthocyanins (25–495 mg g^{-1} FW) [128–130].

The antioxidant activity is higher in wild blueberry species, and part of this activity is conserved in cultivated varieties [131]. The total antioxidant activity of blueberry species ranges from 15.88 to 18.41 μmol Fe^{2+} kg^{-1} FW, using the FRAP reagent [130]. Contreras et al. [132] showed values near to 80% of antioxidant capacity measured by the DPPH method under *in vitro* conditions. The same authors affirmed that antioxidant capacity is related with the content of chlorogenic acid, myricetin, syringic acid, and rutin.

Figure 7. Fruits of *Vaccinium corymbosum* commonly name highbush blueberry.

Plasma antioxidant capacity (PAC) is considered a biomarker for antioxidant status of humans. In this context, Fernández-Panchon et al. [129] indicated that PAC increased following consumption of some foods rich in phenols, which could be related with *in vivo* bioactivity, and its consequent positive effects for human health. More specific biological properties of blueberry have been described, such as anticarcinogenic, antidiabetic, antiinflammatory, antimicrobial, and reducing cholesterol, among other activities [133–135]. The hypoglycemic activity of blueberry is mentioned by several authors. Aktan et al. [135] reported a severe case of hypoglycemia in a patient of 75 years old, who had diagnosed but untreated prediabetes. Just before the episode, this patient consumed about 500 mL juice of blueberry and *Laurocerasus officinalis*, which is also considered hypoglycemic. Similarly, Cheplick et al. [136] affirmed from *in vitro* studies that blueberry fruit has potential for diet-based management of hyperglycemia, especially in the early stages of disease.

Blueberry extracts also have antimicrobial activity, which have interest considering that many microorganisms are pathogenic to humans. In this line, a significant effect of extracts on *Listeria monocytogenes* and *Salmonella enteritidis* was found under laboratory conditions by Shen et al. [137]. In the same conditions (laboratory), other extracts from dried fruits and leaves were tested on contaminant/pathogenic microorganisms. The findings indicate good results in the inhibition of development of *S. aureus* ATCC 29213, *Enterococcus faecalis* ATCC 29212, *E. coli* ATCC 27853, *K. pneumoniae* ATCC10031, *Acetobacter baumanii* ATCC 19609, *S. enteritidis* ATCC 3076, *Salmonella typhimurium* KCCM 11862, *Enterococcus faecium* LGM 11423, *Listeria innocua* NCTC 11286, *Bacillus cereu* ATCC 11778, and *P. aeruginosa* ATCC 27853 [134, 138, 139].

Zhong et al. [140] tested homogenized fresh blueberry juice as a therapy of juvenile idiopathic arthritis. The combined therapy of blueberry juice and etanercept (the typical drug used to treat this condition), improved the therapeutic effect of etanercept in patients with this pathology. Samad et al. [141] confirmed the antiinflammatory activity of extracts of blueberry in an *in vitro* study.

Yi et al. [133] studied the effect of phenolic compounds over colon cancer cell proliferation. Results indicated that these phytochemicals could inhibit the carcinogenic cells. Massarotto et al. [142] demonstrated that anthocyanins and other phenolic compounds have cytotoxic activity, as tested in tumoral cell lines under *in vitro* conditions; thus, blueberry extracts could be useful for future treatment of cancer, as a natural cytotoxic agent. In this context, Tsuda et al. [143] obtained similar results with human leukemia cells and ethanolic extracts of several berries that include blueberry fruits. In all cases, the induction of apoptosis in the cancerous cells may be the mechanism triggered by blueberry.

The antilipidemic and antiatherogenic actions of blueberry have been reported by several authors. Coban et al. [144] indicated that the fresh fruit is food supplements that generate a positive effect over aorta and liver of hypercholesterolemic Guinea pigs [144]. In this respect, Cutler et al. [145] confirmed that berries are a special source of phytochemicals (anthocyanins and other phenolic compounds) and can be exploited as natural phytochemicals to contribute toward the amelioration of several chronic diseases, including those derived from alterations in the lipid profile in vascular systems.

3.4. Maracuyá (*Passiflora edulis* Sims)

Maracuyá (*Passiflora edulis* Sims) is a species that belongs to the *Passifloraceae* family (**Figure 8**) [146, 147] which is native to Brazil, South America. Nevertheless, some authors report that its real origin is Australia and is called *Passiflora edulis* forma flavicarpa [148, 149]. Variability studies have been carried out in South America, mainly in Colombia, as this region is particularly rich in this genus, although a low variability has been reported [150, 151]. Both scientific names *Passiflora edulis* Sims and *Passiflora edulis* forma are indistinctly considered in this chapter. Talcott et al. [152] identified several phenolic acids such as galacturonic acid, p-hydroxybenzoic acid, syringic acid, caffeic acid, p-coumaric acid, tryptophan, flavonoid glycoside, sinapic acid, ferulic acid, o-coumaric acid, and syringic acid in this species. Some compounds such as tryptophan, sinapic acid, and p-coumaric acid are in higher quantity with 733, 626, and 623 μg L^{-1} DW, respectively. Within the latter, total phenolic compounds fluctuated from 342.8 to 382 mg GAE L^{-1} FW [153]; total carotenoids varied between 22.4 and 29.1 mg L^{-1} DW, and the ascorbic acid content from 0.22 to 0.33 g kg^{-1} FW [152, 153]. The anthocyanin concentration in pulps and by product, on the other hand, were 3.48 and 3.7 mg 100 g^{-1} DW, respectively, while the flavonoids in pulps and by product were 60.3 and 40 mg 100 g^{-1} DW, respectively [154]. Regarding the above, Zucolotto et al. [155] indicated that C-glycosyl flavonoids are present in several species of *Passiflora* in South America. Furthermore, Da Silva et al. [154] informed that values for *β*-carotene fluctuated from 57.93 to 1362.07 μg 100 g^{-1} DW for both pulp and by product. It is worth noting, that piceatannol, a compound with an important antioxidant characteristic, is present in peel and seeds of the maracuyá fruit [156]. Moreover, the total antioxidant activity was found to reach values from 409.13 to 805.5 μM TE L^{-1} FW in this fruit [153], while Marcoris et al. [157] indicate values ranging from 1279 to 1460 μM TE L^{-1} FW. Total dietary and soluble fiber is another important characteristic attributed to this species, with values fluctuating from 35.5 to 81.5 g 100 g^{-1} DM. These values are higher in comparison to other tropical fruits such as *Mangifera indica, Ananas comosus,* and *Psidium guajava* [158]. The maracuyá fruit has several special characteristics that are beneficial for human health, described in **Table 2**. Within the latter, the most important properties are the sedative and anxiolytic activities, which are common to several other species of the *Passiflora* genus [159]. Evaluation of aqueous extracts of pericarp fruit on rats concluded that a sedative effect was obtained with an oral administration of 300 mg kg^{-1} [160]. This effect was corroborated by a dose-dependent decrease of the locomotor-activity. Similar studies using ethanolic extracts of maracuyá leaves in rats exhibited sedative effects at 400 mg kg^{-1} [161]. Figueiredo et al. [162] reported that 130 mg kg^{-1} of bark flour of maracuyá fruits showed a sedative effect in rats. On the other hand, antiproliferative properties on cancer cells evaluated in SW480 and SW620 cells lines showed that cell growth in both lines was inhibited with 50–500 μg mL^{-1} of leaf ethanolic extracts and maracuyá fruit juice [163]. The polysaccharide peel of maracuyá fruits was also evaluated for its antiinflammatory effects and antidiabetic properties. In this context, the reduction of the inflammation was associated with the liberation or synthesis of histamine and serotonin, in response to a polysaccharide fraction of the maracuyá fruit [164]. Moreover, flour peel of maracuyá fruits in diabetic rats showed a decline in glucose content in the blood [165], probably associated with the high fiber level in this fruit tissue, which could prevent absorption of carbohydrates [166, 167]. Furthermore, triglycerides levels significantly decreased with 25 mg kg^{-1} of flour peel, however, no changes in total cholesterol levels were observed [165]. Still, Barbalho et al. [167] concluded that maracuyá fruit juice could improve the lipid profile, including the triglycerides, cholesterol, LDL-cholesterol, and HDL-cholesterol levels.

Figure 8. Fruits of *Passiflora edulis* commonly name maracuyá.

4. Conclusion and perspectives

A wealth of information in the field of phytochemical compounds and their impact on human health has been generated. Nowadays, it is possible to affirm that fruits and vegetables must be a part of daily diet. This is not simply a recommendation, but must be treated as an urgent requirement to ameliorate human health, especially in decreasing chronic nontransmissible diseases. We believe that additional efforts of governments and diverse organisms related with human health are necessary in order to highlight the benefits of these food types. Coriander and kiwifruit have remarkable characteristics and are excellent functional foods. We highlight these species for their wide range of benefits in different human diseases and their worldwide distribution. Likewise, further investigation is required to understand the mechanisms associated with several biochemical and physiological processes induced by fruit and vegetable intake in humans. Furthermore, we consider that leeks and artichokes have special potential as functional foods. Although there is a lot of information about the beneficial effects of fruits, we believe it is possible to extend studies to other organs like leaves and stems in artichokes, and roots in leeks, because they can offer additional benefits to human health.

Acknowledgments

We are grateful to FONDECYT 1110726, 1120917, and 1150220 projects and FONDECYT-POSDOCTORAL 3120248, Doctoral Fellowship CONICYT-PCHA/Doctorado Nacional/2016-21160984, as well as the Millennium Nucleus for Plant Synthetic Biology and Systems Biology NC130030. We would like to thank Dr. Michael Handford and Dr. Franko Restovic for language support.

Author details

Ricardo Tighe-Neira[1], Miren Alberdi[2,3], Patricio Arce-Johnson[4], Jesús L. Romero-Romero[4,5,6], Marjorie Reyes-Díaz[2,3] and Claudio Inostroza-Blancheteau[1,7]*

*Address all correspondence to: claudio.inostroza@uct.cl

1 School of Agronomy, Faculty of Natural Resources, Universidad Católica de Temuco, Temuco, Chile

2 Department of Chemical Sciences and Natural Resources, Faculty of Engineering and Sciences, Universidad de La Frontera, Temuco, Chile

3 Center of Plant, Soil Interaction and Natural Resources Biotechnology, Scientific and Technological Bioresource Nucleus (BIOREN), Universidad de La Frontera, Temuco, Chile

4 Department of Molecular Genetics and Microbiology, Faculty of Biological Sciences, Pontificia Universidad Católica de Chile, Santiago, Chile

5 Faculty of Agronomy and Forest Engineering, Pontificia Universidad Católica de Chile, Santiago, Chile

6 Department of Agricultural Biotechnology, Instituto Politécnico Nacional, CIIDIR Sinaloa Unid, Guasave, Mexico

7 Nucleus of Research in Food Production, Faculty of Natural Resources, Universidad Católica de Temuco, Temuco, Chile

References

[1] Hasler CM. Functional foods: their role in disease prevention and health promotion. Food Technology. 1998; 52: 63–147.

[2] Institute of Medicine. Opportunities in the Nutrition and Food Sciences: Research Challenges and the Next Generation of Investigators. Washington, DC: The National Academies Press, 1994. p. 328. DOI:10.17226/2133

[3] Steinmetz KA, Potter JD. Vegetables, fruit, and cancer. II. Mechanisms. Cancer Causes & Control. 1991; 2: 427–442. DOI:10.1007/BF00054304

[4] Palermo M, Pellegrini N, Fogliano V. The effect of cooking on the phytochemical content of vegetables: Effect of cooking on vegetable phytochemicals. Journal of the Science of Food and Agriculture. 2014; 94: 1057–1070. DOI:10.1002/jsfa.6478

[5] Slavin JL, Lloyd B. Health benefits of fruits and vegetables. Advances in Nutrition. 2012; 3: 506–516. DOI:10.3945/an.112.002154

[6] Uchida A, Ogihara J, Kumagai H, Akao M, Matsufuji H, Sakurai H, Sekiguch I. Mechanisms of sulfide components expression and structural determination of substrate precursor in jumbo leek (*Allium ampeloprasum* L.). Nippon Shokuhin Kagaku Kogaku Kaishi. 2009; 56: 280–285. DOI:10.3136/nskkk.56.280

[7] Ceccarelli N, Curadi M, Picciarelli P, Martelloni L, Sbrana C, Giovannetti M. Globe artichoke as a functional food. Mediterranean Journal of Nutrition and Metabolism. 2010; 3: 197–201. DOI:10.1007/s12349–010-0021-z.

[8] Nambiar VS, Daniel M, Guin P. Characterization of polyphenols from coriander leaves (*Coriandrum sativum*), Red amaranthus (*A. paniculatus*) and green amaranthus (*A. frumentaceus*) using paper chromatography and their health implications. Journal of Herbal Medicine and Toxicology. 2010; 4: 173–177.

[9] Simonovska J, Rafajlovska V, Kavrakovski Z, Srbinoska M. Nutritional and bioactive compounds in hot fruits of Capsicum annuum L. from Macedonia. Macedonian Journal of Chemistry and Chemical Engineering. 2014; 33: 95–102.

[10] Bratsch A, Arancibia A. Specialty Crop Profile: Globe Artichoke. College of Agriculture and Life Sciences, Virginia Polytechnic Institute and State University. 2014; Publication 438–108, p. 2–9.

[11] Bianco VV. Present situation and future potential of artichoke in the mediterranean basin. Acta Horticulturae. 2005; 681: 39–58. DOI:10.17660/ActaHortic.2005.681.2

[12] Pennington JAT, Fisher RA. Classification of fruits and vegetables. Journal of Food Composition and Analysis. 2009; 22: S23-S31. DOI:10.1016/j.jfca.2008.11.012

[13] Alamanni MC, Cossu M. Antioxidant activity of the extracts of the edible part of artichoke (*Cynaras colymus* L.) var. spinososardo. Italian Journal of Food Science. 2003; 15: 187–195.

[14] Romani A, Pinelli P, Cantini C, Cimato A, Heimler D. Characterization of Violetto di Toscana, a typical Italian variety of artichoke (*Cynaras colymus* L.). Food Chemistry. 2006; 95: 221–225. DOI:10.1016/j.foodchem.2005.01.013

[15] Karami R, Hosseini M, Mohammadpour T, Ghorbani A, Sadeghnia HR, Rakhshandeh H, Vafaee F, Esmaeilizadeh M. Effects of hydroalcoholic extract of Coriandrum sativum on oxidative damage in pentylenetetrazole-induced seizures in rats. Iranian Journal of Neurology. 2015; 14: 59–66.

[16] Ruiz-Aceituno L, García-Sarrió M J, Alonso-Rodriguez B, Ramos L, Sanz L. Extraction of bioactive carbohydrates from artichoke (*Cynara scolymus* L.) external bracts using microwave assisted extraction and pressurized liquid extraction. Food Chemistry. 2016; 196: 1156–1162.

[17] Hernández-Hernández O, Ruiz-Aceituno L, Sanz M, Martínez-Castro I. Determination of free inositols and other low molecular weight carbohydrates in vegetables. Journal of Agricultural and Food Chemistry. 2011; 6: 2451–2455.

[18] Ostlund RE, Sherman WR. Pinitol and derivatives thereof for the treatment of metabolic disorders (US Patent 5827896). 1998.

[19] Crawford TJ, Crowther CA, Alsweiler J, Brown J. Antenatal dietary supplementation with myo-inositol in women during pregnancy for preventing gestational diabetes. The Cochrane Library. 2015.

[20] Roberfroid, M. Dietary fiber, inulin, and oligofructose: a review comparing their physiological effects. Critical Reviews in Food Science & Nutrition. 1993; 33, 103–148.

[21] Shoaib M, Shehzad A, Omar M, Rakha A, Raza H, Sharif HR, Shakeel A, Ansari A, Niazi S. Inulin: Properties, health benefits and food applications. Carbohydrate Polymers. 2016; 147: 444–454.

[22] Bosscher D, Van Loo J, Franck A. Inulin and oligofructose in lowering the risk of osteoporosis: a major concern for women's health. Agro Food Industry HI-TECH. 2006; 1: 10–12.

[23] Jones HA, Mann LK. Onions and their allies: botany, cultivation, and utilization. Hill L, editor. Cornell University, 1963. p. 286.

[24] Fritsch RM, Friesen N. Evolution, Domestication and Taxonomy, Chapter 1. In: Rabinowitch HD, Currah L, editors. Allium Crop Science: Recent Advances. Wallingford, UK, CABI Publishing; 2002. p. 5–30. DOI:10.1079/9780851995106.0005

[25] Hirschegger P, Galmarini C, Bohanec B. Characterization of a novel form of fertile great headed garlic (*Allium* sp.). Plant Breeding. 2006; 125: 635–637. DOI:10.1111/j.1439–0523.2006.01279.x

[26] Mathew BA. Review of Allium Section Allium. Royal Botanic Gardens. Kew, editors. Chicago USA. Kew Publishing. 1996. p. 182.

[27] Hirschegger P, Jakše J, Trontelj P, Bohanec B. Origins of Allium ampeloprasum horticultural groups and a molecular phylogeny of the section Allium (Allium: Alliaceae). Molecular Phylogenetics and Evolution. 2010; 54: 488–497. DOI:10.1016/j.ympev.2009.08.030

[28] Bernaert N, De Paepe D, Bouten C, De Clercq H, Stewart D, Van Bockstaele E, Van Droogenbroeck B. Antioxidant capacity, total phenolic and ascorbate content as a function of the genetic diversity of leek (*Allium ampeloprasum* var. porrum). Food Chemistry. 2012; 134: 669–677. DOI:10.1016/j.foodchem.2012.02.159

[29] Bernaert N, De Clercq H, Van Bockstaele E, De Loose M, Van Droogenbroeck B. Antioxidant changes during postharvest processing and storage of leek (*Allium ampeloprasum* var. porrum). Postharvest Biology and Technology. 2013; 86: 8–16.

[30] Vandekinderen I, Van Camp J, Devlieghere F, Ragaert P, Veramme K, Bernaert N, Denon Q, De Meulenaer B. Evaluation of the use of decontamination agents during fresh-cut leek processing and quantification of their effect on its total quality by means of a multidisciplinary approach. Innovative Food Science & Emerging Technologies. 2009; 10: 363–373. DOI:10.1016/j.ifset.2009.02.002

[31] Adao CR, da Silva BP, Parente JP. A new steroidal saponin from *Allium ampeloprasum* var. porrum with antiinflammatory and gastroprotective effects. Phytochemistry Letters. 2011a; 4: 306–310. DOI:10.1016/j.phytol.2011.05.009

[32] Adao CR, da Silva BP, Parente JP. A new steroidal saponin with antiinflammatory and antiulcerogenic properties from the bulbs of *Allium ampeloprasum* var. porrum. Fitoterapia. 2011b; 82: 1175–1180. DOI:10.1016/j.fitote.2011.08.003

[33] Sobolewska D, Michalska K, Podolak I, Grabowska K. Steroidal saponins from the genus Allium. Phytochemistry Reviews 2014; 15: 1–35. DOI:10.1007/s11101-014-9381-1

[34] Maidment DCJ, Dembny Z, Watts DI. The anti-bacterial activity of 12 Alliums against *Escherichia coli*. Nutrition & Food Science. 2001; 31: 238–241. DOI:10.1108/EUM0000000005614

[35] Radovanović B, Mladenović J, Radovanović A, Pavlović R, Nikolić V. Phenolic composition, antioxidant, antimicrobial and cytotoxic activities of *Allium porrum* L. (Serbia) extracts. Journal of Food and Nutrition Research. 2015; 3: 564–569. DOI:10.12691/jfnr-3-9-1

[36] Hervert-Hernández D, Sáyago-Ayerdi SG, Goñi I. Bioactive compounds of four hot pepper varieties (*Capsicum annuum* L.), antioxidant capacity, and intestinal bioaccessibility. Journal of Agricultural and Food Chemistry. 2010; 58: 3399–3406. DOI:10.1021/jf904220w

[37] Tighe-Neira R, Navarro RM, Cantergiani GL, Novoa AC. Efecto de dos extractos botánicos en el desarrollo y contenido de polifenoles de ají (*Capsicum annuum* L.). Revista Mexicana de Ciencias Agrícolas. 2013; 5: 115–127

[38] Serrano J; Goñi I, Saura-Calixto F. Determination of β-carotene and lutein available from green leafy vegetables by an in vitro digestion and colonic fermentation method. J. Agric. Food Chem. 2005; 53: 2936–2940. DOI:10.1021/jf0480142

[39] Golubkina NA, Dzhos EA, Pishnaya ON, Mamedov MI, Nadezhkin SM. Biochemical characteristics of hot pepper fruit grown in Moscow region. Russian Agricultural Sciences. 2014; 40: 27–31. DOI:10.3103/S1068367414010042

[40] De M, De Ak, Banerjee AB. Screening of spices for antimicrobial activity. Journal of Spices and Aromatic Crops. 1999; 2.

[41] Soetarno S, Sukrasno S, Yulinah E, Sylvia S. Antimicrobial activities of the ethanol extracts of capsicum fruits with different levels of pungency. Jurnal Matematika & Sains. 1997; 2: 57–63.

[42] Huang X, Huadan TAN, Zheng Y, Zhang J, Yijun ZHOU. Study on antimicrobial activities of ethanol extracts from different capsicum cultivars. 农业科学与技术 (英文版). 2012; 13: 2514–2517.

[43] Mhemdi H, Rodier E, Kechaou N, Fages J. A supercritical tuneable process for the selective extraction of fats and essential oil from coriander seeds. Journal of Food Engineering. 2011; 105: 609–616. DOI:10.1016/j.jfoodeng.2011.03.030

[44] Sahib NG, Anwar F, Gilani AH, Hamid AA, Saari N, Alkharfy KM. Coriander (*Coriandrum sativum* L.): A potential source of high-value components for functional foods and nutraceuticals - A review. Phytotherapy Research 2012; 27: 1439–1456. DOI:10.1002/ptr.4897

[45] Bhuiyan MNI, Begum J, Sultana M. Chemical composition of leaf and seed essential oil of *Coriandrum sativum* L. from Bangladesh. Bangladesh Journal of Pharmacology. 2009; 4: 150–153. DOI:10.3329/bjp.v4i2.2800

[46] Zoubiri S, Baaliouamer A. Essential oil composition of *Coriandrum sativum* seed cultivated in Algeria as food grains protectant. Food Chemistry. 2010; 122: 1226–1228. DOI:10.1016/j.foodchem.2010.03.119

[47] Buthelezi MND, Soundy P, Jifon J, Sivakumar D. Spectral quality of photo-selective nets improves phytochemicals and aroma volatiles in coriander leaves (*Coriandrum sativum* L.) after postharvest storage. Journal of Photochemistry and Photobiology B: Biology. 2016; 161: 328–34. DOI:10.1016/j.jphotobiol.2016.05.032

[48] Kumar RS, Balasubramanian P, Govindaraj P, Krishnaveni T. Preliminary studies on phytochemicals and antimicrobial activity of solvent extracts of *Coriandrum sativum* L. roots (Coriander). Journal of Pharmacognosy and Phytochemistry. 2014; 2: 74–78.

[49] Deepa B, Anuradha CV. Antioxidant potential of *Coriandrum sativum* L. Indian Journal Experimental Biology. 2011; 49: 30–8.

[50] Dhanapakiam P, Joseph JM, Ramaswamy VK, Moorthi M, Kumar AS. The cholesterol lowering property of coriander seeds (*Coriandrum sativum*): mechanism of action. Journal of Environmental Biology. 2008; 29: 53–56.

[51] Saravanan M, Ignacimuthu S. Hypocholesterolemic effect of indian medicinal plants - A review. Medicinal Chemistry. 2015; 5: 040–049. DOI:10.4172/2161-0444.10002

[52] Kharade SM, Gumate DS, Patil VM, Kokane SP, Naikwade NS. Behavioral and biochemical studies of seeds of Coriandrum sativum in various stress models of depression. International Journal of Current Research and Review. 2011; 1003: 4–8.

[53] Zargar-Nattaj SS, Tayyebi P, Zangoori V, Moghadamnia Y, Roodgari H, Jorsaraei SG, Moghadamnia AA. The effect of Coriandrum sativum seed extract on the learning of new born mice by electric shock: interaction with caffeine and diazepam. Psychology Research and Behavior Management. 2011; 4: 13–19. DOI:10.2147/PRBM.S15905

[54] Mahendra P, Bisht S. Anti-anxiety activity of Coriandrum sativum assessed using different experimental anxiety models. Indian Journal of Pharmacology. 2011; 43: 574–577. DOI:10.4103/0253–7613.84975

[55] Fatemeh KS, Vafadar LS, Hosseini M, Naser SM, Ghorbani A, Pourganji M. The analgesic effects of different extracts of aerial parts of Coriandrum Sativum in mice. International Journal of Biomedical Science. 2015; 11: 23–28.

[56] Latha K, Rammohan B, Sunanda BPV, Maheswari MSU, Mohan SK. Evaluation of anxiolytic activity of aqueous extract of Coriandrum sativum Linn. in mice: A preliminary experimental study. Pharmacognosy Research. 2015; 7: (Suppl 1): S47–S51. DOI:10.4103/0974–8490.157996

[57] Soquetta MB, Stefanello FS, Da Mota Huerta K, Monteiro SS, Da Rosa CS, Terra NN. Characterization of physiochemical and microbiological properties, and bioactive compounds, of flour made from the skin and bagasse of kiwi fruit (Actinidia deliciosa). Food chemistry. 2016. 199: 471–478. DOI:10.1016/j.foodchem.2015.12.022

[58] Liang Chou-Fen, Ferguson AR. The botanical nomenclature of the kiwifruit and related taxa. New Zealand Journal of Botany. 1986; 24: 183–184. DOI:10.1080/0028825X.1986.10409728

[59] Drzewiecki J, Latocha P, Leontowicz H, Leontowicz M, Park YS, Najman K,Weisz M, Ezra A, Gorinstein S. Analytical methods applied to characterization of Actinidia arguta, Actinidia deliciosa, and Actinidia eriantha kiwi fruit cultivars. Food Analytical Methods. 2016; 9: 1353–1366. DOI:10.1007/s12161-015-0309-1

[60] D'Evoli L, Moscatello S, Lucarini M, Aguzzi A, Gabrielli P, Proietti S, Battistelli A,Famianic F, Böhmd V, Lombardi-Boccia G. Nutritional traits and antioxidant capacity of kiwifruit (Actinidia deliciosa Planch., cv. Hayward) grown in Italy. Journal of Food Composition and Analysis. 2015; 37: 25–29. DOI:10.1016/j.jfca.2014.06.012

[61] Gammon CS, Kruger R, Minihane AM, Conlon CA, von Hurst PR, Stonehouse W. Kiwifruit consumption favourably affects plasma lipids in a randomised controlled

trial in hypercholesterolaemic men. British Journal of Nutrition. 2013; 109: 2208–2218. DOI:10.1017/S0007114512004400

[62] Hunter DC, Skinner MA, Wolber FM, Booth CL, Loh JM, Wohlers M, Stevenson LM, Kruger MC. Consumption of gold kiwifruit reduces severity and duration of selected upper respiratory tract infection symptoms and increases plasma vitamin C concentration in healthy older adults. British Journal of Nutrition. 2012; 108: 1235–1245. DOI:10.1017/S0007114511006659

[63] Lee I, Im S, Jin CR, Heo HJ, Cho YS, Baik MY, Kim DO. Effect of maturity stage at harvest on antioxidant capacity and total phenolics in kiwifruits (Actinidia spp.) grown in Korea. Horticulture, Environment, and Biotechnology. 2015; 56: 841–848. DOI:10.1007/s13580-015-1085-y.

[64] Carr AC, Bozonet SM, Vissers MCM. A randomised cross-over pharmacokinetic bioavailability study of synthetic versus kiwifruit-derived vitamin C. Nutrients. 2013a; 5: 4451–4461. DOI:10.3390/nu5114451

[65] Carr AC, Bozonet SM, Pullar JM, Simcock JW, Vissers M. A randomized steady-state bioavailability study of synthetic versus natural (Kiwifruit-Derived) vitamin C. Nutrients. 2013b; 5: 3684–3695. DOI:10.3390/nu5093684

[66] Vissers MC, Bozonet SM, Pearson JF, Braithwaite LJ. Dietary ascorbate intake affects steady state tissue concentrations in vitamin C–deficient mice: tissue deficiency after suboptimal intake and superior bioavailability from a food source (kiwifruit). The American journal of clinical nutrition. 2011; 93: 292–301. DOI:10.3945/ajcn.110.004853

[67] Beck K, Conlon C, Kruger R, Coad J, Stonehouse W. The effect of gold kiwifruit consumed with an iron fortified breakfast cereal meal on iron status in women with low iron stores: A 16 week randomised controlled intervention study. BMC Public Health. 2010; 10: 1. DOI:10.1186/1471-2458-10-36

[68] Beck K, Conlon CA, Kruger R, Coad J, Stonehouse W. Gold kiwifruit consumed with an iron-fortified breakfast cereal meal improves iron status in women with low iron stores: a 16-week randomised controlled trial. British Journal of Nutrition. 2011; 105: 101–109. DOI:10.1017/S0007114510003144

[69] Schakel S, Sievert YA, Buzzard IM. In: Dietary fiber values for common foods. The CRC handbook of dietary fiber in human nutrition 3rd ed. Spiller G. Editor. London: CRC Press. 2001.

[70] Chan AO, Leung G, Tong T, Wong NY. Increasing dietary fiber intake in terms of kiwifruit improves constipation in Chinese patients. World journal of gastroenterology. 2007; 13: 4771–4775.

[71] Singletary K. Kiwifruit: overview of potential health benefits. Nutrition Today. 2012; 47: 133–147. DOI:10.1097/NT.0b013e31825744bc

[72] Chang CC, Lin YT, Lu YT, Liu YS, Liu JF. Kiwifruit improves bowel function in patients with irritable bowel syndrome with constipation. Asia Pacific Journal of Clinical Nutrition. 2010; 19: 451–457. DOI: 10.6133/apjcn.2010.19.4.01.

[73] Rutherfurd SM, Montoya CA, Zou ML, Moughan PJ, Drummond LN, Boland, MJ. Effect of actinidin from kiwifruit (Actinidiadeliciosa cv. Hayward) on the digestion of food proteins determined in the growing rat. Food Chemistry. 2011; 129: 1681–1689. DOI:10.1016/j.foodchem.2011.06.031

[74] Kaur L, Rutherfurd SM, Moughan PJ, Drummond L, Boland MJ. Actinidin enhances gastric protein digestion as assessed using an in vitro gastric digestion model. Journal of agricultural and food chemistry. 2010a; 58: 5068–5073. DOI:10.1021/jf903332a

[75] Kaur L, Rutherfurd SM, Moughan PJ, Drummond L, Boland MJ. Actinidin enhances protein digestion in the small intestine as assessed using an in vitro digestion model. Journal of Agricultural and Food Chemistry. 2010b; 58: 5074–5080.DOI:10.1021/jf903835g

[76] Gammon CS, Minihane AM, Kruger R, Conlon CA, von Hurst PR, Jones B, Stonehouse W. TaqIB polymorphism in the cholesteryl ester transfer protein (CETP) gene influences lipid responses to the consumption of kiwifruit in hypercholesterolaemic men. British Journal of Nutrition. 2014a; 111: 1077–1084. DOI:10.1017/S0007114513003437

[77] Gammon CS, Kruger R, Conlon CA, von Hurst PR, Jones B, Stonehouse W. Inflammatory status modulates plasma lipid and inflammatory marker responses to kiwifruit consumption in hypercholesterolaemic men. Nutrition, Metabolism and Cardiovascular Diseases. 2014b; 24: 91–99. DOI:10.1016/j.numecd.2013.06.001

[78] Duttaroy AK, Jørgensen A. Effects of kiwi fruit consumption on platelet aggregation and plasma lipids in healthy human volunteers. Platelets. 2004; 15: 287–292. DOI:10.1080/09537100410001710290

[79] Karlsen A, Svendsen M, Seljeflot I, Laake P, Duttaroy AK, Drevon CA, Arnesen H, Tonstad S, Blomhoff R. Kiwifruit decreases blood pressure and whole-blood platelet aggregation in male smokers. Journal of human hypertension. 2013; 27: 126–130. DOI:10.1038/jhh.2011.116

[80] Svendsen M, Tonstad S, Heggen E, Pedersen TR, Seljeflot I, Bøhn SK, Bastani NE, Blomhoff R, Holme IM, Klemsdal TO. The effect of kiwifruit consumption on blood pressure in subjects with moderately elevated blood pressure: A randomized, controlled study. Blood Pressure. 2014; 24: 48–54. DOI:10.3109/08037051.2014.976979

[81] Brevik A, Gaivão I, Medin T, Jørgenesen A, Piasek A, Elilasson J, Karlsen A, Blomhoff R, Veggan T, Duttaroy AK, Collins AR. Supplementation of a western diet with golden kiwifruits (*Actinidia chinensis* var.'Hort 16A':) effects on biomarkers of oxidation damage and antioxidant protection. Nutrition Journal. 2011; 10: 54.DOI:10.1186/1475-2891-10-54

[82] Hunter DC, Greenwood J, Zhang J, Skinner MA. Antioxidant and "natural protective" properties of kiwifruit. Current Topics in Medicinal Chemistry. 2011; 11: 1811–1820. DOI:10.2174/156802611796235134

[83] Skinner MA. Wellness foods based on the health benefits of fruit: gold kiwifruit for immune support and reducing symptoms of colds and influenza. J. Food Drug Anal. 2012; 20 (Suppl 1): 261–264.

[84] Calder PC, Kew S. The immune system: a target for functional British Journal of Nutrition. 2002; 88: S165-S176. DOI:10.1079/BJN2002682

[85] Collins BH, Horská A, Hotten PM, Riddoch C, Collins AR. Kiwifruit protects against oxidative DNA damage in human cells and in vitro. Nutrition and Cancer. 2001; 39: 148–153. DOI:10.1207/S15327914nc39120

[86] Fiorentino A, D'Abrosca B, Pacifico S, Mastellone C, Scognamiglio M, Monaco P. Identification and assessment of antioxidant capacity of phytochemicals from kiwi fruits. Journal of Agricultural and Food Chemistry. 2009; 57: 4148–4155. DOI:10.1021/jf900210z

[87] Collins AR, Harrington V, Drew J, Melvin R. Nutritional modulation of DNA repair in a human intervention study. Carcinogenesis. 2003; 24: 511–515.DOI:10.1093/carcin/24.3.511

[88] Moore GA. Oranges and lemons: clues to the taxonomy of Citrus from molecular markers. Trends in Genetics. 2001; 17: 536–540. DOI:10.1016/S0168-9525(01)02442-8

[89] Mallick N, Khan RA. Effect of *Citrus paradisi* and *Citrus sinensis* on glycemic control in rats. African Journal of Pharmacy and Pharmacology. 2015; 9: 60–64. DOI:10.5897/AJPP2014. 4185

[90] Gmitter FG, Hu X. The possible role of Yunnan, China, in the origin of contemporary Citrus species (Rutaceae).Economic Botany. 1990; 44: 267–277. DOI:10.1007/BF02860491

[91] Bayer RJ, Mabberley DJ, Morton C, Miller CH, Sharma IK, Pfeil B E, Rich S, Hitchcock R, Sykes S. A molecular phylogeny of the orange subfamily (Rutaceae: Aurantioideae) using nine cpDNA sequences. American Journal of Botany. 2009; 96: 668–685. DOI:10.3732/ajb.0800341

[92] Xu Q, Chen LL, Ruan X, Chen D, Zhu A, Chen C, Bertrand D, Jiao Wen-Biao, HaoBao-Hai, Lyon MP, Chen J, Gao S, Xing F, Lan H, Chang Ji-Wei, Ge X, Lei Y, Hu Q, Miao Y, Wang L, Xiao S, Biswas MK, Zeng W, Guo F, Cao H, Yang X, Xu Xi-Wen, Cheng Yun-Jiang, Xu J, Liu Ji-Hong, Luo OJ, Tang Z, Guo Wen-Wu, Kuang H, Zhang Hong-Yu, Roose ML, Nagarajan N, Deng Xiu-Xin, Ruan Y. The draft genome of sweet orange (*Citrus sinensis*). Nature genetics.2013; 45: 59–66. DOI:10.1038/ng.2472

[93] Grosso G, Galvano F, Mistretta A, Marventano S, Nolfo F, Calabrese G, Buscemi S, Drago F, Veronesi U, Scuderi A. Red orange: Experimental models and epidemiological evidence of its benefits on human health. Oxidative Medicine and Cellular Longevity. 2013; Article ID 157240: 11. DOI:10.1155/2013/157240

[94] Roussos PA. Phytochemicals and antioxidant capacity of orange (*Citrus sinensis* (L.) Osbeck cv. Salustiana) juice produced under organic and integrated farming system in Greece. Scientia horticulturae. 2011; 129: 253–258. DOI:10.1016/j.scienta.2011.03.040

[95] Goulas V, Manganaris GA. Exploring the phytochemical content and the antioxidant potential of Citrus fruits grown in Cyprus. Food Chemistry. 2012; 131: 39–47. DOI:10.1016/j.foodchem.2011.08.007

[96] Aschoff JK, Kaufmann S, Kalkan O, Neidhart S, Carle R, Schweiggert RM. In vitro bioaccessibility of carotenoids, flavonoids, and vitamin C from differently processed oranges and orange juices [Citrus sinensis (L.) Osbeck]. Journal of Agricultural and Food Chemistry. 2015; 63: 578–587. DOI:10.1021/jf505297t

[97] Letaief H, Zemni H, Mliki A, Chebil S. Composition of Citrus sinensis (L.) Osbeck cv «Maltaise demi-sanguine» juice. A comparison between organic and conventional farming.Food Chemistry. 2016; 194: 290–295. DOI:10.1016/j.foodchem.2015.08.025

[98] Mattila P, Astola J, Kumpulainen J. Determination of flavonoids in plant material by HPLC with diode-array and electro-array detections. Journal of Agricultural and Food Chemistry. 2000; 48: 5834–5841. DOI:10.1021/jf000661f

[99] Neveu V, Perez-Jimenez J, Vos F, Crespy V, du Chaffaut L, Mennen L., Knox C, Eisner R, Cruz J, Wishart D, Scalbert A. Phenol-Explorer: an online comprehensive database on polyphenol contents in foods. Database. 2010. DOI:10.1093/database/bap024

[100] Perche O, Vergnaud-Gauduchon J, Morand C, Dubray C, Mazur A, Vasson, MP. Orange juice and its major polyphenol hesperidin consumption do not induce immunomodulation in healthy well-nourished humans. Clinical Nutrition. 2014; 33: 130–135. DOI:10.1016/j.clnu.2013.03.012

[101] Adu F, Apenteng JA, Kuntworbe N, Akanwariwiak WG, Appiah T, Mintah DN. Antimicrobial and antioxidant activities of Citrus sinensis var. late Valencia fruits at various stages of development. African Journal of Microbiology Research. 2016; 10: 73–78. DOI:10.5897/AJMR2015.7725

[102] Molan AL, Ismail MH, Nsaif RH. Phenolic contents and antioxidant activity of peels and seeds of orange (Citrus sinensis) cultivated in Iraq. 2016. DOI:10.20959/wjpps20164-6478

[103] Fallico B, Ballistreri G, Arena E, Brighina S, Rapisarda P. Bioactive compounds in blood oranges (Citrus sinensis (L.) Osbeck): Level and intake. Food Chemistry. 2017; 215: 67–75. DOI:10.1016/j.foodchem.2016.07.142

[104] Ahamad SR. Phytochemical, heavy metal and metabolomic anti-diabetic analysis of Citrus sinensis. Bulletin of Environment, Parmacology and Life Sciences. 2015; 5: 22–27.

[105] Kumar PRZA, Bhaskar A. Evaluation of antihyperglycaemic and antihyperlipidemic activity of Citrus sinensis peel extract on streptozotocin-induced diabetic rats. International Journal of Diabetes in Developing Countries. 2015; 35: 448–453. DOI:10.1007/s13410-015-0310-7

[106] Aptekmann NP, Cesar TB. Orange juice improved lipid profile and blood lactate of overweight middle-aged women subjected to aerobic training. Maturitas. 2010; 67: 343–347. DOI:10.1016/j.maturitas.2010.07.009

[107] Aptekmann NP, Cesar TB. Long-term orange juice consumption is associated with low LDL-cholesterol and apolipoprotein B in normal and moderately hypercholesterolemic subjects. Lipids in health and disease. 2013; 12: 1. DOI:10.1186/1476-511X-12-119

[108] Choi JS, Yokozawa T, Oura H. Antihyperlipidemic effect of flavonoids from *Prunus davidiana*. Journal of Natural Products. 1991; 54: 218–224. DOI:10.1021/np50073a022

[109] Cesar TB, Aptekmann NP, Araujo MP, Vinagre CC, Maranhão R C. Orange juice decreases low-density lipoprotein cholesterol in hypercholesterolemic subjects and improves lipid transfer to high-density lipoprotein in normal and hypercholesterolemic subjects. Nutrition Research. 2010; 30: 689–694. DOI:10.1016/j.nutres.2010.09.006

[110] Kurowska EM, Spence JD, Jordan J, Wetmore S, Freeman DJ, Piché LA, Serratore P. HDL-cholesterol-raising effect of orange juice in subjects with hypercholesterolemia. The American Journal of Clinical Nutrition. 2000; 72: 1095–1100.

[111] Mallick N, Khan RA. Antihyperlipidemic effects of *Citrus sinensis, Citrus paradisi,* and their combinations. Journal of Pharmacy & Bioallied Sciences. 2016; 8: 112.

[112] Napoleone E, Cutrone A, Zurlo F, Di Castelnuovo A, D'Imperio M, Giordano L., De Curtis A, Iacoviello L, Rotilio D, Cerletti C, de Gaetano G, Donati MB, Lorenzet R. Both red and blond orange juice intake decreases the procoagulant activity of whole blood in healthy volunteers. Thrombosis Research. 2013; 132: 288–292. DOI:10.1016/j.thromres.2013.06.022

[113] Morand C, Dubray C, Milenkovic D, Lioger D, Martin JF, Scalbert A, Mazur A. Hesperidin contributes to the vascular protective effects of orange juice: a randomized crossover study in healthy volunteers. The American Journal of Clinical Nutrition. 2011; 93: 73–80. DOI:10.3945/ajcn.110.004`+945

[114] Rangel-Huerta OD, Aguilera CM, Martin MV, Soto MJ, Rico MC, Vallejo F, Tomas-Barberan F, Perez-de-la-Cruz JA, Gil A, Mesa MD. Normal or high polyphenol concentration in orange juice affects antioxidant activity, blood pressure, and body weight in obese or overweight adults. The Journal of Nutrition 2015; 145: 1808–1816. DOI:10.3945/jn.115.213660

[115] Schär MY, Curtis PJ, Hazim S, Ostertag LM, Kay CD. Potter J Cassidy A. Orange juice–derived flavanone and phenolic metabolites do not acutely affect cardiovascular risk biomarkers: a randomized, placebo-controlled, crossover trial in men at moderate risk of cardiovascular disease. The American Journal of Clinical Nutrition. 2015; 101: 931–938. DOI:10.3945/ajcn.114.104364

[116] Giordano L, Coletta W, Tamburrelli C, D'Imperio M, Crescente M, Silvestri C, Rapisarda P, Reforgiato Recupero G, De Curtis A, Iacoviello L, de Gaetano G, Rotilio D, Cerletti C, Donati MB. Four-week ingestion of blood orange juice results in measurable anthocyanin urinary levels but does not affect cellular markers related to cardiovascular risk: a randomized cross-over study in healthy volunteers. European Journal of Nutrition. 2012; 51: 541–548. DOI:10.1007/s00394–011-0237–9

[117] Silveira JQ, Dourado GK, Cesar TB. Red-fleshed sweet orange juice improves the risk factors for metabolic syndrome. International Journal of Food Sciences and Nutrition. 2015; 66: 830–836. DOI:10.3109/09637486.2015.1093610

[118] Mohanty P, Hamouda W, Garg R, Aljada A, Ghanim H, Dandona P. Glucose challenge stimulates reactive oxygen species (ROS) generation by leucocytes. The Journal of Clinical Endocrinology & Metabolism. 2000; 85: 2970–2973.

[119] Ghanim H, Sia CL, Upadhyay M, Korzeniewski K, Viswanathan P, Abuaysheh S, Mohanty P, Dandona P. Orange juice neutralizes the proinflammatory effect of a high-fat, high-carbohydrate meal and prevents endotoxin increase and Toll-like receptor expression. The American Journal of Clinical Nutrition. 2010; 91: 940–949. DOI:10.3945/ajcn.2009.28584

[120] Khan RA, Mallick N, Feroz Z. Anti-inflammatory effects of *Citrus sinensis* L., *Citrus paradisi* L. and their combinations. Pakistan Journal of Pharmaceutical Sciences. 2016; 29: 843–852.

[121] Javed S, Javaid A, Nawaz S, Saeed MK, Mahmood Z, Siddiqui SZ, Ahmad R. Phytochemistry, GC-MS analysis, antioxidant and antimicrobial potential of essential oil from five Citrus species. Journal of Agricultural Science. 2014; 6: 201–208. DOI:10.5539/jas.v6n3p201

[122] Dixon, R. A., Dey, P. M. and Lamb, C. J. Phytoalexins: Enzymology and Molecular Biology, in Advances in Enzymology and Related Areas of Molecular Biology. 1983; 55: (ed A. Meister), John Wiley & Sons, Inc., Hoboken, NJ, USA. DOI:10.1002/9780470123010

[123] Carvalho M, Matos M, Carnide V. Fingerprinting of *Vaccinium corymbosum* cultivars using DNA of fruits. Horticultural Science. 2014; 41: 175–184.

[124] Luby JJ, Ballington JR, Draper AD, Pliszka K, Austin ME. Blueberries and cranberries (*Vaccinium*).Genetic Resources of Temperate Fruit and Nut Crops. 1991; 290: 393–458. DOI:10.17660/ActaHortic.1991.290.9

[125] Galleta, GJ, Ballington JR. Blueberries, cranberries and lingonberries In: Janick J, Moore JN, Editors. Fruit breeding. Vol 2. New York: Wiley; 1996. p. 1–107.

[126] Levi A, Rowland LJ. Identifying blueberry cultivars and evaluating their genetic relationships using randomly amplified polymorphic DNA (RAPD) and simple sequence repeat-(SSR-) anchored primers. Journal of the American Society for Horticultural Science. 1997; 122: 74–78.

[127] Figueira ME, Oliveira M, Direito R, Rocha J, Alves P, Serra AT, Duarte C, Bronze R, Fernandes A, Brites D, Freitas M, Fernandes E, Freitas M. Protective effects of a blueberry extract in acute inflammation and collagen-induced arthritis in the rat. Biomedicine & Pharmacotherapy. 2016; 83: 1191–1202. DOI:10.1016/j.biopha.2016.08.040

[128] Skupien K, Oszmiański J. Comparison of six cultivars of strawberries (Fragaria x ananassa Duch.) grown in northwest Poland. European Food Research and Technology. 2004; 219: 66–70. DOI:10.1007/s00217–004-0918–1

[129] Fernández-Panchon MS, Villano D, Troncoso AM, Garcia-Parrilla MC. Antioxidant activity of phenolic compounds: from in vitro results to in vivo evidence. Critical Reviews in Food Science and Nutrition. 2008; 48: 649–671. DOI:10.1080/10408390701761845

[130] Chiabrando V, Giacalone G. Anthocyanins, phenolics and antioxidant capacity after fresh storage of blueberry treated with edible coatings. International Journal of Food Sciences and Nutrition. 2015; 66: 248–253. DOI:10.3109/09637486.2014.986075

[131] Braga PC, Antonacci R, Wang YY, Lattuada N, Dal Sasso M, Marabini L, Fibiani M, Lo Scalzo R. Comparative antioxidant activity of cultivated and wild *Vaccinium* species investigated by EPR, human neutrophil burst and COMET assay. European Review for Medical and Pharmacological Sciences. 2013; 17: 1987–1999.

[132] Contreras RA, Köhler H, Pizarro M, Zúiga GE. In vitro cultivars of *Vaccinium corymbosum* L. (Ericaceae) are a source of antioxidant phenolics. Antioxidants. 2015; 4:281–292. DOI:10.3390/antiox4020281

[133] Yi W, Fischer J, Krewer G, Akoh CC. Phenolic compounds from blueberries can inhibit colon cancer cell proliferation and induce apoptosis. Journal of agricultural and food chemistry 2005; 53: 7320–7329. DOI:10.1021/jf051333o

[134] Pervin M, Hasnat MA, Lim BO. Antibacterial and antioxidant activities of *Vaccinium corymbosum* L. leaf extract. Asian Pacific Journal of Tropical Disease. 2013; 3: 444–453. DOI:10.1016/S2222-1808(13)60099-7

[135] Aktan AH, Ozcelik A, Cure E, Cure MC, Yuce S. Profound hypoglycemia-ύnduced by *Vaccinium corymbosum* juice and laurocerasus fruit. Indian Journal of Pharmacology. 2014; 46: 446–447. DOI:10.4103/0253-7613.135963

[136] Cheplick S, Sarkar D, Bhowmik P, Shetty K. Phenolic bioactives from developmental stages of highbush blueberry (*Vaccinium corymbosum*) for hyperglycemia management using in vitro models. Canadian Journal of Plant Science. 2015; 95: 653–662. DOI:10.4141/CJPS-2014-352

[137] Shen X, Sun X, Xie Q, Liu H, Zhao Y, Pan Y, Cheng-An H, Wu VC. Antimicrobial effect of blueberry (*Vaccinium corymbosum* L.) extracts against the growth of *Listeria monocytogenes* and *Salmonella enteritidis*. Food Control. 2014; 35: 159–165. DOI:10.1016/j.foodcont.2013.06.040

[138] Silva S, Costa EM. Pereira MF, Costa MR, Pintado ME. Evaluation of the antimicrobial activity of aqueous extracts from dry *Vaccinium corymbosum* extracts upon food microorganism. Food Control. 2013; 34: 645–650. DOI:10.1016/j.foodcont.2013.06.012

[139] Silva S, Costa EM, Costa MR, Pereira MF, Pereira JO, Soares, JC, Pintado MM. Aqueous extracts of *Vaccinium corymbosum* as inhibitors of *Staphylococcus aureus*. Food Control. 2015; 51: 314–320. DOI:10.1016/j.foodcont.2014.11.040

[140] Zhong Y, Wang Y, Guo J, Chu H, Gao Y, Pang L. Blueberry improves the therapeutic effect of etanercept on patients with juvenile idiopathic arthritis: Phase III Study. The Tohoku Journal of Experimental Medicine. 2015; 237:183–191. DOI:10.1620/tjem.237.183

[141] Samad NB, Debnath T, Ye M, Hasnat MA, Lim BO. In vitro antioxidant and anti–inflammatory activities of Korean blueberry (*Vaccinium corymbosum* L.) extracts. Asian Pacific Journal of Tropical Biomedicine. 2014; 4: 807–815. DOI:10.12980/APJTB.4.2014C1008

[142] Massarotto G, Barcellos T, Garcia CSC, Brandalize APC, Moura S, Schwambach J, Henriquez JAP, Roesch-Ely M. Chemical characterization and cytotoxic activity of blueberry extracts (cv. Misty) cultivated in Brazil. Journal of Food Science. 2016; 81. DOI:10.1111/1750-3841.13385

[143] Tsuda H, Kunitake H, Kawasaki-Takaki R, Nishiyama K, Yamasaki M, Komatsu H, Yukizaki C. Antioxidant activities and anti-cancer cell proliferation properties of natsuhaze (*Vaccinium oldhamii* Miq.), shashanbo (*V. bracteatum* Thunb.) and blueberry cultivars. Plants. 2013; 2: 57–71. DOI:10.3390/plants2010057

[144] Coban J, Evran B, Özkan F, Çevik A, Doğru-Abbasoğlu S, Uysal M. Effect of blueberry feeding on lipids and oxidative stress in the serum, liver and aorta of guinea pigs fed on a high-cholesterol diet. Bioscience, Biotechnology, and Biochemistry. 2013; 77: 389–391. DOI:10.1271/bbb.120722

[145] Cutler BR, Petersen C, AnandhBabu PV. Mechanistic insights into the vascular effects of blueberries: Evidence from recent studies. Molecular Nutrition and Food Research. 2016. DOI:10.1002/mnfr.201600271

[146] Oliveira GAF, Pádua JG, Costa JL, Jesus OND, Carvalho FMD, Oliveira EJD. Cross-species amplification of microsatellite loci developed for *Passiflora edulis* Sims. in related Passiflora Species. Brazilian Archives of Biology and Technology. 2013; 5: 785–792.

[147] Vanderplank J, Passion flowers. 2nd ed. MIT Press, Cambridge, Mass; 1996.

[148] Vitta FA, Bernacci LC A new species and two overlooked species of Passiflora (*Passifloraceae*) from Brazil. Brittonia. 2004; 56: 89–95.

[149] Bernacci LC, Soares-Scott MD, Junqueira NTV, Passos IRDS, Meletti LMM. *Passiflora edulis* Sims: the correct taxonomic way to cite the yellow passion fruit (and of others colors). Revista Brasileira de Fruticultura. 2008; 2: 566–576.

[150] Ocampo J, Coppens D'G, Jarvi A. Distribution of the genus *Passiflora* L. diversity in Colombia and its potential as an indicator for biodiversity management in the coffee growing zone. Diversity. 2010; 11: 1158–1180.

[151] Ortiz DC, Bohórquez A, Duque MC, Tohme J, Cuéllar D, Vásquez TM. Evaluating purple passion fruit (*Passiflora edulis* Sims f. edulis) genetic variability in individuals from commercial plantations in Colombia. Genetic Resources and Crop Evolution. 2012; 6: 1089–1099.

[152] Talcott ST, Percival SS, Pittet-Moore J, Celoria C. Phytochemical composition and antioxidant stability of fortified yellow passion fruit (*Passiflora edulis*). Journal of Agricultural and Food Chemistry. 2003; 4: 935–941.

[153] Ramaiya S, Bujang JS, Zakaria MH, King WS, Sahrir S, Arif M. Sugars, ascorbic acid, total phenolic content and total antioxidant activity in passion fruit (Passiflora) cultivars. Journal of the Science of Food and Agriculture. 2012; 5: 1198–1205.

[154] Da Silva LMR, De Figueiredo EAT, Ricardo NMPS. Vieira IGP, De Figueiredo RW, Brasil IM, Gomes CL. Quantification of bioactive compounds in pulps and by-products of tropical fruits from Brazil. Food Chemistry. 2014; 143: 398–404.

[155] Zucolotto SM, Fagunde C, Reginatto FH, Ramos FA, Castellanos L, Duque C, Schenkel EP. Analysis of C-glycosyl Flavonoids from South American Passiflora Species by HPLC-DAD and HPLC-MS. Phytochemical Analysis. 2012, 3; 232–239.

[156] Maruki-Uchida H, Kuritak, Sugiyam, K, Sai M, Maeda, K., yo Ito, T. (2013). The protective effects of piceatannol from passion fruit (*Passiflora edulis*) seeds2 in UVB-irradiated keratinocytes. Biological and Pharmaceutical Bulletin, 36(5), 845–849.

[157] Marcoris MS, De Marchi R, Janzantti N, Monteiro M. The influence of ripening stage and cultivation system on the total antioxidant activity and total phenolic compounds of yellow passion fruit pulp. Journal of the Science of Food and Agriculture. 2012; 9: 1886–1891.

[158] Martínez R, Torres P, Meneses MA, Figueroa JG, Pérez-Álvarez JA, Viuda-Martos M. Chemical, technological and *in vitro* antioxidant properties of mango, guava, pineapple and passion fruit dietary fibre concentrate. Food Chemistry. 2012; 3: 1520–1526.

[159] Sato CA, Andrade AS, Brito VM, Sampaio MAUM, Maffei H de AF, Oliva LVM. Effects of compounds from *Passiflora edulis* Sims f. flavicarpa juice on blood coagulation and on proteolytic enzymes. Protein and Peptide Letters. 2012; 5: 501–508.

[160] Klein N, Gazola AC, Lima TCM, Schenkel E, Nieber K, Butterweck V. Assessment of sedative effects of *Passiflora edulis* f. flavicarpa and *Passiflora alata* extracts in mice, measured by telemetry. Phytotherapy Research. 2014; 5: 706–713.

[161] Li H, Zhou P, Yang Q, Shen Y, Deng J, Li L, Zhao D. Comparative studies on anxiolytic activities and flavonoid compositions of *Passiflora edulis* 'edulis' and *Passiflora edulis* 'flavicarpa'. Journal of Ethnopharmacology. 2011; 3: 1085–1090.

[162] Figueiredo DA, Pordeus L, Paulo LL, Braga RM, Fonsêca DV, Sousa BS, Costa MJC, Goncalves MCR, Oliveir KH. Effects of bark flour of *Passiflora edulis* on food intake, body weight and behavioral response of rats. Revista Brasileira de Farmacognosia. 2016; 5: 595–600.

[163] Aguillón OJ, Maldonado ME, Loango CN, Arango VSS, Landázuri P. (2013). Antioxidant and antiproliferative activity of ethanolic and aqueous extracts from leaves and fruits juice *Passiflora edulis*. Perspectivas en Nutrición Humana. 2013; 1: 13–25.

[164] Silva RO, Damasceno SR, Brito TV, Dias JM, Fontenele AM, Braúna IS, Junior JSC, Miaciel JS, Paulac RCM, Ribeiroa RA, Souza HLP, Freitas ALP, Medeiros Jand-Venes, Silva DC, Barbosa ALR, Polysaccharide fraction isolated from *Passiflora edulis* inhibits the inflammatory response and the oxidative stress in mice. Journal of Pharmacy and Pharmacology. 2015; 7: 1017–1027.

[165] Silva DC, Freitas AL, Pessoa CD, Paula RC, Mesquita JX, Leal L K, Brito GAC, Goncalves OD, Viana GS. Pectin from *Passiflora edulis* shows anti-inflammatory action as well as hypoglycemic and hypotriglyceridemic properties in diabetic rats. Journal of Medicinal Food. 2011; 10: 1118–1126.

[166] Salgado JM, Bombarde TAD, Mansi DN, Piedade SMDS, Meletti LMM. Effects of different concentrations of passion fruit peel (*Passiflora edulis*) on the glicemic control in diabetic rat. Food Science and Technology (Campinas), 2010; 3: 784–789.

[167] Barbalho SM, Damasceno DC, Spada APM, Lima IEDR, Araújo AC, Guiguer EL, Martuchi KA,Oshiiwa M, Mendes CG. Effects of Passiflora edulis on the metabolic profile of diabetic wistar rat offspring. Journal of Medicinal Food. 2011; 12; 1490–1495.

[168] Kirchhoff R, Beckers C, Kirchhoff GM, Trinczek-Gärtner H, Petrowicz O, Reimann HJ. Increase in choleresis by means of artichoke extract. Phytomedicine. 1994; 1: 107–115. DOI:10.1016/S0944-7113(11)80027-9

[169] Gebhardt R. Antioxidative and protective properties of extracts from leaves of the artichoke (Cynara scolymus L.) against hydroperoxide-induced oxidative stress in cultured rat hepatocytes. Toxicology and Applied Pharmacology. 1997; 144: 279–286. DOI:10.1006/taap.1997.8130

[170] Lattanzio V, Kroon PA, Linsalata V, Cardinali A. Globe artichoke: a functional food and source of nutraceutical ingredients. Journal of Functional Foods. 2009; 1: 131–144. DOI:10.1016/j.jff.2009.01.002

[171] Gebhardt R. Inhibition of cholesterol biosynthesis in HepG2 cells by artichoke extracts is reinforced by glucosidase pretreatment. Phytotherapy Research. 2002; 16: 368–372. DOI:10.1002/ptr.960

[172] Movahedian A, Sadeghi H, Ghannadi A, Gharavi M, Azarpajooh S. Hypolipidemic activity of *Allium porrum* L. in cholesterol-fed rabbits. Journal of Medicinal Food. 2006; 9: 98–101. DOI:10.1089/jmf.2006.9.98

[173] Roghani M, Aghaei M. The effect *Allium ampeloprasum* feeding on serum level of glucose, triglyceride, and total cholesterol of diabetic rats. Journal of Semnan University of Medical Science. 2007; 8: 73–78.

[174] Del Rio D, Rodriguez-Mateos A, Spencer JP, Tognolini M, Borges G, Crozier A. Dietary (poly) phenolics in human health: structures, bioavailability, and evidence of protective effects against chronic diseases. Antioxidants & Redox Signaling. 2013; 18: 1818–1892. DOI:10.1089/ars.2012.4581

[175] Pak N. La fibra dietética en la alimentación humana, importancia en la salud. Anales de la Universidad de Chile. 2000. DOI:10.5354/0717-8883.2000.2504

[176] Bhat S, Kaushal P, Kaur M, Sharma HK. Coriander (*Coriandrum sativum* L.): Processing, nutritional and functional aspects. African Journal of Plant Science. 2014; 8: 25–33. DOI:10.5897/AJPS2013.1118

[177] Forastiere F, Pistelli R, Sestini P, Fortes C, Renzoni E, Rusconi F, Dell' Orco V, Ciccone G, Bisanti L. Consumption of fresh fruit rich in vitamin C and wheezing symptoms in children. Thorax. 2000; 55: 283–288. DOI:10.1136/thorax.55.4.283

[178] Del Bo C, Riso P, Campolo J, Møller P, Loft S, Klimis-Zacas D, Brambilla A, Rizzolo A, Porrini M. A single portion of blueberry (*Vaccinium corymbosum* L) improves protection against DNA damage but not vascular function in healthy male volunteers. Nutrition Research. 2013; 33: 220–227. DOI:10.1016/j.nutres.2012.12.009

[179] Del Bo C, Porrini M, Fracassetti D, Campolo J, Klimis-Zacas D, Riso P. A single serving of blueberry (*V. corymbosum*) modulates peripheral arterial dysfunction induced by acute cigarette smoking in young volunteers: a randomized-controlled trial. Food and Function. 2014; 5: 3107–3116. DOI:10.1039/c4fo00570h

9

Selenium Biofortification and the Problem of its Safety

Renato de Mello Prado,

Flávio José Rodrigues Cruz and

Raphael Leone da Cruz Ferreira

Abstract

Selenium is an essential mineral element to humans and animals because it is an important component of selenoproteins that are important for functioning of the metabolism. Because of poor soil conditions in various regions of the world, the enrichment of edible plants with selenium via the biofortification strategy has been implemented. However, selenium in the context of plant mineral nutrition appears twofold due to its biofortifying character at low concentrations and toxicity at high concentrations. In this sense, understanding of the functional mechanisms in which selenium is involved is important, ranging from its absorption and assimilation in organic compounds to its beneficial or harmful effects, considering its role in food security and human health. Therefore, this chapter addresses the key aspects related to selenium in the soil-plant-man environment and the narrow limit between biofortification and toxicity, as well as the main scientific findings on this mineral element in the biochemical, physiology and plant nutrition contexts.

Keywords: selenate, selenite, sulfur, toxicity, metabolism

1. Introduction

The discovery of selenium (Se) by Swedish chemist Jons Jacob Berzelius about 200 years ago initiated the start of several studies with this chemical element in living organisms [1,2]. However, it was [3] who identified Se as an essential nutrient for the growth and development of bacteria, mammals and birds. It is currently considered as an essential micronutrient for humans and animals, as well as for certain lower organisms such as algae, fungi and bacteria [4].

Se is a trace element widespread in various geochemical environments [5], and its average content in the earth's crust is estimated at 0.05 mg kg^{-1}. However, concentrations exceeding 0.5 mg kg^{-1} can be found. Selenium levels are slightly higher in mafic rocks; however, they rarely exceed 0.1 mg kg^{-1}. In sedimentary rocks, Se is associated with the clay fraction, and thus, its abundance is higher in clayey sediments compared to sandstone and limestone. Selenium exists in four valence states, of which the state −2 predominates in organic compounds. The forms selenite (Se^{4+}) and selenate (Se^{6+}) do not form stable compounds in geochemical environments and are preferably adsorbed by minerals, particularly clay minerals, and oxides and hydroxides of Fe and Mn [6]. In the soil-plant system, the plant has the important role of recycling and releasing Se forms from the soil to the food chain. In this sense, the Se concentration in agricultural products and forage depends on the Se content in the soil and its availability [7–9]. However, Se availability in the soil is restricted, and its content is relatively low due to weathering and acidity [6, 7, 9, 10].

Because of the chemical similarity between Se and sulfur (S), the metabolism of Se in higher plants is closely related to S. Therefore, its absorption by the roots of plants is governed by the solubility of adsorbed forms and the transformation of organic Se forms (**Figure 1**). Due to the ability of plants to absorb Se, they can be classified according to Se accumulation in their tissues, including the classes of hyperaccumulators (1000–15,000 mg Se kg^{-1} dry weight), accumulators (>100 mg Se kg^{-1} dry weight) and non-accumulators (<100 mg Se kg^{-1} dry weight) [11]. Many species of the genera *Astragulus, Xylorrhyza* and *Stanleya* are typically Se accumulators because they are capable of growing in selenium-rich soil without showing any negative effect of Se toxicity, accumulating Se contents in the range of 20–40 mg Se g^{-1} DM while presenting no toxic effect [12, 13].

Because the range of Se concentrations in the soil varies, for example, from 8000 mg kg^{-1} in soils of the Russian city of Tuva to 0.005 mg kg^{-1} in soils of Finland and China [15], there is close relationship between Se deficiency in soils and the appearance of disease symptoms related to low Se intake in humans and mammals. Various efforts have thus been made to enrich agricultural feed crops with Se via fertilization or genetic breeding in a strategy called biofortification [16, 17]. According to this strategy, the supply of Se to the plant promotes a positive response in human health, preventing the onset of diseases related to low intake of this nutrient. In this sense, several studies report the positive effect of using Se in biofortification of lettuce [18], tomato [19], broccoli [20, 21], cucumber [22] and carrot crops [23].

In this context, biofortification of edible plants appears to be important from a nutritional point of view because selenium participates in constitution of selenoproteins, for example, the active site of antioxidant enzymes from the group of glutathione peroxidases (GSH-Px), which are reactive oxygen species detoxifying enzymes [24, 25]. There are several isoenzymes of the glutathione peroxidase family (with 4 g atom per mole of protein): cellular glutathione peroxidase (GSH-Px 1); glutathione peroxidase of the intestinal epithelium (GSH-Px 2); plasma glutathione peroxidase (GSH-Px 3); hydroperoxide phospholipid glutathione peroxidase (GSH-Px 4); and glutathione peroxidase present in the sperm (GSH-Px 5). Of these, GSH-Px 1 is the most abundant in mammalian selenoproteins [26]. Additionally, Se is of great importance to human health because of its presence in chemical constitution of the iodothyronine deiodinase enzyme, which is involved in metabolism of the thyroid hormone [27].

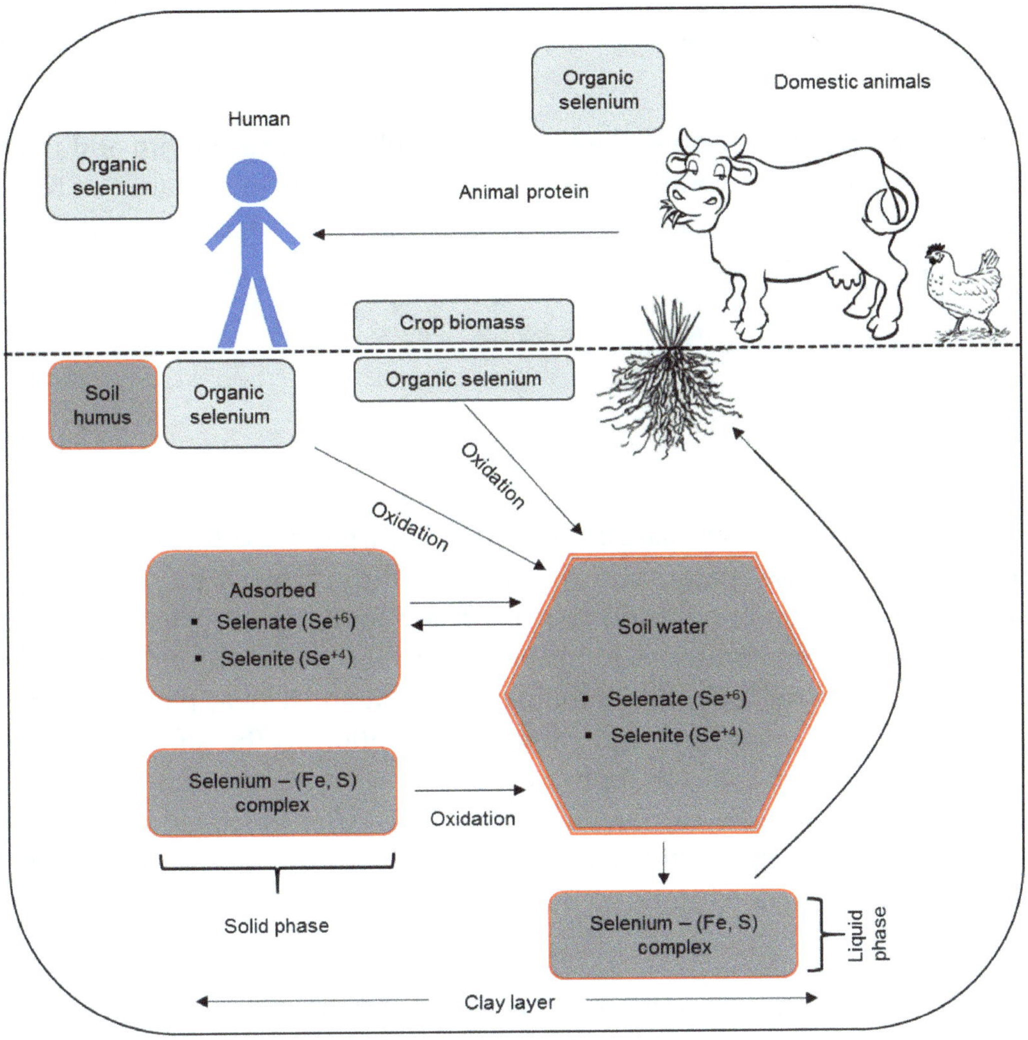

Figure 1. Selenium in the soil-plant-atmosphere system. Plants absorb selenate or selenite from the soil solution. Se concentration in the soil solution depends on the solubility of Se forms present and the biological transformation of organic Se forms. Adapted with permission from [14].

Despite the biofortifying character of Se in plants, it also has a toxic effect when provided to plants in high concentrations. This toxicity results from the substitution of S for Se in cysteine and methionine amino acids, reducing the number of disulfide bonds and this altering the structure and functionality of proteins, causing a negative impact on plant growth. Furthermore, another toxic effect of Se during its assimilation into organic compounds results in depletion of the nonenzymatic antioxidant, glutathione. Therefore, there is an imbalance between detoxification and formation of free radical species, which results in a significant oxidative burst and consequent reduction in plant growth [28].

It should be emphasized that biofortification/toxicity of Se is well reported in literature [18, 21–23, 29]; however, the transition between biofortification and toxicity by Se is narrow and depends on the concentration and source, as well as the plant genotype. Despite this, literature does not report Se levels in plants based on determination of critical Se concentrations with regard to leaf content and Se concentrations in the culture medium. This shortcoming complicates the adoption of one Se concentration or a narrow range of concentrations that promote plant growth at the expense of biochemical, physiological and nutritional disorders promoted by toxic Se levels.

In this chapter, we sought to address the key aspects inherent to selenium and its functional relationships in the plant environment, as well as the intrinsic importance of food security regarding this nutrient, considering the main current scientific findings in the biochemical, physiological and nutritional fields of plants.

2. Selenium metabolism

In previous studies, it was shown that Se was absorbed by passive diffusion [30, 31]; however, it was recently shown that selenate is absorbed by the sulfur carrier, while selenite is absorbed by the phosphate carrier, and both processes are dependent on energy expenditures [32, 33]. Among the inorganic (selenate and selenite) and organic forms of selenium [selenomethionine (SeMet) and selenocysteine], selenomethionine in canola and wheat plants is the form that present the highest absorption rate and rapid translocation to the shoots [34]. Due to the chemical similarity between S and Se, both present the same route of absorption and assimilation, competing for the same carrier membrane [16]. After absorbed through the sulfur carrier (Sultr) and translocated to the shoot, selenate can be assimilated in chloroplasts and reduced to selenite in a reaction catalyzed by the enzyme ATP sulfurylase (APS) and then into selenide [16].

Non-accumulating plants can concentrate Se because APS has limited catalytic activity. On the other hand, accumulating plants overexpress APS resulting in accumulation and tolerance to high Se concentrations. Selenide may be incorporated into the S-amino acid similar to selenocysteine which can be converted to selenomethionine (SeMet) in three enzymatic steps. Incorrect insertion of the amino acid selenomethionine/selenocysteine in proteins can cause the formation of protein aggregates that promote disruption of important cellular functions [35, 28]. The incorporation of Se in proteins may occur when Se is converted to less toxic forms, because some plant species present nonprotein organic compounds containing Se such as methylselenocysteine (MeSeCys), γ-glutamyl-MeSeCys and/or selenocysteine [36]. Se can be volatilized by plants through the dimetilselenide or dimetildiselenide compounds, synthesized from selenomethionine and methylselenocysteine, respectively [16]. Se accumulating plants have the synthesis of methyl-SeCys catalyzed by the enzyme SeCys methyltransferase (SMT), accumulating methyl-SeCys, a nonprotein amino acid. Furthermore, methyl-SeCys may be converted into dimetildiselenide, a volatile compound. Expression of the enzyme SMT in non Se accumulating plants increases accumulation of Se in the form of methyl-SeCys, and its activity is related to tolerance to Se accumulation [16].

In the biochemical field, the metabolic pathways of plants are interconnected by means of some compounds. In the case of Se and nitrogen (N), the metabolism of these inorganic elements is interconnected by means of the O-acetylserine compound. Therefore, alterations to the S metabolism induced by Se interfere with that of N with respect to the metabolism of amino acids and proteins [16], considering that the amino acids methionine, phenylalanine, tyrosine and tryptophan are precursors of glucosinolate, while phenylalanine is a precursor of phenolic compounds. Thus, variations in the synthesis of these amino acids influence the synthesis of nutraceutical compounds such as glucosinolate and phenolic compounds [16].

Several studies report the positive impact of Se on the plant metabolism, particularly due to its abiotic stress mitigating effect. In this sense, Se plays an important role in increasing the activity of antioxidant enzymes to contribute to the detoxification of reactive oxygen species, considering that this mineral element participates in the active site of these enzymes. These enzymes, called glutathione peroxidases, appear quite active in plants subjected to various abiotic stresses such as drought stress [37, 38], salinity [39] and heavy metal toxicity [40], conferring stress tolerance to plants. This effect of Se is evident, because when supplied at concentrations of 10 and 50 μM of selenate beneficial effects were observed in wheat plants grown under appropriate and reduced N availability. In this study, Se promoted a better response of the parameters fluorescence and gas exchange, with a positive impact on the growth of wheat plants [41].

On the other hand, Se in toxic concentrations may compromise energy synthesis by redox reactions (*i.e.*, photosynthesis and respiration) due to substitution of S for Se in the cysteine amino acid residue. Cysteine constitutes an important site for binding and stabilization of Fe-S metal centers, heme groups and ions participating in the flow of electrons in the mitochondria and chloroplasts. In this regard, it is speculated that substitution of the cysteine amino acid residue by selenocysteine in proteins rich in Fe-S metal centers disturbs the flow of electrons in the mitochondria and chloroplasts [28]. This fact implies the reduction of energy synthesis, and consequently reduced plant growth.

Moreover, the reduction of selenite to selenate via the S metabolic pathway demands a great glutathione input, a biochemical component involved in important redox reactions in cellular homeostasis [42]. This fact explains the decrease in root growth of plants induced by selenate when it is assimilated into organic compounds, since there is a depletion in the cellular glutathione content [28]. This was observed in brassica plants of *Stanleya albescens* sensitive to Se toxicity, which when exposed to toxic concentrations of this element had their growth compromised due to oxidative stress caused by the increased leaf accumulation of hydrogen peroxide (H_2O_2) and superoxide anion (O_2^-) in contrast to the more tolerant genotype *Stanleya pinnata*, which has a high glutathione content [43].

The metabolisms of nitrogen (N) and S are interconnected by means of the compound O-acetylserine [16]. Thus, the supply of Se to plants may interfere with the N metabolism. It was indicated that the supply of Se to barley plants reduced the nitrate assimilatory process because of reduced activity of the nitrate and nitrite reductase enzymes in leaf and root tissues. However, intensity of the reduction was greater when Se was supplied in the form of selenite [44].

It was recently demonstrated that toxic concentrations of Se promote reduction of plant growth in *Arabidopsis thaliana* due to incomplete mobilization of starch reserves overnight, reduced expression of genes encoding the synthesis of endotransglucosilase/endohidrolase enzymes and expansins, as well as nutritional disorders [45]. Therefore, despite the benefits of low Se concentrations, it has a large negative impact on the plant metabolism when provided in high concentrations on plant growth by affecting metabolic processes of energy acquisition, cell expansion, and absorption and assimilation of essential nutrients.

3. Selenium biofortification

Selenium is an essential inorganic element for humans and animals, and one of the organic forms of Se, methylselenocysteine, appears to be an effective food source of Se [16]. Se is incorporated into a range of selenoproteins involved in several important metabolic activities such as synthesis of thyroid hormones and antioxidative activity [46, 47]. Selenium is an important inorganic component for the antioxidant metabolism of enzymes, making up part of the active site of enzymes from the group of glutathione peroxidases (GSH-Px) which plays an important role in detoxification of free radicals. The GSH-Px catalyze the reduction of hydroperoxide radicals (H_2O_2 for instance) by the oxidation of glutathione (GSH), a nonenzymatic component of the antioxidative metabolism [48].

It is noted that the accumulation of Se in foods is closely related to the content of this nutrient in the soil. However, consumption of foods poor in Se or low ingestion of foods containing Se is associated with the emergence of numerous diseases such as cancer, type II diabetes, heart disease, pulmonary dysfunction, seizures in children, impaired development and cerebral functions, as well as pregnancy and conception [49–51].

Currently, Se deficiency affects about 1 billion people worldwide due to soils lacking this mineral nutrient in some countries [52]. This edaphic characteristic was registered in countries such as Sweden, Finland, USA and China [16, 53, 54]. Because there is a close relationship between plant mineral nutrition and human health, food fortification with Se, via a strategy known as biofortification, is an effective way to add Se to human food and prevent the emergence of diseases related to deficient Se intake. This strategy proved to be effective due to the fact that Se presents chemical similarity to S, and both have the same carrier membranes and biochemical pathway of assimilation [28, 33].

Because vegetables are considered an important source of bioactive compounds that contain polyunsaturated fatty acids, phytochemicals such as flavonoids and glucosinolates, many of which can inhibit cell proliferation, induce apoptosis and act synergistically when combined in foods [55], some vegetable groups are more suitable for biofortification because they are natural accumulators of Se such as brassicas [16]. This group of plants has significant levels of glucosinolates, a substance of great nutraceutical interest. In this context, recent studies confirm the biofortifying effect of Se in edible plants in the group of brassicas. For example, in a major study [21] showed the positive impact of Se biofortification in broccoli. This study found that the broccoli extract showed high levels of phenolic compounds, significant antioxidant effect and anti-proliferative activity of tumor cells on the effect of Se in the synthesis of glucosinolates and phenolic compounds, and anticancer activity, see [16, 56].

Although Se biofortification is a strategy undertaken in edible and domesticated plants, it can occur naturally, as in the Brazilian Amazon, where important food sources of Se are the shrub species of the family Lecythidaceae, *Bertholletia excelsa* Humb. and Bonpl and *Lecythis usitata* Miers, which produce nuts presenting high Se concentrations [57, 58]. For these species, biofortification is a natural phenomenon, because the average levels of Se in their nuts range from 0.03 to 512 $\mu g\ g^{-1}$ [59, 60]. Thus, consumption of only one nut per day appears to be sufficient to meet the daily Se requirement of an adult, because according to the Scientific Committee on Foods of the European Commission, the human daily intake requirement of Se is around 55 μg dia^{-1} [27]. However, the production of nuts is seasonal and the season of low production affects the nutrition of people near regions of natural occurrence of these two species [61, 62].

Among domesticated plants of food interest, the biofortification strategy can be carried out by providing Se to the soil, nutrient solution or leaf. However, considering large-scale plantations, the crop cycle and the economic value of the Se sources can make biofortification expensive, with the need for less costly and more practical strategies. One example is pelletizing the seeds of Se accumulator species, like radish, because of its short life cycle. Another strategy is to add Se sources to fertilizers used in agriculture. This strategy was adopted by Whelan and Barrow [63] in cultivation of the grass *Trifolium subterraneum* L in a soil classified as podzol laterite. These authors used a fertilizer containing 1% Se, composed of the sources Na_2SeO_4 and $BaSeO_4$ in a 1:1 ratio, which is characterized by slow release of Se in the soil. This strategy proved to be interesting due to the fact that there is a synchrony between plant growth and Se liberation, a fact which favors Se absorption and culture biofortification based on soil-plant interaction.

4. Toxicity of selenium

From the point of view of plant nutrition, the effects of Se are twofold, since at low concentrations it can promote biofortification but at high concentrations it triggers toxicity in plants [64]. This duality of effects on plant growth is related to the narrow range between biofortification and toxicity, which in literature is not well established based on determination of critical Se levels. It is understood that the critical toxicity level in the soil or nutrient solution is that which implies 10% reduction in plant growth, since the benefit of fertilization reaches its maximum between 90 and 95% of relative growth of the culture [65]. In plants, the effect of Se toxicity is dependent on the Se concentration and source, as well as plant genotype. However, for the same Se concentration, selenite appears to be more toxic than selenate. This fact is justified by the rapid incorporation of selenite into organic compounds still in the root system [28, 66]. In general, symptoms of Se toxicity in plants are characterized by reduced growth, as observed in lettuce [67], spinach [68], cucumber [64] and pea [29]. Reduced plant growth, as a symptom of Se toxicity, is based on two biochemical mechanisms related to Se absorption and assimilation in organic compounds.

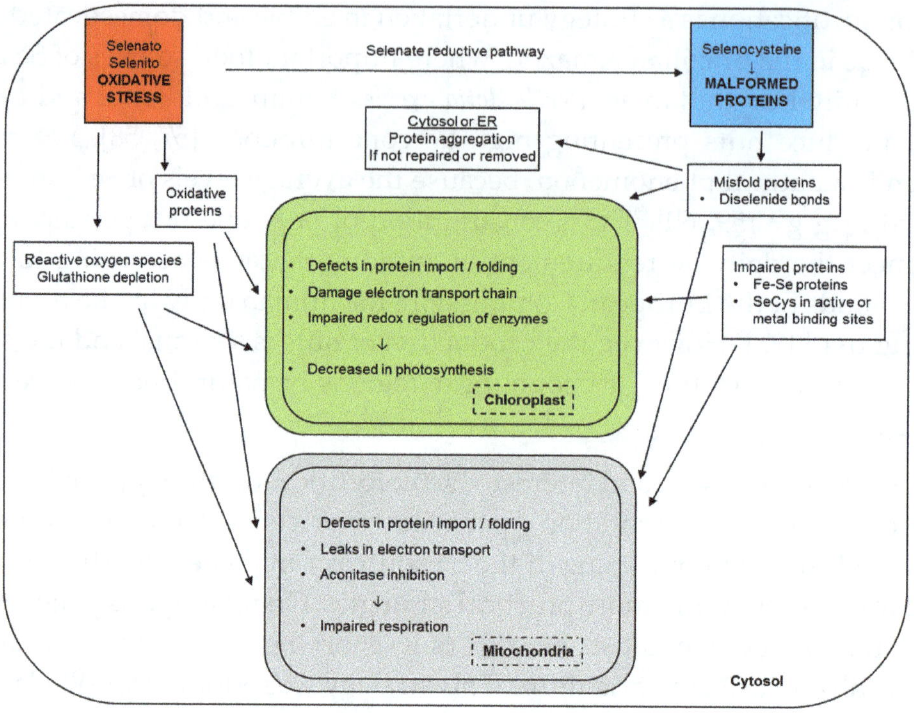

Figure 2. Model describing the two distinct mechanisms of Se toxicity in plants. Inorganic Se contributes to oxidative stress, while its reduction to selenocysteine may inadvertently substitute the amino acid cysteine and create malformed selenoproteins. The possible targets and ramifications of oxidative stress induced by Se and nonspecific selenoproteins are proposed. Adapted with permission from [28]. These symptoms were observed in the study of toxicity to Se sources and concentrations in the lettuce cultivar Vera, conducted under hydroponic conditions in a study of Ref. [71]. These symptoms were most intense in the source selenite at Se concentrations >8 μM (**Figure 3**). In this study (unpublished data), two response patterns were evident for lettuce plants with respect to Se sources and concentrations in the shoot, root and leaf area growth.

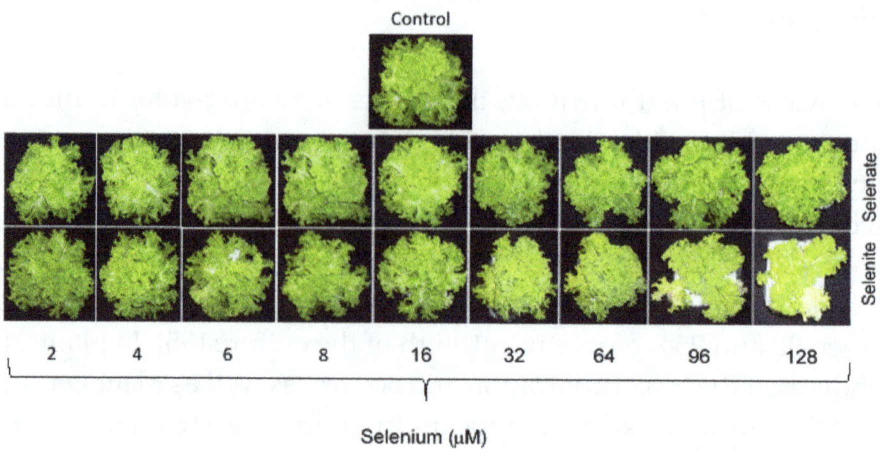

Figure 3. Visual aspect of applying increasing concentrations of selenite and sodium selenate to lettuce plants. Sources [71].

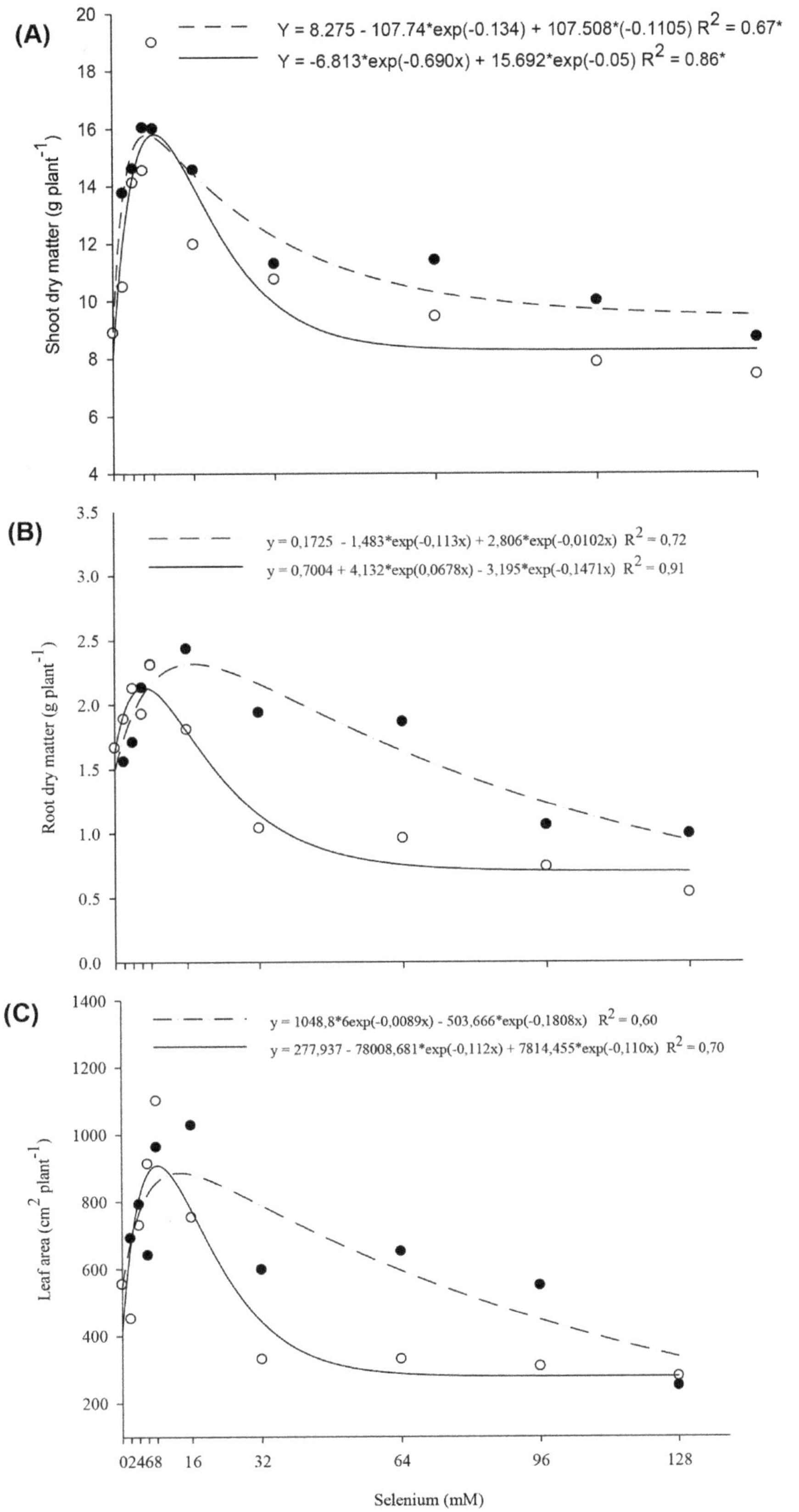

Figure 4. Leaf area (A), shoot (B) and root dry mass (C) of plants and the lettuce cultivar Vera submitted to selenium sources and concentrations in hydroponic cultivation. Selenite (–°–); Selenate (--•--). Source: [71].

The first mechanism is related to substitution of S for Se in the amino acids cysteine and methionine. This exchange of S for Se implies substitution of the amino acids cysteine and methionine for their analogs selenomethionine and selenocysteine during protein synthesis. This substitution is more deleterious in relation to cysteine, because this amino acid residue is of great importance in the structure and function of proteins and formation of disulfide bonds, enzymatic catalysis, metal bonding sites and redox state regulation. It is also suggested that iron as a metal cofactor complexed to selenocysteine can interrupt the flow of electrons in chloroplasts and mitochondria (**Figure 2**), since these cytoplasmic organelles have electron transport systems with supramolecular organization [28]. Therefore, any replacement of S by Se impedes electron flow due to changes in the protein conformational structure, and hence energy synthesis coordinated through the electron transport chain. The second mechanism is related to the participation of glutathione (GSH), a tripeptide active in cellular redox homeostasis regulation [69] and in the selenite reduction stage [28]. The participation of GSH in this step is crucial, since under Se toxicity conditions there may be a functional imbalance of GSH, since there is increased GSH demand for selenite reduction compared to hydroperoxide reduction in reactions catalyzed by glutathione peroxidase (**Figure 2**).

This imbalance in favor of selenite reduction triggers an oxidative burst that results in reduced plant growth [28]. However, plants considered accumulators or hyperaccumulators have biochemical mechanisms that prevent Se incorporation in proteins, by adding the methyl radical to the intermediate compound selenocysteine which is volatilized [16, 28]. From a symptomatology point of view, Se toxicity is characterized by reduced growth and chlorosis of plants [70].

In the lower concentrations of selenite or selenate, there was greater shoot, root and leaf area growth (**Figure 4A–C**). However, concentrations >14 and 16 μM of Se (selenite and selenate) for the shoots, and 9.7 and 30 μM (selenite and selenate) for the root resulted in growth reduction in these organs and more intense chlorosis in plants grown in the presence of selenite (**Figure 4A–C**). These responses were similar to those observed in lettuce [18] and cucumber [64] grown under sources and increasing Se concentrations.

5. Limits between selenium biofortification and toxicity

Plant growth is affected by excess Se, and the sensitivity of plants to toxicity of this mineral element is dependent on its concentration and source. However, in literature, the limits between Se biofortification and toxicity are not evident and based on critical toxicity levels which consider 10% reduction in plant growth [65, 72]. For example, Ríos et al. [18] observed that the concentration of 40 μmol L^{-1} selenate, in contrast to concentrations of selenite, was optimal for lettuce growth and biofortification. However, these authors reported a reduction of plant growth when the selenate concentration was >40 μmol L^{-1}.

It should be noted that the above results reported by the limits between Se biofortification and toxicity are not evident and based on critical toxicity levels which consider 10% reduction in plant growth [65, 72]. For example, Ríos et al. [18] contradict those presented by

Hawrylak-Nowak [67], which places the limit of lettuce biofortification at concentrations 15 and 20 μmol L^{-1} for the sources of selenite and selenate, respectively, by the fact that no negative impact was observed on growth and the content of photosynthetic pigments at these Se concentrations. To investigate the effect of Se in lettuce, Ríos et al. [73] observed that the provision of selenate was not toxic up to the concentration of 80 μmol L^{-1}. On the other hand, selenite concentrations >5 μmol L^{-1} showed symptoms of toxicity related to reduced growth. These authors attributed growth reduction in plants to the nutritional imbalance of macro- and micronutrients, as well as the oxidative stress caused by toxic levels of selenite used in the study.

In cucumbers, the application of selenate and selenite showed toxic effects with in concentrations >80 and 20 μmol L^{-1}, respectively [64]. In this study, leaf area, leaf and root fresh weight were severely reduced at selenite concentrations >20 μmol L^{-1}. These authors showed that the concentration of chlorophyllian pigments and fluorescence of chlorophyll were more negatively affected by selenite compared to selenate.

It is known that the human health and plant nutrition are closely linked to soil fertility, and diseases associated with Se deficiency are documented in areas where the content of this element in soil is low, as in China, Denmark, Finland, New Zealand and central and eastern Siberia [74, 75]. However, there is no record of human intoxication by Se due to consumption of plant species of food interest (*e.g.*, corn, rice, beans, wheat, barley, etc.), presumably due to the fact that these plants are not Se accumulators. An interesting fact was recorded on a farm in the state of South Dakota, United States, where the owners recorded low hatchability of chicken eggs or chicks born with deformities and soon died. When investigating the fact, the United States Department of Agriculture identified high levels of Se in wheat grains which were used to feed the chickens [76]. Although literature relates the intake of foods rich in Se to the lower risk of emergence of diseases such as colon, stomach, prostate and lung cancer [51], there are studies that relate Se supplementation to increased risk of death due to prostate cancer [77]. However, the population along the Tapajos River in the Brazilian state of Pará presents high levels of Se in their blood in function of the considerable consumption of nuts from *Bethollethia excelsa* and *Lecythis usitata*, which have high Se levels. In these nuts, the Se concentration is quite variable, within the range from 0.03 to 512 mg g^{-1} [59], well above the daily intake recommended by the Scientific Committee on Foods of the European Commission, which recommends a daily intake of approximately 55 μg day^{-1} [27]. Nevertheless, in this population, the consumption of nuts rich in Se appears to protect against the emergence of diseases resulting from advancing age such as cataract and motor dysfunction [78]. The recommended daily Se intake (**Table 1**) varies in function of age, gender, pregnancy and lactation. Among the sexes, there is no difference with regard to ingestion demands, except when women are pregnant or lactating, where intake should be increased by 9 and 27%, respectively [79]. However, the maximum tolerable daily Se limit at which there occurs initial toxicity symptoms (**Table 2**) characterized by hair loss, and the appearance of brittle nails is similarly dependent on sex, age, pregnancy and lactation [79].

Age	Male	Female	Pregnancy	Lactation
Birth to 6 months	45 mcg*	45 mcg*		
7–12 months	60 mcg*	60 mcg*		
1–3 years	90 mcg	90 mcg		
4–8 years	150 mcg	150 mcg		
9–13 years	280 mcg	280 mcg		
14–18 years	400 mcg	400 mcg	400 mcg	400 mcg
19* years	400 mcg	400 mcg	400 mcg	400 mcg

Source: [79].
*Adequate Intake (AI).

Table 1. Recommended dietary allowances for selenium.

Age	Male	Female	Pregnancy	Lactation
Birth to 6 months	15 mcg*	15 mcg*		
7–12 months	20 mcg*	20 mcg*		
1–3 years	20 mcg	20 mcg		
4–8 years	30 mcg	30 mcg		
9–13 years	40 mcg	40 mcg		
14–18 years	55 mcg	55 mcg	60 mcg	70 mcg
19–50 years	55 mcg	55 mcg	60 mcg	70 mcg
51+ years	55 mcg	55 mcg		

Source: [79].
*Breast milk, formula, and food should be the only sources of selenium for infants.

Table 2. Tolerable upper intake levels for selenium.

6. Conclusions and future perspectives

This chapter addressed the main issues referring to the mineral element selenium, its history of discovery, its presence in the soil-plant system and its main functions in mammals and humans, always based on recent scientific findings that can guide students, teachers and researchers in their studies. Also assessed were the mechanisms of absorption, transportation and assimilation of selenium in organic compounds, as well as its biofortifying and toxic effect in plants. Moreover, its essentiality in mammals has been addressed in order to emphasize its importance in food and human health due to its function in the human metabolism. It was therefore sought to present the importance of this nutrient in the plant-human system, and therefore provide information for future studies seeking to clarify other functional aspects of selenium in plants.

Acknowledgements

To the Plant Nutrition Study Group (GenPlant) of the Universidade Estadual Paulista (UNESP), Campus Jaboticabal—São Paulo—Brazil, led by Professor Renato de Mello Prado, to UNESP for providing the new doctor scholarship to post-doctoral fellow Flávio José Rodrigues Cruz.

Author details

Renato de Mello Prado, Flávio José Rodrigues Cruz* and Raphael Leone da Cruz Ferreira

*Address all correspondence to: fjrc@bol.com.br

Department of Soils and Fertilizers, Faculty of Agricultural and Veterinary Sciences, Plant Nutrition Study Group-GenPlant, University State of São Paulo "Júlio de Mesquita Filho", São Paulo, Brazil

References

[1] Kieliszek M, Błazejak S. 2016. Current knowledge on the importance of selenium in food for living organisms: a review. Molecules, 21:609.

[2] Reich HJ, Hondal RJ. 2016. Why nature chose selenium. ACS Chemical Biology, 11:821–841.

[3] Schwarz K, Foltz CM. 1957. Selenium as an integral part of cofactor 3 against dietary necrotic liver degradation. Journal of the American Chemical Society, 79:3292–3293.

[4] Ros GH, van Rotterdam AMD, Bussink DW, Bindraban PS. 2016. Selenium fertilization strategies for bio-fortification of food: an agro-ecosystem approach. Plant and Soil, 404:99–112.

[5] Gabos MB, Alleini LRF, Abreu CA. 2014. Background levels of selenium in some selected Brazilian tropical soils. Journal of Geochemical Exploration, 145:35–39.

[6] Kabata-Pendias A, Pendias H. Trace elements in soils and plants. Boca Raton: CRC, 2001. 315p.

[7] Sippola J. 1979. Selenium content of soils and timothy (*Plheum pretense* L.) in Filand. Annales Agriculturae Fenniae, 18:182–187.

[8] Koivistoinen P. 1980. Mineral element composition of Finnish foods: N, K, Ca, Mg, P, S, Fe, Cu, Mn, Zn, Mo, Co, Ni, Cr, F, Se, Si, Rb, Al, B, Br, Hg, As, Cd, Pb, and ash. *Acta Agriculturae Scandinavica*. Supplementum 22. Stockholm, 171p.

[9] Yläranta T. 1983. Sorption of Selenite and selenate added in the soil. Annales Agriculturae Fenniae, 22:29–39.

[10] Koljonen T. 1975. The behavior of selenium in finnish soils. Annales Agriculturae Fenniae, 14:240–247.

[11] White PJ. 2016. Selenium accumulation by plants. Annals of Botany, 117:217–235.

[12] Rosenfeld I, Beath OA. Selenium. Geobotany, biochemistry, toxicity and nutrition. New York: Academic Press, 1964.

[13] Brown TA, Shrift A. 1982. Selenium: toxicity and tolerance in higher plants. Biologial Reviews, 57:59–84.

[14] Burau RG. 1985. Environmental chemitry of selenium. California Agriculture, 39:16–18.

[15] Mora ML, Durán P, Acuña AJ, Cartes P, Demanet R, Gianfreda L. 2015. Improving selenium status in plant nutrition and quality. Journal of Soil Science and Plant Nutrition, 15:486–503.

[16] Malagoli M, Schiavon M, Dall'acqua S, Pilon-Smits EAH. 2015. Effects of selenium biofortification on crop nutritional quality. Frontiers in Plant Science, 6:1–5.

[17] Pinto E, Ferreira IMPLVO. 2015. Cation transporters/channels in plants: tools for nutrient biofortification. Journal of Plant Physiology, 179:64–82.

[18] Ríos JJ, Rosales MA, Blasco B, Cervilla LM, Romero L, Ruiz JM. 2008. Biofortification of Se and induction of the antioxidant capacity in lettuce plants. Scientia Horticulturae, 116:248–255.

[19] Pezzarossa B, Rosellini I, Borghesi E, Tonutti P, Malorgio F. 2014. Effects of Se enrichment on yield, fruit composition and ripening of tomato (*Solanum lycopersicum*) plants grown in hydroponics. Scientia Horticulturae, 165:106–110.

[20] Bañuelos GS, Arroyo I, Pickering IJ, Yang SI, Freeman JL. 2015. Selenium biofortification of broccoli and carrots grown in soil amended with Se-enriched hyperaccumulator *Stanleya pinnata*. Food Chemistry, 166:603–608.

[21] Bachiega P, Salgado JM, Ernesto de Carvalho J, Ruiz ALTG, Schwarz K, Tezotto T, Morzelle MC. 2016. Antioxidant and antiproliferative activities in different maturation stages of broccoli (*Brassica oleracea Italica*) biofortified with selenium. Food Chemistry, 190:771–776.

[22] Businelli D, D'Amato R, Onofri A, Tedeschini E, Tei F. 2015. Se-enrichment of cucumber (*Cucumis sativus* L.), lettuce (*Lactuca sativa* L.) and tomato (*Solanum lycopersicum* L. Karst) through fortification in pre-transplanting. Scientia Horticulturae, 197:697–704.

[23] Smoleń S, Skoczylas L, Ledwozyw-Smolen I, Rakoczy R, Kopec A, Ewa Piatkowska E, Biezanowska-Kopec R, Koronowicz A, Kapusta-Duch J. 2016. Biofortification of carrot

(*Daucus carota* L.) with iodine and selenium in a field experiment. Frontiers in Plant Science, 7:1–17.

[24] Flohé L, Gunzler WA, Schock HH. 1973. Glutathione peroxidase: a selenium enzyme. FEBS Letters, 32:132–134.

[25] Rotruck E, Pope AL, Ganther HE, Swason AB, Hafeman DG, Hoestra WG. 1973. Selenium: biochemical role as a component of glutathione peroxidase. Science, 179:558–560.

[26] Redestorff M, Krol A, Lescure A. 2005. Understanding the importance of selenium and selenoproteins in muscle function. Cellular and Molecular Life Science, 63:52–59.

[27] Duntas LH, Benvenga S. 2015. Selenium: an element for life. Endocrine, 48:756–775.

[28] Van Hoewyk D. 2013. A tale of two toxicities: malformed selenoproteins and oxidative stress both contribute to selenium stress in plants. Annals of Botany, 112:965–972.

[29] Lehotai N, Lyubenova L, Schröder P, Feigl G, Ördög A, Szilágyi K, Erdei L, Kolbert Z. 2016. Nitro-oxidative stress contributes to selenite toxicity in pea (*Pisum sativum* L). Plant and Soil, 400:107–122.

[30] Shrift A, Ulrich JM. 1969. Transport of selenate and selenite into Astragalus roots. Plant Physiology, 44:893–896.

[31] Arvy MP. 1993. Selenate and selenite uptake and translocation in bean plants (*Phaseolus vulgaris*). Journal of Experimental Botany, 44:1083–1087.

[32] Maruyama-Nakashita A, Nakamura Y, Yamaya T, Takahashi H. 2004. A novel regulatory pathway of sulfate uptake in Arabidopsis roots: implication of CRE1/WOL/AHK4-mediated cytokinin-dependent regulation. Plant Journal, 38:779–789.

[33] Li HF, Mcgrath SP, Zhao FJ. 2008. Selenium uptake, translocation and speciation in wheat supplied with selenate or selenite. New Phytologist, 178:92–102.

[34] Kikkert J, Berkelaar E. 2013. Plant uptake and translocation of inorganic and organic forms of selenium. Archives of Environmental Contamination and Toxicology, 65:458–465.

[35] Sabbagh M, Van Hoewyk D. 2012. Malformed selenoproteins are removed by the ubiquitin–proteasome pathway in *Stanleya pinnata*. Plant and Cell Physiology, 53:555–564.

[36] Terry N, Zayed AM, Desouza MP, Tarun AS. 2000. Selenium in greater plants. Annual Review of Plant Physiology, 51:401–432.

[37] Wang CQ. 2011. Water-stress mitigation by selenium in *Trifolium repens* L. Journal of Plant Nutrition and Soil Science, 174:276–282.

[38] Durán P, Acuña JJ, Armada E, López-Castillo OM, Cornejo P, Mora ML, Azcón RR. 2016. Inoculation with selenobacteria and arbuscular mycorrhizal fungi to enhance selenium content in lettuce plants and improve tolerance against drought stress. Journal of Soil Science and Plant Nutrition, 16:201–225.

[39] Diao M, Ma L, Wang J, Cui J, Fu A, Liu HY. 2014. Selenium promotes the growth and photosynthesis of tomato seedlings under salt stress by enhancing chloroplast antioxidant defense system. Journal of Plant Growth Regulation, 33:671–682.

[40] Hasanuzzaman M, Hossain MA, Fujita M. 2012. Exogenous selenium pretreatment protects rapeseed seedlings from cadmium-induced oxidative stress by upregulating antioxidant defense and methylglyoxal detoxification systems. Biological Trace Element Research, 149:248–261.

[41] Hajiboland R, Sadeghzade N. 2014. Effect of selenium on CO_2 and NO_3^- assimilation under low and adequate nitrogen supply in wheat (*Triticum aestivum* L.). Photosynthetica, 52:501–510.

[42] Koprivova A, Mugford ST, Kopriva S. 2010. Arabidopsis root growth dependence on glutathione is linked to auxin transport. Plant Cell Reports, 29:1157–1167.

[43] Freeman JL, Tamaoki M, Stushnoff C, Quinn CF, Cappa JJ, Devonshire J, Fakra SC, Marcus MA, McGrath SP, Van Hoewyk D, Pilon-Smits EAH. 2010. Molecular mechanisms of selenium tolerance and hyperaccumulation in *Stanleya pinnata*. Plant Physiology, 153:1630–1652.

[44] Aslam M, Harbit KB, Huffaker RC. 1990. Comparative effects of selenite and selenate on nitrate assimilation in barley seedlings. Plant, Cell & Environment, 13:773–782.

[45] Ribeiro DM, Silva Júnior DD, Cardoso FB, Martins AO, Silva WA, Nascimento VL, Araújo WL. 2016. Growth inhibition by selenium is associated with changes in primary metabolism and nutrient levels in *Arabidopsis thaliana*. Plant, Cell & Environment, 39:2235–2246.

[46] Rayman MP. 2008. Food-chain selenium and human health: emphasis on intake. British Journal Nutrition, 100:254–268.

[47] Rayman, MP. 2012. Selenium and human health. Lancet, 379:1256–1268.

[48] Wang X, Hai C. 2016. Novel insights into redox system and the mechanism of redox regulation. Molecular Biology Reports, 43:607–628.

[49] Seven M, Basaran SY, Cengiz M, Unal S, Yuksel A. 2013. Deficiency of selenium and zinc as a causative factor for idiopathic intractable epilepsy. Epilepsy Research, 104:35–39.

[50] Pitts MW, Byrns CN, Ogawa-Wong AN, Kremer P, Berr MJ. 2014. Selenoproteins in nervous system development and function. Biological Trace Element Research, 161:231–245.

[51] Pieczyńska J, Grajeta H. 2015. The role of selenium in human conception and pregnancy. Journal of Trace Elements in Medicine and Biology, 29:31–38.

[52] Nothstein AK, Eiche E, Riemann M, Nick P, Winkel LHE, Göttlicher J, Steininger R, Brendel R, Brasch MV, Konrad G, Neumann T. 2016. Tracking se assimilation and speciation through the rice plant–nutrient competition, toxicity and distribution. PLoS One, 26;1–15.

[53] Gissel-Nielson G. 1971. Influence of pH and texture of the soil on plant uptake of added seleniuk. Journal of Agricultural and Food Chemistry, 19:1165–1167.

[54] Gupta U, Gupta S. 2000. Selenium in soils and crops, its deficiencies in livestock and humans: implications for management. Communications in Soil Science and Plant Analysis, 31:1791–1807.

[55] Davis CD. 2007. Nutritional interactions: credentialing of molecular targets for cancer prevention. Experimental Biology and Medicine, 232:176–183.

[56] Finley JW. 2007. Increased intakes of selenium-enriched foods may benefit human health. Journal of the Science of Food and Agriculture, 87:1620–1629.

[57] Chang JC, Gutenmann WH, Reid CM, Lisk DJ. 1995. Selenium content of Brazil nuts from two geographic locations in Brazil. Chemosphere, 30:801–802.

[58] Andrade EHA, Maia JGS, Streich R, Marx F. 1999. Seed composition of amazonian lecythidaceae species: part 3 in the series "studies of edible amazonian plants". Journal of food composition and analysis, 12:37–51.

[59] Lisk DJ, Bache CA, Essick LA, Reid CM, Rutzke M, Crown K. 1988. Absorption and excretion of selenium and barium in humans from consumption of Brazil nuts. Nutrition Reports International, 38:183–91.

[60] Welna M, Klimpel M, Zyrnicki W. 2008. Investigation of major and trace elements and their distributions between lipid and non-lipid fractions in Brazil nuts by inductively coupled plasma atomic optical spectrometry. Food Chemistry, 111:1012–1015.

[61] Lemire M, Mergler D, Fillion M, Passos CJS, Guimarães JRD, Davidson R, Lucotte M. 2006. Elevated blood selenium levels in the Brazilian Amazon. Science of the Total Environment, 366:101–111.

[62] Lemire M, Philibert A, Fillion M, Passos CJS, Guimarães JRD, Juniorr FB, Mergler D. 2012. No evidence of selenosis from a selenium-rich diet in the Brazilian Amazon. Environment International, 40:128–136.

[63] Whelan R, Barrow NJ. 1994. Slow-release selenium fertilizers to correct selenium sheep in Western Australia. Fertilizer Research, 38:183–188.

[64] Hawrylak-Nowak B, Matraszek R, Pogorzelec M. 2015. The dual effects of two inorganic selenium forms on the growth, selected physiological parameters and macronutrients accumulation in cucumber plants. Acta Physiologiae Plantarum, 37:41.

[65] Fageria NK. Maximizing crop yields. New York: Marcel Dekker, 1992. 274p.

[66] Geoffroy L, Gilbin R, Simon O, Floriani M, Adam C, Pradines C, Cournac L, Garnier-Laplace J. 2007. Effect of selenate on growth and photosynthesis of *Chlamydomonas reinhardtii*. Aquatic Toxicology, 83:149–158.

[67] Hawrylak-Nowak B. 2013. Comparative effects of selenite and selenate on growth and selenium accumulation in lettuce plants under hydroponic conditions. Plant Growth Regulation, 70:149–157.

[68] Saffaryazdi A, Lahouti M, Ganjeali A, Bayat H. 2012. Impact of selenium supplementation on growth and selenium accumulation on spinach (*Spinacia oleracea* L.) plants. Notulae Scientia Biologicae, 4:95–100.

[69] Jozefczak M, Remans T, Vangronsveld J, Ann CA. 2012. Glutathione is a key player in metal-induced oxidative stress defenses. International Journal of Molecular Sciences, 13:3145–3175.

[70] Malavolta E. Elements of plant mineral nutrition. São Paulo: Ed. Agronômica Ceres, 1980, pp.211-212.

[71] GenPlant-Plant Nutrition Study Group. 2016. Available from: http://www.nutricaode-plantas.agr.br/site/extensao/lab_np.php (Accessed August 2016).

[72] Davis RD, Beckett PHT, Wollan E. 1978. Critical levels of twenty potentially toxic elements in young spring barley. Plant and Soil, 49:395–408.

[73] Ríos JJ, Blasco B, Leyva R, Sanchez-Rodriguez E, Rubio-Wilhelmi MM, Romero L, Ruiz JM. 2013. Nutritional balance changes in lettuce plant grown under different doses and forms of selenium. Journal of Plant Nutrition, 36:1344–1354.

[74] Whanger PD. 1989. China, a country with both selenium deficiency and toxicity: some thoughts and impressions. The Journal of Nutrition, 119:1236–1239.

[75] Spadoni M, Voltaggio M, Carcea M, Coni E, Raggi A, Cubadda F. 2007. Bioaccessible selenium in Italian agricultural soils: comparison of the biogeochemical approach with a regression model based on geochemical and pedoclimatic variables. Science of the Total Environment, 376:160–177.

[76] Franke KW. 1934. A new toxicant occurring naturally in certain samples of plant foodstuffs. 1. Results obtained in preliminary feeding trials. Journal of Nutrition, 8:597–609.

[77] Kenfield SA, Van Blarigan EL, DuPre N, Stampfer MJ, Giovannucci EL, Chan JM. 2015. Selenium supplementation and prostate cancer mortality. Journal of National Cancer Institute, 107:dju360.

[78] Lemire M, Fillion M, Barbosa Junior B, Guimarães JRD, Mergler D. 2010. Elevated levels of selenium in the typical diet of Amazonian riverside populations. Science of the Total Environment, 408:4076–4084.

[79] Institute of Medicine, Food and Nutrition Board. Dietary Reference Intakes: Vitamin C, Vitamin E, Selenium, and Carotenoids. National Academy Press, Washington, DC, 2000. Available from: https://ods.od.nih.gov/factsheets/Selenium-HealthProfessional/ (Accessed August 2016).

10

Functional Properties and Maillard Reaction Product Formation in Rye-Buckwheat Ginger Cakes Enhanced with Rutin

Małgorzata Starowicz and Henryk Zieliński

Abstract

A study of functional properties and Maillard reaction progress in rye-buckwheat ginger cakes supplemented with low and high rutin amount was conducted. The cakes were formulated on rye flour substituted by flour from husked buckwheat or flour from roasted buckwheat groats at 30 % level. The dough was either spontaneously fermented for 72 h at 21 °C, or the fermentation step was omitted. The cakes were baked at 180 °C for 18 min. Fortification of rye-buckwheat ginger cakes by low and high level of rutin was associated with the beneficial progress of the Maillard reaction towards the formation of melanoidins, while furosine formation at the early stage of Maillard reaction was decreased. However, loss of the nutritional value due to the formation of high amount of fluorescent (carboxymethyllysine (CML)) and fluorescent-intermediate compounds was observed. It has also been proved that rye-buckwheat ginger cakes fortified with low and high rutin doses are a rich source of biologically active compounds. Therefore, the cakes showed a high ability to inhibit the formation of advanced glycation end products (AGEs) in vitro and revealed high antioxidant potential. These findings may be important factors in complete evaluation of functional properties of ginger cakes. Stronger influence of rutin enrichment was observed in cakes baked from fermented-like dough than without this process.

Keywords: maillard reaction products, antioxidant potential, AGE inhibition, bioactive compounds analysis, buckwheat-based product, ginger cakes

1. Introduction

Ginger cakes are traditional pastries from Central and Eastern Europe. The very important feature of ginger cakes is their ability to remain fresh and savoury for a long time. Currently, ginger cakes are baked on the basis of wheat and rye flour. However, a traditional recipe for ginger cakes was based on the use of rye flour. In Poland, rye (*Secale cereale* L.) is an important grain for bread making and cookies production; therefore, in 2012 18 % of cereal products were made of rye [1]. Rye grain is considered to be a good source of biologically active compounds like antioxidants [2]. Referring to the up-to-date literature, it is highlighted that there is a lack of wider use of buckwheat in pastry.

The buckwheat is a rich source vitamin B1 and B2, lysine, protein with balanced amino acid composition [3], flavonoids [4], phytosterols [5], soluble carbohydrates, D-chiro-inositol and other fagopyritols [6] and thiamin-binding proteins [7]. Buckwheat is also rich in antioxidant compounds such as flavonoids, phenolic acids, tocopherols, reduced glutathione, inositol phosphates and melatonin [8]. Furthermore, buckwheat contains a high amount of rutin (quercetin-3-rutinoside) and has antioxidant, anti-inflammatory and anticarcinogenic properties [9]. According to various chemical compositions, buckwheat-based products were found to display several biological activities, including the increasing number of lactic acid bacteria in rat intestine, treatment of allergic inflammation, reducing the serum glucose level, suppressing cholesterol level, inhibiting protease and scavenging free radicals [10, 11]. These healthy and dietary benefits of buckwheat are main aspects in determining the usage of buckwheat to produce functional products.

The wide spectrum of buckwheat-based bakery and pastry products, e.g. bread, biscuits, crackers, cookies or muffins, was designed by researchers [12]. Mancebo et al. [13] observed that consumers' rating of cookies prepared from buckwheat did not reach high quality score which was mainly related to unpleasant and pungent taste of buckwheat. Therefore, Filipčev et al. [14] noted that 30 % of buckwheat flour is appropriate to create buckwheat-based product with high sensorial acceptability. Chlopicka et al. [15] also showed that 30 % addition of buckwheat flour is highly acceptable, and moreover buckwheat bread has a high antioxidant potential. Moreover, while analysing gluten-free products' sensory profiles, Loredana et al. [16] suggested that the optimum buckwheat flour addition is different for cake, cookies and muffins. The optimum amount in cake was established on the level of 30 % and 10 % for cookies, while for muffins 20 %. Not only the optimisation of the recipes but also technological process parameters for buckwheat-based product preparation have acquired an increasing interest. Lee [17] achieved high overall acceptance for steam bread (wheat with 3 % of buckwheat flour). However, the addition of buckwheat flour was not as high as in the previous studies yet, during the steaming process, not as high amount of an undesirable Maillard reaction products may be formed. Moreover, it is said that dough fermentation step can lead to nutritive and antinutritive compound formation, but some studies also suggested that fermentation process negatively influenced sensory properties of Turkish bread yufka supplemented with 10 % of buckwheat flour [18, 19].

The baking process is inherent in Maillard reaction product formation. The Maillard reaction is a reaction which might have nutritional and toxicological effects on processed food. The Maillard reaction is initiated by the reaction between the carbonyl group of a reduced sugar and a free amino group of proteins, and then subsequent and parallel reactions go. The progress of the reaction can be considered in the context of early, advanced and final Maillard reaction product formation such as furosine (ε-N-2-furoylmethyl-L-lysine) [20], fluorescent intermediary compounds (FIC) formed at the advanced stage [21], carboxymethyllysine (CML) [22] and melanoidins [23]. The latter are responsible for product's colour formation, nevertheless possessing the ability to scavenge free radicals. The degradation of proteins is usually expressed as FAST index [24], based on the measurement of the fluorescence of tryptophan and formation of intermediatory compounds. Moreover, in thermally treated food *via* Maillard reaction, dietary advanced glycation end products (dAGEs) can be formed [25]. dAGEs are an important contributor to the total pool of AGEs formed in the living organism and can induce oxidant stress and inflammation resulting in increasing risk of diabetes and cardiovascular diseases [26]. The AGE accumulation in the body can be regulated by low dAGE diet or by consuming food with natural AGE inhibitors such as plant extracts and plant products, which are good source of antioxidant polyphenols [27, 28]. Thereby, the high antiglycation potential of wheat-buckwheat bread extract has been recently reported [29].

Gathering all the above-mentioned information led to enriching the traditional ginger cakes recipe with 30 % of light buckwheat flour or flour obtained from roasted buckwheat groats. The new recipe for buckwheat ginger cakes has been created, which fits well with contemporary trends in the bakery. Furthermore, characterisation of the changes occurring during fermentation and baking processing and effect of rutin supplementation should be valuable for the understanding of quality and safety of buckwheat ginger cakes. To achieve the aim, the evaluation of the total phenolics; rutin; products of early, advanced and final Maillard reaction stages; and antioxidative capacity of rye-buckwheat ginger cakes enriched with rutin was addressed in this study.

2. Research methods

2.1. The rye-buckwheat ginger cake preparation

The ginger cakes were prepared using a mix of rye flour (70 %) and light buckwheat flour/flour from roasted buckwheat groats (30 %). Then the rye-buckwheat ginger cake recipe was modified by the addition of low (50 mg/of rutin/100 g of flour mix) and high rutin (100 mg/of rutin/100 g of flour mix) dosage. The rutin dosage added to buckwheat ginger cakes was adjusted to rutin content in one tablet of OTC drugs. The control cake was prepared from rye flour. The ingredient list used for rye-buckwheat ginger cake preparation is included in **Table 1**. All the ingredients were well mixed, and then half of dough was set aside, and the other half was cut into regular discs and baked at 180 °C for 18 min. The first half of dough was spontaneously fermented for 72 h at 21 °C in fermented chamber. Then, the fermented dough was prepared as the previous one. The cakes were freeze-dried and powdered after baking and cooling. The powdered samples were stored at −20 °C until analysis of functional properties and Maillard reaction product formation.

Ingredients	RGC	BERGC-1	BERGC -1L	BERGC -1H	BERGC -2	BERGC -2L	BERGC -2H
Rye flour [g]	100	70	70	70	70	70	70
Light buckwheat flour [g]	–	30	30	30	–	–	–
Flour from roasted buckwheat groats [g]	–	–	–	–	30	30	30
Buckwheat honey [g]	50	50	50	50	50	50	50
Sugar [g]	20	20	20	20	20	20	20
Baking soda [g]	3	3	3	3	3	3	3
Butter [g]	25	25	25	25	25	25	25
Spice mix for ginger cakes [g]	2	2	2	2	2	2	2
Rutin [mg]	0	0	50	100	0	50	100

Table 1. The list of ingredients used for rye-buckwheat ginger cake formulation.

Sample description: RGC (control), rye ginger cake; BERGC-1, buckwheat-enhanced rye ginger cake formulated on (1) light buckwheat flour; BERGC-1L, buckwheat-enhanced rye ginger cake with low rutin dose; BERGC-1H, buckwheat-enhanced rye ginger cake with high rutin dose; BERGC-2, buckwheat-enhanced rye ginger cake formulated on (2) flour from roasted buckwheat groats; BERGC-2L, buckwheat-enhanced rye ginger cake with low rutin dose; and BERGC-2H, buckwheat-enhanced rye ginger cake with high rutin dose.

2.2. The determination of total phenolic, rutin contents and antioxidant capacity in ginger cakes

The initial step included preparation of ginger cake extracts. Therefore, 100 mg of powdered samples was extracted with 1 ml of 80 % (v/v) methanol solution. Then, the mixture was treated by ultrasounds (30 s) and vortexed (30 s) three times. After centrifugation (6860 rpm at controlled temperature 4 °C, 5 min) the supernatant was collected into 5-ml flask. That step was repeated five times to achieve the final extract concentration 20 mg/ml.

The total phenolic content (TPC) was measured using Folin-Ciocalteu reagent according to Przygodzka et al. [30], whereas the rutin content was determined with HPLC with UV detector (330 nm), reported by Zielińska [31]. The antioxidant properties were determined by measurement of scavenging ability against ABTS radical cation, DPPH radical and superoxide anion radical ($O_2^{\bullet-}$) [30]. The measurements were carried out using spectrophotometer UV-160 1PC (Shimadzu, Japan) and Photochem® apparatus (Analytical Jena, Germany). The Trolox was used as a standard.

2.3. The Maillard reaction product determination: furosine, fluorescent intermediary compounds (FIC), carboxymethyllysine (CML) and melanoidins

In these buckwheat ginger cakes, markers of early (furosine), advanced (carboxymethyllysine (CML), total fluorescent intermediate compounds) and final (melanoidins) Maillard reaction compounds were identified and quantified.

The furosine content was determined as described by Delgado-Andrade et al. [32], using HPLC (Shimadzu, Japan) with PDA detector setup at 280 nm. To determine the furosine quantity, the external standard of furosine at concentration range 0.2–9 μg/ml was applied. In the next step, the fluorescent intermediary compounds (FIC) was measured. The total sum of intermediatory compounds in buckwheat ginger cake extracts was determined according to procedure described by Delgado-Andrade et al. [21]. The analysis was followed by enzymatic hydrolysis step using pronase E to break the bindings between intermediatory compounds and proteins. The fluorescent readings were registered at extinction wavelength 347 nm and emission, 415 nm, using a luminescent spectrofluorometer (LS-50B, PerkinElmer, USA). The results are expressed as fluorescence intensity (FI) per milligram of dry matter.

The degradation of proteins (nutritional value) was expressed as FAST index according to Damjanovic Desic and Birlouez-Aragon's procedure [24]. The FAST index was calculated as a ratio of the fluorescence of intermediatory compounds, measured at extinction wavelength 347 nm and emission, 415 nm, using a luminescent spectrofluorometer (LS-50B, PerkinElmer, USA), to fluorescence of tryptophan (extinction ,290 nm, and emission, 340 nm) and described as a percentage.

The carboxymethyllysine (CML), one of the intermediatory compounds, was quantified and determined by HPLC method. The CML extraction was followed by a detailed description of Peng et al. [9]. The OPA reagent solution, which is a mixture of 10 mg of *o*-phthaldialdehyde (OPA) in 2 ml of methanol and the CML determination, was evaluated by HPLC (Dionex, USA) with fluorescent detector (SFLD-3400RS, Dionex, USA). The detector settings were established as the excitation wavelength 455 nm and emission, 340 nm, whereas the oven temperature was adjusted at 35 °C and flow rate 0.2 ml/min. The CML was separated on Luna® 3 μm C18 column (Phenomenex, USA) and eluted in isocratic gradient by water with 0.05 % of *o*-phosphoric acid and acetonitrile with 0.05 % of *o*-phosphoric acid. For quantitative analysis, calibration curve of CML standard was prepared in the range from 2.5 to 20 μM. The results were expressed in μg per gram of dry matter.

The formation of final Maillard reaction products was estimated as reported previously by Zieliński et al. [33]. The absorbance of buckwheat ginger cake methanolic extracts was measured at 410 nm using UV-Vis spectrophotometer (Shimadzu, Japan). Final results were expressed as the absorbance units (AU).

2.4. The evaluation of buckwheat ginger cake inhibitory activity against advanced glycation end-product formation

The inhibitory effect on formation of advanced glycation end products (AGEs) in ginger cakes was studied according to the procedure described by Szawara-Nowak et al. [29] in two in vitro model systems: bovine serum albumin-glucose (BSA-glu) and bovine serum albumin-methylglyoxal (BSA-MGO). The fluorescence intensity was measured at the excitation wavelength 330 nm and emission 410 nm using a luminescent spectrofluorometer (LS-50B, PerkinElmer, USA). The results are expressed in percentage inhibition of AGE formation. Aminoguanidine solution (1 mmol/l) was used as a positive control in this experiment.

3. Results and discussion

3.1. The results of total phenolic and rutin contents and antioxidant capacity in buckwheat ginger cakes

Table 2 shows total phenolic content (TPC) in rye-buckwheat ginger cakes prepared with and without dough fermentation-like step, respectively. The substitution of rye flour by buckwheat flours at level of 30 % w/w on total flour basis resulted in higher TPC values in rye-buckwheat ginger cake BERGC-1 and BERGC-2 obtained without dough fermentation-like step than RGC, 52 and 85 %. In contrast, this effect was not seen in BERGC-1 and BERGC-2 cakes after dough fermentation-like preparation. Furthermore, the addition of low and high rutin doses increased the TPC. The highest TPC was noted in BERGC-2H cake after dough fermentation-like process, whereas the result was almost two times higher than in BERGC-2. Higher TPC values were observed in ginger cakes after 72 h at constant 21 °C fermentation-like process. It can be related to the positive influence of fermentation process of dough on effective formation of antioxidants [34]. Moreover, all types of rye-buckwheat ginger cakes reached higher TPC values than did ginger nut biscuits [14]. It means that this innovative buckwheat-based product is a good source of phenolic compounds.

It is generally known that rutin is the main bioactive compound in buckwheat-based products [9], and for this reason, the rutin content in rye-buckwheat ginger cakes was analysed. The results of rutin content are presented in **Figure 1**. The rutin identification in RGC, obtained both with or without dough fermentation-like process, was related to the buckwheat honey usage in the recipe. Except for buckwheat honey, another crucial source of this flavonoid was light buckwheat flour and flour from roasted buckwheat groats. Therefore, rutin content increased almost twice in BERGC-1 and BERGC-2 cakes due to the 30 % substitution of rye flour by buckwheat flours. The highest rutin content was noted in BERGC-1H cake after dough fermentation-like process; this result was almost 25 times higher than in BERGC-1. In the case of cakes with omitted fermentation-like step, it was noted that BERGC-1 and BERGC-2 with low and high rutin doses have 7 times higher rutin content in BERGC-1L and 15 times in BERGC-1H. The analogous tendency was observed in BERGC-2L and in BERGC-2H, 12 times and 15 times, respectively. Moreover, a similar trend was noted in BERG-1 and BERG-2 obtained after dough fermentation-like process. Generally, the fermentation-like process caused some changes in dough, resulting in higher rutin content in cakes after this process. Compared to Filipčev et al. [14], the rutin content in rye-buckwheat ginger cakes (BERGC-1 and BERGC-2 with dough fermentation-like step) was higher than it was noted in ginger nut biscuits (wheat-buckwheat flour, 70:30).

Type of ginger cakes	TPC	DPPH	ABTS	PCL
Dough *without* fermentation-like preparation				
RGC	2.52 ± 0.10eG	5.15 ± 0.65fI	17.96 ± 0.87cE	3.45 ± 0.07fG
BERGC-1	3.85 ± 0.27dF	7.53 ± 1.14e	28.12 ± 2.08aA	7.95 ± 0.71bcC
BERGC-1L	3.79 ± 0.16dF	9.16 ± 0.10dE	23.94 ± 1.12bC	7.64 ± 0.07cdC
BERGC-1H	4.97 ± 0.16cD	9.37 ± 0.24dE	27.18 ± 0.74aA	10.36 ± 0.08aA
BERGC-2	4.66 ± 0.32cD	8.74 ± 0.77eF	23.48 ± 0.70bB	6.93 ± 0.12dD
BERGC-2L	5.52 ± 0.36bC	12.15 ± 0.36bC	8.64 ± 0.28dG	5.49 ± 0.56eE
BERGC-2H	7.20 ± 0.46aB	17.74 ± 0.29aA	24.49 ± 0.53bB	8.65 ± 0.79bB
Dough *with* fermentation-like preparation				
RGC	4.80 ± 0.14dD	6.87 ± 0.05fH	17.41 ± 0.87cE	4.84 ± 0.06fF
BERGC-1	4.91 ± 0.23cdD	7.20 ± 0.21efG	20.19 ± 1.16abCD	5.93 ± 0.80cE
BERGC-1L	5.71 ± 0.49bC	7.78 ± 0.28dG	21.25 ± 0.37abC	5.96 ± 0.53cE
BERGC-1H	5.80 ± 0.42bC	8.18 ± 0.08cF	21.62 ± 0.90aC	7.32 ± 0.42bC
BERGC-2	4.26 ± 0.14eE	9.40 ± 0.35cE	19.72 ± 0.36cD	4.91 ± 0.79cdF
BERGC-2L	4.63 ± 0.36deD	11.02 ± 0.45bD	14.23 ± 0.65bF	5.97 ± 0.62cE
BERGC-2H	8.43 ± 0.37aA	14.55 ± 0.68aB	21.03 ± 1.41abC	8.97 ± 1.17aB

Table 2. The total phenolic content (TPC) and antioxidant capacity of rye-buckwheat ginger cakes determined by DPPH, ABTS and PCL methods.

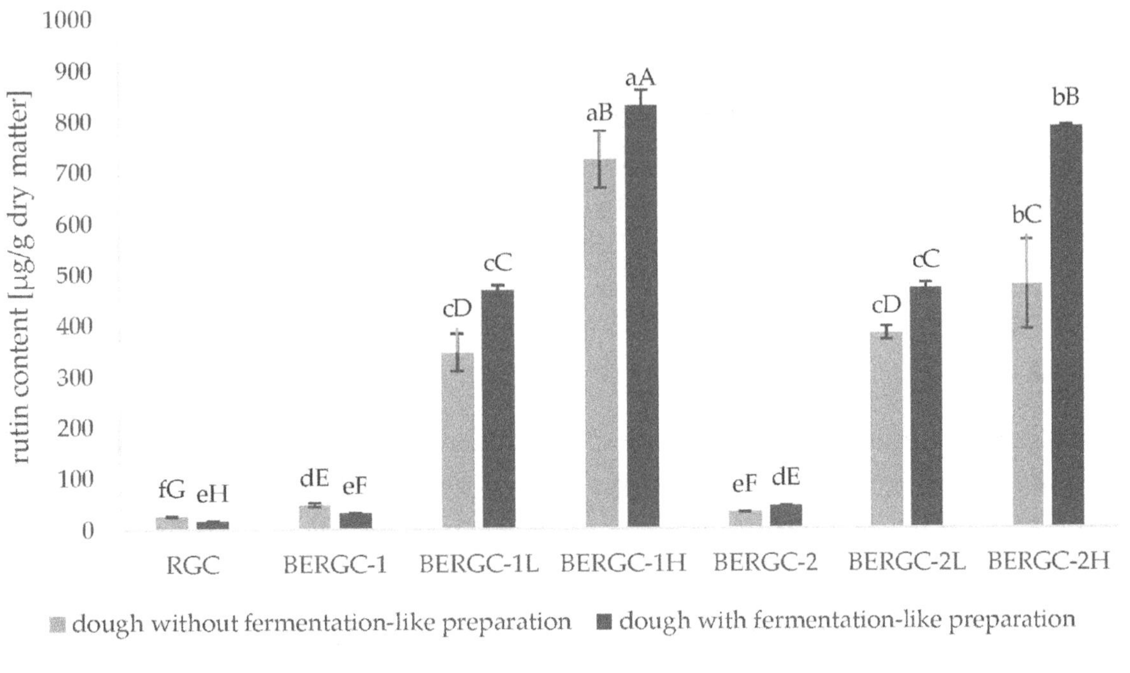

Figure 1. Rutin content in rye-buckwheat ginger cakes from dough obtained with or without fermentation-like preparation. Sample description under **Table 1**.

The 80 % methanol-water extracts of rye-buckwheat ginger cakes were examined for their free radical scavenging activity against DPPH$^{\bullet}$, ABTS$^{+\bullet}$ and $O_2^{\bullet-}$ radicals. The results are collected in **Table 2**. In DPPH method, the highest scavenging ability was noted in BERGC-2H without dough fermentation-like process. In general, higher antioxidant capacity was noted for ginger cakes made of flour from roasted buckwheat groats (BERGC-2) and in cakes from dough without fermentation-like treatment. These results are in accordance with Sedej et al. [35], in which authors maintained that buckwheat groat possesses the strongest DPPH scavenging ability. The 16 % increase of antioxidant activity values was noted in BERGC-2 compared to BERGC-1, 89 % in BERGC-2H to BERGC-1H. A similar increasing trend of antioxidant capacity vs. rutin addition was found for rye-buckwheat ginger cakes with dough after fermentation-like step. The antioxidant scavenging ability of rye-buckwheat ginger cakes against DPPH$^{\bullet}$ was higher than in ginger cakes evaluated from rye flours [33]. For the ABTS method (**Table 2**), the highest scavenging ability was noted in BERGC-1H without dough fermentation-like process. In this case, higher antioxidant capacity as 11 % was noted for ginger cakes made of light buckwheat flour (BERGC-1) and, as previously in DPPH method, in cakes from dough without fermentation-like treatment. The increase of almost 56 % in antioxidant activity values was noted in cakes after buckwheat flour incorporation to achieve BERGC-1 and 31% to BERGC-2 without fermentation-like usage. Then the influence of low and high rutin application was not observed. For cakes after dough fermentation-like process, the 16 % increase of antioxidant activity values was noted in cakes after buckwheat flour incorporation to achieve BERGC-1 and 13%—BERGC-2. The results were similar for BERGC-1H and BERGC-2H. At least, in PCL method the highest scavenging ability was noted, as well as for ABTS method, in BERGC-1H without dough fermentation-like process. Therefore, higher antioxidant capacity was noted for ginger cakes made of light buckwheat flour (BERGC-1) and, as previously in DPPH and ABTS method, in cakes from dough without fermentation-like treatment. The antioxidant activity values have 2.3 times increase in cakes after buckwheat flour incorporation to achieve BERGC-1 and two times in BERGC-2 (dough without fermentation-like step). Then the influence of low and high rutin application was not observed. For cakes after dough fermentation-like process, no spectacular increase of antioxidant activity values was noted in cakes after buckwheat flour incorporation BERGC-1 and BERGC-2. Furthermore, a significant increase of ability to scavenge superoxide anion radicals was observed after addition of high rutin dose, 23 % in BERGC-1H compared to BERGC-1 and 83 % in BERGC-2H in comparison to BERGC-2.

Sample description under **Table 1**. TPC is expressed in mg of rutin eq. per gram of dry matter. Antioxidant capacity is expressed in μmol of Trolox eq. per gram of dry matter. Values are means and standard deviations (n = 3).

3.2. The results of Maillard reaction products formed during buckwheat ginger cake baking

In **Table 3**, data of furosine, fluorescent intermediary compounds (FIC), carboxymethyllysine (CML) and melanoidin contents formed *via* Maillard reaction are summarised.

Type of ginger cakes	Furosine (mg/g DM)	Total FIC (FI/mg DM)	CML (μg/g DM)	Melanoidins (AU)
Dough *without* fermentation-like preparation				
RGC	0.94 ± 0.08aA	132.8 ± 0.08eI	12.70 ± 1.20dE	0.40 ± 0.01eI
BERGC-1	0.54 ± 0.01bC	166.6 ± 3.05dG	15.60 ± 1.00cD	0.75 ± 0.01bB
BERGC-1L	0.58 ± 0.01bC	208.2 ± 5.08aC	15.90 ± 2.00bcD	0.76 ± 0.00bB
BERGC-1H	0.53 ± 0.07bC	187.6 ± 3.33cE	11.90 ± 0.90dE	0.91 ± 0.01aA
BERGC-2	0.52 ± 0.02bC	200.8 ± 11.53abCD	17.90 ± 1.20abC	0.42 ± 0.01eI
BERGC-2L	0.29 ± 0.02cD	202.2 ± 4.51abD	19.10 ± 0.60aC	0.50 ± 0.01dE
BERGC-2H	0.27 ± 0.01cD	198.2 ± 0.19bD	17.90 ± 1.20aC	0.62 ± 0.01cC
Dough *with* fermentation-like preparation				
RGC	0.49 ± 0.10cC	125.7 ± 5.62eJ	12.69 ± 1.15dE	0.40 ± 0.01fI
BERGC-1	0.80 ± 0.06bB	151.3 ± 2.09dH	27.63 ± 1.95aA	0.45 ± 0.00dG
BERGC-1L	0.14 ± 0.03dE	250.9 ± 1.29aA	18.67 ± 1.66cC	0.35 ± 0.00gJ
BERGC-1H	0.17 ± 0.01dE	230.7 ± 2.49bB	17.98 ± 0.88cC	0.49 ± 0.00bE
BERGC-2	0.76 ± 0.08bB	172.0 ± 0.01cF	24.36 ± 1.31bB	0.43 ± 0.00eH
BERGC-2L	0.98 ± 0.04aA	174.2 ± 6.62cF	23.84 ± 0.80bB	0.48 ± 0.00cF
BERGC-2H	0.47 ± 0.04cC	175.6 ± 6.25cF	22.96 ± 0.90bB	0.55 ± 0.00aD

Table 3. Data of furosine, fluorescent intermediary compounds (FIC), carboxymethyllysine (CML) and melanoidin contents formed via Maillard reaction.

At the early stage of Maillard reaction, furosine was analysed due to its influence on nutritional protein damage in thermally treated food products [36]. According to obtained results, furosine was formed in all types of ginger cakes. In cakes from non-fermented-like dough, light buckwheat flour or flour from roasted buckwheat groats, limited formation of furosine was observed from 0.94 in RGC to 0.54 mg/g of dry matter in BERGC-1 and to 0.52 mg/g of dry matter in BERGC-2. Furthermore, the highest reduction of furosine, being about twofold, was noted in BERGC-2L and BERGC-2H. In cakes with dough after fermentation-like preparation, 63 and 55 % higher formation of furosine was noted in BERGC-1 and BERGC-2 than in RGC. These findings are in accordance with other studies where high protein content of light buckwheat flours and from roasted buckwheat groats was confirmed [38]. Then, the protective effect of rutin application was observed. In contrast, in BERGC-2L the polyphenolic protective effect was not observed. The most effective furosine decrease 4.7-fold and 1.6-fold was noted for BERGC-1H and BERGC-2H, respectively. Moreover, the lowest furosine content was determined in BERGC-1H with fermentation-like step during dough preparation. The furosine content formed in rye-buckwheat ginger cakes reported in this study was at the same level as previously described in enteral formula [37] and rye ginger cakes [33].

In the next step, the total fluorescence of intermediatory compounds (total FIC) formed at the advanced stage of Maillard reaction was studied. The formation of FIC is related to nutritional loss due to the formation of new molecules from lysine-free amino residues and reducing sugars [21]. These results are presented in **Table 3**. The total FIC were determined in all types of cakes within the range of 132.8–208.2 FI/mg in rye-buckwheat ginger cakes from non-fermented-like dough and 125.7–250.9 FI/mg in cakes from fermented-like dough. The addition of buckwheat flours to ginger cake recipe from dough after fermentation-like processing influenced the increasing total amount of FIC formation around 1.25 times and 1.51 times in BERGC-1 and BERGC-2, respectively. The lower formation of total FIC was noted after incorporation of high rutin dose compared to ginger cakes with low rutin dose. The most significant decrease 1.1-fold in BERGC-1H was noted. Similar findings were noted in ginger cakes prepared from dough after fermentation-like process. The buckwheat flour incorporation increased FIC formation 1.2-fold and 1.36-fold in BERGC-1 and BERGC-2, respectively. Then the application of low and high rutin content increased total FIC value up to 65 % in BERGC-1L. These results may suggest that FI compounds are more likely to be formed after the addition of phenolic compounds such as rutin. The FIC values remain at the same level after rutin application to cakes made of flour from roasted buckwheat groats (BERGC-2L and BERGC-2H).

In **Table 3**, the results of carboxymethyllysine (CML) content in rye-buckwheat ginger cakes enriched with rutin, made in two dough preparation procedures, are collected. The CML is known as a nonfluorescent intermediatory compound and characteristic marker of advanced Maillard reaction stage [38]. The CML was identified in all ginger cake samples. The addition of buckwheat flours to ginger cakes made of dough without fermentation-like step influenced increasing CML content around 1.22 times and 1.41 times in BERGC-1 and BERGC-2, respectively. Then, enrichment with rutin proceeded to achieve 24 % lower amount of CML in BERGC-1H than in BERGC-1, whereas no change was observed for BERGC-2H. In cakes evaluated from dough after fermentation-like step, the buckwheat flour addition increased around twofold CML content in BERGC-1 and BERGC-2. Then, addition of rutin decreased CML amount to 1.5 times in BERGC-1H. However, in cakes made of flour from roasted buckwheat groats, no effect was observed. In general, the lowest CML content was found in GC-1H (without dough fermentation-like preparation) 11.9 μg/g of dry matter. According to high values of CML content, it may be observed that fermentation-like preparation of dough negatively influenced intensified formation of CML. The restricted parameters of fermentation process, e.g. using specific bacterial strain and temperature, are required to control CML formation in further studies. Moreover, CML formation was linked to furosine content (r = 0.63), suggesting that loss of nutritional quality of elaborated rye-buckwheat ginger cakes was caused by Maillard reaction progressing at the advanced stage.

Sample description under **Table 1**. Furosine is expressed as mg/g of dry matter. Total FIC is expressed in fluorescence intensity (FI) per mg of dry matter. Melanoidin content is expressed as arbitrary units (AU). Values are means and standard deviations (n = 3).

In **Figure 2**, the results of FAST index are displayed. In ginger cakes prepared from fermented-like dough, the FAST values ranged from 461 to 1309 % and in cakes without previous fermentation-like preparation from 271 to 1477 %. The obtained results were at least twice lower than those described for ginger cakes made from rye flour [33]. The addition of buckwheat flours to ginger cakes made of dough without fermentation-like process showed FAST index value on the same level for BERGC-1 as in RGC, while for BERGC-2 FAST value was above twice higher. Then, enrichment with rutin proceeds to achieve 2.8 higher FAST in BERGC-1H than in BERGC-1, whereas no change was observed for BERGC-2H. In cakes evaluated from dough after fermentation-like step, the buckwheat flour addition decreased around 2.6-fold and 1.4-fold FAST in BERGC-1 and BERGC-2, respectively. Then, addition of rutin increased FAST values up to 5.4 times in BERGC-1H. Moreover, in cakes made of flour from roasted buckwheat groats, no significant effect of rutin supplementation was observed. An increase of FAST values was observed in using light buckwheat flour in ginger cake recipe. In contrast, in ginger cakes baked from flour from roasted buckwheat groats, no significant effect was observed. However, their FAST values were significantly higher than in raw buckwheat groats [33].

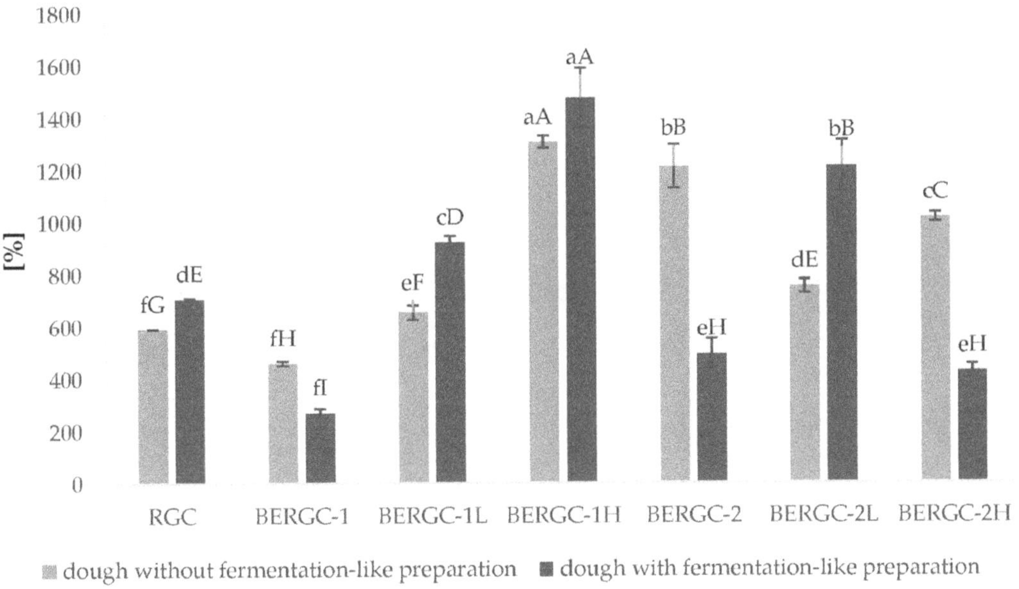

Figure 2. FAST index evaluation in rye-buckwheat ginger cakes obtained from dough with or without fermentation-like process.

The results of brown pigment formations as melanoidin, as the markers of the last stage of Maillard reaction, are shown in **Table 3**. As it may be observed, these brown polymers were formed in RGC as well as in new elaborated rye-buckwheat ginger cakes enriched with rutin. The addition of buckwheat flours to ginger cakes made of dough without fermentation-like process influenced on increasing melanoidin content by 88 % and 5 % in BERGC-1 and BERGC-2, respectively. Additional increase up to 21 % (BERGC-1) and 48 % (BERGC-2) was observed after rutin substitution. Therefore, the most advanced melanoidin formation process was observed in BERGC-1H. In cakes evaluated from dough after fermentation-like step, the results were similar. The obtained values noted in our study were slightly higher than those previously found in ginger cakes from rye flour, but they were twice as those in ginger cakes formulated on rye and wheat flours mix [39]. Moreover, melanoidin formation was found to be positively correlated with antioxidant capacity measured by ABTS test (r = 0.61) and PCL assay (r = 0.84) and DPPH (r = 0.94) as well as TPC and rutin content in cakes without fermentation-like process (r = 0.97 and 0.64). The slightly lower correlation coefficients for ginger cakes prepared from fermented-like dough were noted. Our findings are in accordance with previous studies, where positive correlation between melanoidin formation and antioxidant activity was proved [40].

3.3. The inhibitory activity of buckwheat ginger cakes against advanced glycation end-product (AGE) formation

The inhibitory ability of rye-buckwheat ginger cake extracts, prepared from dough without or with fermentation-like step, against AGE formation was evaluated using in vitro BSA-MGO and BSA-glu model systems. The obtained results are presented in **Figures 3** and **4**.

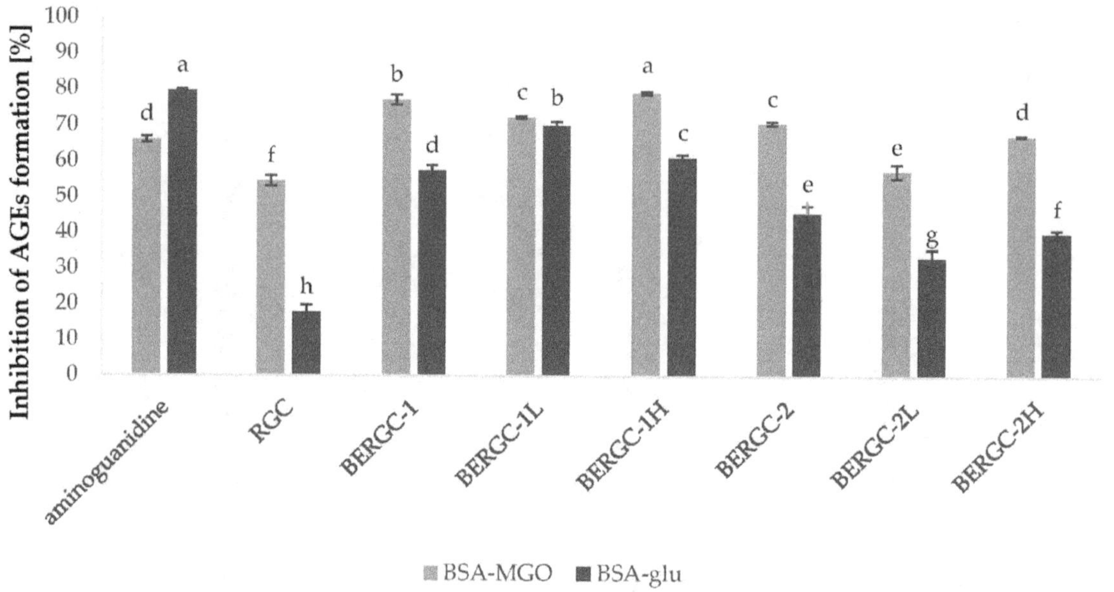

Figure 3. Inhibition of AGE formation in rye-buckwheat ginger cakes from dough without fermentation-like preparation. Sample description under **Table 1**.

Firstly, it was found that in BSA-MGO model in cakes, prepared without fermentation-like preparation, the AGE inhibitory activity increased after usage of buckwheat flours almost 42 and 30 % in BERGC-1 and BERGC-2, respectively (**Figure 3**). Their inhibitory activity values were higher than for aminoguanidine, a medicine used during medical treatment against diseases related to AGE accumulation in human tissues [29]. The enrichment of rye-buckwheat ginger cakes with rutin increased inhibitory activity and the highest value was noted in cakes with high dose of rutin (BERGC-1H). The application of light buckwheat flour in ginger cake formula (BERGC-1) offered also higher values of AGE inhibitory activity than in flour from roasted buckwheat groat incorporation (BERGC-2). In BSA-glu model, also the usage of buckwheat flours almost 23 and 56 % increased inhibitory activity of BERGC-1 and BERGC-2, respectively. However, in this model system, inhibitory effect of aminoguanidine was higher reaching 80 %. Also, the addition of low and high rutin dosages did not increase the inhibitory ability. Then, in BSA-MGO model system but in cakes baked from fermented-like dough, 30 % of light buckwheat flour addition increased the AGE inhibitory ability from 59 to 65 %, while no change was obtained after flour from roasted buckwheat groats addition. Therefore, in BSA-MGO model system, the highest values of inhibitory activity of ginger cakes were noted for BERGC-2H. The same results were obtained in BSA-glu model system, whereas the highest inhibitory activity was achieved in BERGC-2H (48 %). According to these results, it has been confirmed that the antiglycation activity was strongly correlated with polyphenolic compound content and the scavenging ability measured for rye-buckwheat ginger cakes obtained from fermented-like dough. It may be summarised that the effect of rutin enrichment was clearly seen in cakes obtained with fermented-like dough, even if the inhibitory activity was slightly lower than those cakes produced from non-fermented dough. Our findings are confirmed in previous studies of antiglycation and antioxidative activities of buckwheat breads [29]. These breads substituted by light buckwheat flour inhibited at 40 % the AGE formation and breads baked from flour from roasted buckwheat groats at 60 % measured in BSA-glu model system, in BSA-MGO at 20 % and 40 %, respectively [29]. Moreover, the high AGE inhibitory potential of coriander, which is an ingredient of spice mix, may contribute to the total antiglycation ability of rye-buckwheat ginger cakes [41].

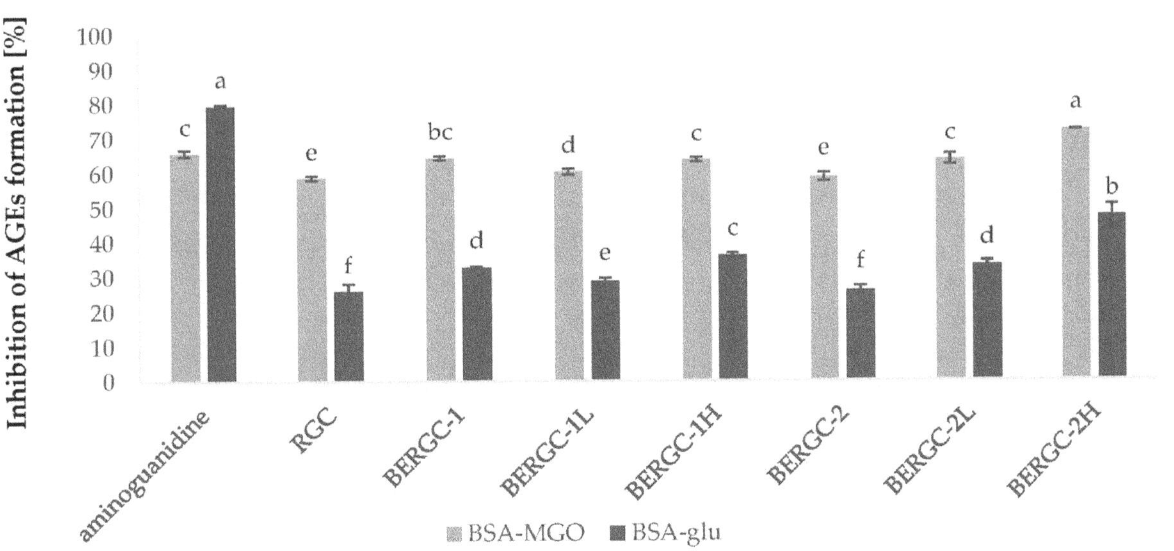

Figure 4. Inhibition of AGE formation in rye-buckwheat ginger cakes from dough obtained with fermentation-like preparation. Sample description under **Table 1**.

4. Conclusions

The new functional product as rye-buckwheat ginger cakes enriched with rutin has been elaborated, and data of total phenolics; rutin contents; antioxidative capacity measured by ABTS, DPPH and PCL methods; and characterisation of Maillard reaction markers have been provided. Moreover, the inhibitory activity against AGE formation using in vitro model systems BSA-MGO and BSA-glu has been studied. The enrichment of rye-buckwheat ginger cakes with rutin improved their increased total phenolic content and antioxidant properties. The protective effect on furosine formation has been observed, whereas melanoidin formation has been stimulated. In contrast, the loss of nutritional quality of rye-buckwheat ginger cakes enriched with rutin has been noted due to the formation of CML and FI compounds at the advanced stage of MR. Moreover, high antiglycation potential of rye-buckwheat ginger cakes enriched with rutin has been confirmed. The relationship between antiglycation ability and rutin content and antioxidant capacity has been found. The addition of buckwheat flours as well as rutin supplementation in ginger cakes has influenced the increase of AGE inhibitory potential.

Author details

Małgorzata Starowicz* and Henryk Zieliński

*Address all correspondence to: m.przygodzka@pan.olsztyn.pl

Institute of Animal Reproduction and Food Research, Olsztyn, Poland

References

[1] Rynek zbóż w Polsce. Raport Agencji Rynku Rolnego. 2013. The cereals market in Poland. The Agricultural Market Agency Report. 2013. http://www.gov.pl/data/1900321/rynek_zboz_2013_pl.pdf

[2] Michalska A, Ceglińska A, Amarowicz R, Piskuła MK, Szarawa-Nowak D, Zieliński H. Antioxidant contents and antioxidative properties of traditional rye bread. Journal of Agricultural and Food Chemistry. 2007; 55: 734-740.

[3] Li S, Zhang OH. Advance in the development of functional foods from buckwheat. Critical Review in Food Science and Nutrition. 2001;41(6):451-464.

[4] Jiang P., Burczynski F, Campbell C, Pierce G, Austria JA, Briggs CJ. Rutin and flavonoid contents in three buckwheat species *Fagopyrum esculentum, F. tataricum* and *F. homotropicum* and their protective effects against lipid peroxidation. Food Research International. 2007;40:356-364. DOI: 10.10/j.foodres.2006.009

[5] Horbowicz M, Obendorf RF. Changes in sterols and fatty acids of buckwheat endosperm and embryo during seed development. Journal of Agricultural and Food Chemistry. 1992;40(5):742-752.

[6] Steadman KJ, Burgoon MS, Schuster RL, Lewis BA, Edwardson SE, Obendorf RL. Fagopyritols, D-chiro-inositol and other soluble carbohydrates in buckwheat

seed milling fractions. Journal of Agricultural and Food Chemistry. 2000;48:2843-2847.

[7] Watanabe K, Shimizu M, Adachi T, Mitsunaga T. Characterization of thiamine-binding protein from buckwheat seeds. Journal of Nutrition Science and Vitaminology. 1999;44(2):323-328.

[8] Zieliński H, Michalska A, Piskuła MK, Kozłowska H. Antioxidants in thermally treated buckwheat groats. Molecular Nutrition and Food Research. 2006;50:824-832.

[9] Peng X, Zheng Z, Cheng K-W, Shan F, Ren G-X, Chen F, Wang M. Inhibitory effect of mung bean extract and its constituents vitexin and isovitexin on the formation of advanced glycation endproducts. Food Chemistry. 2008;106:475-481.

[10] Kawa JM, Taylor CG, Przybylski R . Buckwheat concentrate reduces serum glucose in streptozotocin-diabetic rats. Journal of Agricultural and Food Chemistry. 2003;51:7287-7291.

[11] Kim CD, Lee WK, No KO, Park SK, Lee MH, Lim SR. Anti-allergic action of buckwheat (*Fagopyrum esculentum* Moench) grain extract. International Immunopharmacology. 2003;3:129-136.

[12] Giménez-Bastida JA, Piskuła MK, Zieliński H. Recent advances in processing and development of buckwheat derived bakery and non-bakery products—a review. Journal of Food and Nutrition Science. 2015;65:9-20. DOI: 10.1515/pjfns-2015-005 (Polish)

[13] Mancebo CM, Picon J, Gomez M. Effect of flour properties on the quality characteristics of gluten free sugar-snap cookies. LWT—Food Science and Technology. 2015;64:264-269.

[14] Filipčev B, Šimurina O, Sakač M, Sedej I, Jovanov P, Pestorić M, Bodroža-Solarov M. Feasibility of use of buckwheat flour as an ingredient in ginger nut biscuit formulation. Food Chemistry. 2011;125:164-170.

[15] Chlopicka J, Pasko P, Gorinstein S, Jedryas A, Zagrodzki P. Total phenolic and total flavonoid content, antioxidant activity and sensory evaluation of pseudocereal breads. LWT—Food Science and Technology. 2012;46:548-555.

[16] Loredana IM, Petru R B, Daniela S, Ioan TT, Monica N. Sensory evaluation of some sweet gluten-free bakery products based on rice and buckwheat flour. Journal of Biotechnology. 2015;208:5-120.

[17] Lee JH. Influence of buckwheat flour on physiochemical properties and consumer acceptance of steamed breads. Journal of Food and Science Nutrition. 2010;15:329-334.

[18] Dordević TM, Šiler-Marinković S, Dimitrijević-Branković SI. Effect of fermentation on antioxidant properties of some cereals and pseudo cereals. Food Chemistry. 2010;119:957-963. DOI: DOI: 10.1016/j.foodchem.2009.07.049

[19] Levent H, Bilgiçli N. Evaluation of physical, chemical and sensory properties of Turkish flat breads (Bazlama and Yufka) supplemented with lupin, buckwheat and oat flours. International Journal of Food Science and Nutrition Engineering. 2012;2:89-95.

[20] Giannetti V, Boccacci Mariani M, Mannino P. Furosine as a pasta quality marker: evaluation by an innovative and fast chromatographic approach. Journal of Food Science. 2013;78:C994-C999.

[21] Delgado-Andrade C, Rufian-Henares JA, Morales FJ . Study on fluorescence of Maillard reaction compounds in breakfast cereals. Molecular Nutrition and Food Research. 2006;50:799-804.

[22] Bosch L, Sanz ML, Montilla A, Alegría A, Farré R, del Castillo MD. Simultaneous analysis of lysine, Nε-carboxymethyllysine and lysinoalanine from proteins. Journal of Chromatography B. 2007;860:69-77.

[23] Wang H-Y, Qian H, Yao W-R. Melanoidins produced by the Maillard reaction: Structure and biological activity. Food Chemistry. 2011;128:573-584.

[24] Damjanovic Desic S, Birlouez-Aragon I. The FAST index—A highly sensitive indicator of the heat impact on infant formula model. Food Chemistry. 2011;124:1043-1049.

[25] Markowicz Bastos DH, Gugliucci A. Contemporary and controversial aspects of the Maillard reaction products. Current Opinion in Food Science. 2015;1:13-20.

[26] Ames JM. Dietary Maillard reaction products: implications for human health and disease. Czech Journal of Food Science. 2009;27:S66-S69.

[27] Vlassara H, Uribarri J. Advanced Glycation End Products (AGE) and diabetes: cause, effect, or both?. Current Diabietes Reports. 2014;14(1):453.

[28] Peng X, Ma J, Chen F, Wang M. Naturally occurring inhibitors against the formation of advanced glycation end-products. Food Functionality. 2011;2(6):289-301.

[29] Szawara-Nowak D, Koutsidis G, Wiczkowski W, Zieliński H. Evaluation of the in vitro inhibitory effects of buckwheat enhanced wheat bread extracts on the formation of advanced glycation end-products (AGEs). LWT—Food Science and Technology. 2013; 50:1–8.

[30] Przygodzka M, Zielińska D, Ciesarová Z, Kukurová K, Zieliński H. Comparison of methods for evaluation of the antioxidant capacity and phenolic compounds in common spices. LWT—Food Science and Technology. 2014;58:321-326. DOI: 10.1016/j.lwt.2013.09.019

[31] Zielińska D, Szawara-Nowak D, Zieliński H. Antioxidative ant anti-glycation activity of buckwheat hull tea infusion. International Journal of Food Properties. 2013;16:228–239.

[32] Delgado-Andrade C, Seiquer I, Navarro MP, Morales FJ. Maillard reaction indicators in diets usually consumed by adolescent people. Molecular Nutrition & Food Research. 2007;51:341-351. DOI: 10.10/mnfr.200500249.

[33] Zieliński H, del Castillo MD, Przygodzka M, Ciesorová Z, Kukurová K, Zielińska D. Changes in chemical composition and antioxidative properties of rye ginger cakes during their shelf-life. Food Chemistry. 2012;135:2965-2973. DOI: 10.1016/j.foodchem.2012.07.009.

[34] Hur SJ, Lee SY, Kim Y-C, Choi I, Kim G-B. Effect of fermentation on the antioxidant activity in plant-based foods. Food Chemistry. 2014;160:346-356.

[35] Sedej I, Sakač M, Mandić A, Mišan A, Tumbas V, Čanadanović-Brunet J. Buckwheat (*Fagopyrum esculentum* Moench) grain and fractions: antioxidant compounds and activities. Journal of Food Science. 2012;77(9). DOI: 10.1111/j.1750-3841.2012.02867.

[36] Erbersdobler HF, Somoza V. Forty years of furosine- forty years of using Maillard reaction products as indicators of the nutritional quality of foods. Molecular Nutrition and Food Research. 2007;51:423-430. DOI:10.1002/mnfr.200600154.

[37] Rufian-Henares JA, Guerra-Hernández E, García-Villanova B. Maillard reaction in enteral formula processing: furosine, loss of *o*-phthaldialdehyde reactivity, and fluorescence. Food Research International. 2002;35:527–533.

[38] He J, Zeng M, Zheng Z, He Z, Chen J. Simultaneous determination of N^{ε}-(carboxymethyl)lysine and N^{ε}-(carboxyethyl)lysine in cereal foods by LC-MS/MS. European Food Research and Technology. 2014;238:367-374. DOI:10.1007/s00217-013-2085-8.

[39] Zieliński H, Amigo-Benavent M, del Castillo MD, Horszwald A, Zielińska D. Formulation and baking process affect Maillard reaction development and antioxidant capacity of ginger cakes. Journal of Food and Nutrition Research. 2010;49:140-148.

[40] Manzocco L, Calligaris S, Mastrocola D, Nicoli MC, Lerici CR. Review of non-enzymatic browning and antioxidant capacity in processed foods. Trends in Food and Science Technology. 2000;11: 340-346.

[41] Ramkissoon JS, Mahomoodally MF, Ahmed N, Subratty AH. Antioxidant and anti-glycation activities correlates with phenolic composition of tropical medicinal herbs. Asian Pacific Journal of Tropical Medicine. 2013;6:562-569.

Permissions

All chapters in this book were first published in SFFAOTPU, by InTech Open; hereby published with permission under the Creative Commons Attribution License or equivalent. Every chapter published in this book has been scrutinized by our experts. Their significance has been extensively debated. The topics covered herein carry significant findings which will fuel the growth of the discipline. They may even be implemented as practical applications or may be referred to as a beginning point for another development.

The contributors of this book come from diverse backgrounds, making this book a truly international effort. This book will bring forth new frontiers with its revolutionizing research information and detailed analysis of the nascent developments around the world.

We would like to thank all the contributing authors for lending their expertise to make the book truly unique. They have played a crucial role in the development of this book. Without their invaluable contributions this book wouldn't have been possible. They have made vital efforts to compile up to date information on the varied aspects of this subject to make this book a valuable addition to the collection of many professionals and students.

This book was conceptualized with the vision of imparting up-to-date information and advanced data in this field. To ensure the same, a matchless editorial board was set up. Every individual on the board went through rigorous rounds of assessment to prove their worth. After which they invested a large part of their time researching and compiling the most relevant data for our readers.

The editorial board has been involved in producing this book since its inception. They have spent rigorous hours researching and exploring the diverse topics which have resulted in the successful publishing of this book. They have passed on their knowledge of decades through this book. To expedite this challenging task, the publisher supported the team at every step. A small team of assistant editors was also appointed to further simplify the editing procedure and attain best results for the readers.

Apart from the editorial board, the designing team has also invested a significant amount of their time in understanding the subject and creating the most relevant covers. They scrutinized every image to scout for the most suitable representation of the subject and create an appropriate cover for the book.

The publishing team has been an ardent support to the editorial, designing and production team. Their endless efforts to recruit the best for this project, has resulted in the accomplishment of this book. They are a veteran in the field of academics and their pool of knowledge is as vast as their experience in printing. Their expertise and guidance has proved useful at every step. Their uncompromising quality standards have made this book an exceptional effort. Their encouragement from time to time has been an inspiration for everyone.

The publisher and the editorial board hope that this book will prove to be a valuable piece of knowledge for researchers, students, practitioners and scholars across the globe.

List of Contributors

Karuane Saturnino da Silva Araújo
Federal University of Maranhão (UFMA), Imperatriz, Brazil

Dark Luzia dos Santos Neto and Sandra Maria Botelho Mariano
Federal University of Tocantins (UFT), Palmas, Brazil

Banu Yucel
Ege University, Faculty of Agriculture, Department of Animal Science, Bornova, Izmir, Turkey

Erkan Topal and Mustafa Kosoglu
Aegean Agricultural Research Institute, Beekeeping Unit, Menemen, Izmir, Turkey

Felipe Galeti Migue, Nathalia Baptista, Andresa Piacezzi Nascimento, Franciane Marquele-Oliveira1, Juliana Issa Hori and Raul Ferreira
Apis Flora Indl. Coml. Ltda., Ribeirão Preto, Brazil

Caroline Arruda and Jairo Kenupp Bastos
Faculty of Pharmaceutical Sciences of Ribeirão Preto, University of São Paulo, Ribeirão Preto, Brazil

Andresa A. Berretta
Apis Flora Indl. Coml. Ltda., Ribeirão Preto, Brazil
Faculty of Pharmaceutical Sciences of Ribeirão Preto, University of São Paulo, Ribeirão Preto, Brazil
Brazilian Association of Honey Exporters, ABEMEL, Rio Claro, Brazil

Hernane da Silva Barud
Uniara, University of Araraquara, Araraquara, Brazil

Bianca Damaso and César Ramos
Brazilian Association of Honey Exporters, ABEMEL, Rio Claro, Brazil
Natucentro Própolis, Bambuí, Brazil

Elizabeth M. Ulloa-Inostroza, Daniela Peña-Sanhueza and Jorge González-Villagra
Center of Plant, Soil Interaction and Natural Resources Biotechnology, Scientific and Technological Bioresource Nucleus (BIOREN), Universidad de La Frontera, Temuco, Chile

Eric G. Ulloa-Inostroza
Department of Internal Medicine, Faculty of Medicine, Universidad de La Frontera, Temuco, Chile

Miren Alberdi and Marjorie Reyes-Díaz
Center of Plant, Soil Interaction and Natural Resources Biotechnology, Scientific and Technological Bioresource Nucleus (BIOREN), Universidad de La Frontera, Temuco, Chile
Department of Chemical Sciences and Natural Resources, Faculty of Engineering and Sciences, Universidad de La Frontera, Temuco, Chile

Laura Jaakola
Department of Arctic and Marine Biology UiT, The Arctic University of Norway, Tromsø, Norway

Patricia Velásquez
Department of Chemistry and Bioprocess, Faculty of Engineering, Pontifical Catholic University of Chile, Santiago, Chile

Gloria Montenegro
Department of Plant Science, Faculty of Agricultural and Forestry Sciences, Pontifical Catholic University of Chile, Santiago, Chile

Wanessa Almeida da Costa, Mozaniel Santana de Oliveira, Marcilene Paiva da Silva, Vânia Maria Borges Cunha, Rafael Henrique Holanda Pinto, Fernanda Wariss Figueiredo Bezerra and Raul Nunes de Carvalho Junior
LABEX/FEA, Faculty of Food Engineering, Federal University of Pará, Belém, Brazil

Ayse Tulin Oz
Food Engineering Department, Engineering Faculty, Osmaniye Korkut Ata University, Osmaniye, Turkey

Ebru Kafkas
Horticulture Department, Agriculture Faculty, Cukurova University, Adana, Turkey

Ricardo Tighe-Neira
School of Agronomy, Faculty of Natural Resources, Universidad Católica de Temuco, Temuco, Chile

Miren Alberdi and Marjorie Reyes-Díaz
Department of Chemical Sciences and Natural Resources, Faculty of Engineering and Sciences, Universidad de La Frontera, Temuco, Chile
Center of Plant, Soil Interaction and Natural Resources Biotechnology, Scientific and Technological Bioresource Nucleus (BIOREN), Universidad de La Frontera, Temuco, Chile

Patricio Arce-Johnson
Department of Molecular Genetics and Microbiology, Faculty of Biological Sciences, Pontificia Universidad Católica de Chile, Santiago, Chile

Jesús L. Romero-Romero
Department of Molecular Genetics and Microbiology, Faculty of Biological Sciences, Pontificia Universidad Católica de Chile Santiago, Chile
Faculty of Agronomy and Forest Engineering, Pontificia Universidad Católica de Chile, Santiago, Chile
Department of Agricultural Biotechnology, Instituto Politécnico Nacional, CIIDIR Sinaloa Unid, Guasave, Mexico

Claudio Inostroza-Blancheteau
School of Agronomy, Faculty of Natural Resources, Universidad Católica de Temuco, Temuco, Chile
Nucleus of Research in Food Production, Faculty of Natural Resources, Universidad Católica de Temuco, Temuco, Chile

Renato de Mello Prado, Flávio José Rodrigues Cruz and Raphael Leone da Cruz Ferreira
Department of Soils and Fertilizers, Faculty of Agricultural and Veterinary Sciences, Plant Nutrition Study Group-GenPlant, University State of São Paulo "Júlio de Mesquita Filho", São Paulo, Brazil

Małgorzata Starowicz and Henryk Zieliński
Institute of Animal Reproduction and Food Research, Olsztyn, Poland

Index

O

P

Q

R

S

T

www.ingramcontent.com/pod-product-compliance
Lightning Source LLC
LaVergne TN
LVHW080726210926
848485LV00003B/523